CHEMISTRY AND GEOTHERMAL SYSTEMS

ENERGY SCIENCE AND ENGINEERING:
RESOURCES, TECHNOLOGY, MANAGEMENT
An International Series

EDITOR
JESSE DENTON
Belton, Texas

LARRY L. ANDERSON and DAVID A. TILLMAN (eds.), Fuels from Waste, 1977

A. J. ELLIS and W. A. J. MAHON, Chemistry and Geothermal Systems, 1977

In preparation

FRANCIS G. SHINSKEY, Energy Conservation through Control

Chemistry and Geothermal Systems

A. J. ELLIS

Department of Scientific and Industrial Research
Chemistry Division
Petone, New Zealand

W. A. J. MAHON

Department of Scientific and Industrial Research
Chemistry Division
Taupo, New Zealand

ACADEMIC PRESS   New York   San Francisco   London   1977
A Subsidiary of Harcourt Brace Jovanovich, Publishers

ACADEMIC PRESS, INC.
111 Fifth Avenue, New York, New York 10003

United Kingdom Edition published by
ACADEMIC PRESS, INC. (LONDON) LTD.
24/28 Oval Road, London NW1

Library of Congress Cataloging in Publication Data

Ellis, Albert James,   Date
      Chemistry and geothermal systems.

      Bibliography: p.
          1.   Geothermal resources.   2.   Geochemistry.
3.   Water chemistry.   I.   Mahon, W. A. J., joint
author.   II.   Title.
GB1199.5.E43   553'.79   76-52715
ISBN 0–12–237450–9

CONTENTS

Chapter 4   HYDROTHERMAL SOLUTIONS

Chapter 5   SAMPLING AND DATA COLLECTION

Chapter 6   PREDEVELOPMENT INVESTIGATIONS

Chapter 7   DATA PROCESSING AND CHANGES WITH TIME

Chapter 8   SOME SPECIFIC CHEMICAL TOPICS

Chapter 9 PHYSICAL AND CHEMICAL CHARACTERISTICS
OF NEW ZEALAND GEOTHERMAL AREAS

PREFACE

With the threat of depletion of conventional sources of energy there
has been increased interest in harnessing geothermal energy for man's
use in many countries. However, there are comparatively few people
with wide experience in the science and technology of developing this
new energy resource. Geothermal training programs are therefore
being organized by many governments, universities, and industries,
and there is a need for literature that summarizes the current state of
knowledge in various facets of the work.

We have drawn upon our background as geochemists with experience in many New Zealand and overseas geothermal areas, to provide
information on the theories and methods used in the geochemical
aspects of geothermal work. We also describe and seek to place in
context the contributions from other sciences in a geothermal development program and provide, as a background, descriptions of the
nature and occurrence of geothermal areas. While aimed at geothermal
specialist training, the book may also provide useful material in
graduate earth science courses, as well as be of general interest to
geologists, geophysicists, chemists, and chemical engineers.

The first two chapters review the extent of geothermal development
throughout the world and outline the geological and hydrological
features of geothermal areas. With this introduction, the chemistry of
geothermal fluids is then described in detail, and the reasons for the
various fluid compositions are interpreted through the chemical and
mineralogical reactions that occur in the geothermal systems. A brief
description of the physical and chemical nature of high-temperature
water solutions follows to aid this understanding. Procedures of
geochemical field and laboratory work are considered in detail in
Chapters 5 and 6 with particular reference to New Zealand experiences. The basic information, theory, and data required for the
chemical work are provided. The approach is practical with many

examples. Recommended sequences are given for the geological, geophysical, and geochemical work required in investigating a new geothermal area. Chapter 7 gives examples of the calculations necessary to interpret geochemical results, and methods of storing, retrieving, and interpreting data. The use of chemistry to monitor changes in a geothermal field is described.

Chapter 8 gives brief coverage of several chemical, chemical engineering, and metallurgical problems associated with geothermal development work, e.g., corrosion of materials, scale formation from waters, environmental effects, effluent treatment, and chemical recoveries. Examples are given from many fields.

The final chapter reviews scientific investigations and drilling in many New Zealand geothermal areas that are typical of the types of system found in many recent volcanic zones.

Frequent reference is made to the publications resulting from Symposia on the Development and Use of Geothermal Resources, organized by the United Nations and held at Pisa in September 1970 and at San Francisco in May 1975. The Proceedings of these symposia should be used to obtain more extensive and specific information on areas, developments, and techniques.

ACKNOWLEDGMENTS

Many of the geochemical techniques described, and much of the earlier chemical knowledge of the New Zealand geothermal areas are due to the work of S. H. J. Wilson. Both of the authors worked with him in the early stages of their careers and are grateful for his enthusiastic introduction to geothermal chemistry.

The authors also acknowledge with thanks the advice, assistance, and comments of many of their colleagues while preparing this book, including W. F. Giggenbach, R. B. Glover, J. R. Hulston, R. James, T. M. Seward, and B. G. Weissberg.

CHEMISTRY AND GEOTHERMAL SYSTEMS

Chapter 1

INTRODUCTION

1.1 DEFINITIONS

The term *geothermal energy* is used in this book in a restricted way, to refer to the potentially useful energy stored as hot water or steam in favorable geological situations within the top few kilometers of the earth's crust. It is contained in a *geothermal area* or *geothermal field* with finite surface boundaries and in a particular rock–hydrological situation (*geothermal system*). The specialized use of these terms may be thought unfortunate, since hot water or steam forms a very small fraction of the total geothermal energy contained within the earth's interior. Nevertheless, for the present it is the only fraction which can be utilized economically. Terms such as *geothermal energy, geothermal area,* and *geothermal system* now carry through accepted usage the suggestion of utilization of the heat resource.

Geothermal development refers to the harnessing of steam and hot water flows, either natural or induced through drilling into the rock system. The utilization may simply involve reticulation of hot water to buildings or glasshouses; steam or hot water may be used directly in industrial processes, or the thermal energy of steam may be converted into electricity.

Hydrothermal system is a general term used when discussing a rock–water system containing high-temperature water, in the laboratory or in the field.

The various types of geothermal system are discussed in Chapter 2, but for preliminary reading it should be noted that wells drilled into geothermal systems may in some areas intersect steam to considerable depths, whereas in other areas high-temperature liquid water is encountered. In the former situation the wells produce a single steam

1

phase; in hot water geothermal systems the wells produce a boiling mixture of steam and water in a proportion which depends on the original underground water temperature.

Throughout the book, temperatures are given in degrees Celsius (°C). Unless otherwise noted, pressures given are gage pressures, not absolute pressures.

1.2 GENERAL BACKGROUND

It is convenient to classify the energy sources used by man under three headings: solar, atomic, and geothermal. Traditional energy sources, which provide most of the power requirements of the world, stem ultimately from solar energy; examples are the burning of present-day and fossil organic materials and the harnessing of wind and water flows. There is great interest in the direct use of solar energy but so far it is of minor importance. Nuclear power can potentially supply many times the energy stored in the world's fossil fuels, but the construction and operation of nuclear power stations is at present limited to countries with a high level of scientific and technological development. Also, the safety of nuclear power plant designs and the problem of long-term disposal of radioactive wastes is currently being debated.

Except for the thin layer of the earth's surface that is affected by seasonal variations of temperature and by groundwater, the earth's temperature increases gradually with depth and the thermal gradient usually ranges from about 5° to 70° per kilometer. The outward flow of heat by conduction per unit of surface area and of time is given by the product of the vertical thermal gradient and the thermal conductivity coefficient characteristic of the rock. For example, if a gradient of 30° per kilometer occurs in a rock having a thermal conductivity of 0.0047 cal cm^{-1} sec^{-1} deg^{-1}, the conducted heat flow is 1.4×10^{-6} cal cm^{-2} sec^{-1} or 44 cal each year per square centimeter. This is the mean value of the earth's heat flow (Bullard, 1973).

The earth's heat is most obvious at the surface in volcanic zones of the world where lava eruptions, fumaroles, and hot water flows are well known. For example, Shimozura (1968) gave the energy released by volcanic activity in Japan as 2×10^{16} cal/yr. The lava pool created by the eruption of Kilauea Iki, Hawaii, in 1959 contained nearly 100 million tons of molten lava with about 3×10^{16} cal or 3×10^{10} kWh of energy. A molten rock body of this magnitude is a significant natural energy resource, but it is difficult to extract the energy at a usable rate because the low thermal conductivity of the rock and its high viscosity on cooling make any rapid, continuing heat exchange process almost impossible. The present utilization of volcanic heat in geothermal areas relies on natural slow heat exchange processes which have transferred heat from the rocks into deeply circulating waters.

Natural hot waters have been used since earliest times for cooking, washing, and bathing, but only in the last 30 years has geothermal energy been harnessed to provide a significant proportion of the energy requirements of any country. The pioneering countries in this field, Italy, Iceland, Japan, New Zealand, and the western United States, contain zones of recent volcanism, and until recently the search for further areas for development has been centered largely on this type of area. However, there are warm springs in most countries. Within nonvolcanic areas they may occur where water circulates or is stored at deep levels in the rocks, not necessarily in zones of elevated geothermal gradient. Investigations in Russia (Tikhonov and Dvorov, 1970) and Hungary (Boldizar, 1970) show that vast storage of moderate-temperature (50–150°) water may occur at depths down to 3500 m in deep sedimentary basins. These can provide important resources for city and agricultural heating.

The modern trends in the utilization of geothermal energy resources can be traced back to the middle of the last century, and to the areas of boraciferous steam and waters in the Tuscan hills in Italy. In the Larderello area the jets of steam (*soffioni*) and the pools of water which were formed in small craters and maintained at boiling temperatures by the passage of natural steam (*lagoni*) were known for many centuries, but were regarded by the local peasants as being due to unknown and unfriendly powers. The discovery of boric acid in them by Hoeffer and Mascagni in 1777 led to the establishment of a successful boric acid industry by the Conte Francesco de Larderel in the early 1800s. Natural steam was used to evaporate the boracic waters. Later, to increase the flow of steam, shallow wells were drilled (Nasini, 1930).

The first attempt to harness the energy of Larderello was made in 1897 by using natural steam to heat a boiler and produce pure steam for a reciprocating engine. On 4 July 1904 the first electric power was generated from geothermal steam by engineers under the direction of Prince Ginori Conti, when a small motor was operated by natural steam to drive a dynamo connected to some electric lamps.

In 1912 the first turbine generator unit powered by underground steam was installed at Larderello. The condensing turbine, driving an alternator, was supplied with purified steam obtained from evaporators, since the high gas content of the natural steam made it difficult to maintain a vacuum in the condenser. The turbine rating was 0.25 MW (Villa, 1961). During 1914 three larger turbine generator units (each 2.75 MW) were installed.

By 1944 the installed electric-generating capacity in the Larderello area was 127 MW, but the plant was destroyed during the later stages of the war. Postwar reconstruction of the Italian geothermal plants has resulted in an efficient power complex with a present generating capacity of about 400 MW.

In the 1920s geothermal power investigations were also made in California, Japan, and Iceland but there were no significant power developments there at that time.

1.3   GEOTHERMAL DEVELOPMENTS

Many countries are now actively involved in geothermal development programs. Although it is not possible to give an extensive account of the various projects, the following is a brief outline, country by country, of some developments which are known to the authors. At present a geothermal area with water or steam temperatures of at least $180°$–$200°$ is required for large-scale electricity generation, but in the future, binary cycle systems using heat transfer media of lower boiling point than water, such as freon or organic fluids, may allow more extensive development of lower-temperature systems for this purpose. Such developments at present are experimental. For space heating and agriculture, water of only moderate temperatures may be used. The emphasis in discussion is on the higher-temperature hydrothermal areas and on the larger uses of warm waters. Major references are Facca (1970), and the papers of the Second United Nations Symposium on the Development and Use of Geothermal Resources, San Francisco, 1975.

Figure 1.1 is a map of the world giving the location of some major geothermal development areas. Table 1.1 summarizes the major geothermal electric power developments in various countries.

Chile

A United Nations–Chilean government project has investigated the geothermal fields in northern Chile. At El Tatio, an area located high

Fig. 1.1   World map showing some of the main centers of geothermal investigation and development.

in the Andes near the Bolivian border in Antofagasta province, 13 wells from 600 m to 1820 m deep have been drilled. Water at temperatures of 180°–265° was found in ignimbrites and lavas. About 18 MW equivalent of steam is available from production wells (Lahsen and Trujillo, 1975) and a pilot plant is operating. Power will be used to aid mining activities in the province and a desalination plant working from geothermal heat and water is also planned.

Investigations have also been made of the Polloquere and Puchuldisa fields farther north.

Costa Rica

At Miravalles volcano a strong thermal and resistivity anomaly of some 10 km² occurs. Wells have been drilled to some 350 m and there are plans for a geothermal power plant in the early 1980s (M. Corrales, Symposium on Geothermal Energy in Latin America, Guatemala City, October 1976).

El Salvador

The initial studies of the geothermal resources in El Salvador were made in the period 1953–1959. Surface surveys throughout the active zone were followed by the drilling of shallow wells in the Ahuachapan area of Pleistocene andesites. Some years later further investigations

TABLE 1.1

Geothermal Electric Power Schemes in Operation or at Advanced Stage of Development

Area	Av. (max.) well depth (m)	Av. (max.) temp. (°C)	Discharge type[a]	Total generating capacity (MW)	
				Installed	Addition planned
Chile					
El Tatio	1000 (1820)	230 (265)	S + W	—	15
El Salvador					
Ahuachapan	1000 (1400)	230 (250)	S + W	90	increase
Iceland					
Namafjall	1000 (1400)	250 (280)	S + W	3	—
Krafla				55	—
Indonesia					
Kawah Kamojang	650 (761)	220 (238)	S + (W)	—	15
Italy					
Larderello region	650 (1600)	170–220 (240)	S	360	—
Monte Amiata region	780 (1500)	140–170 (190)	S + (W)	25	—
Japan					
Matsukawa	1000 (1500)	220 (280)	S	20	increase
Otake	500 (1500)	230 (250)	S + W	13	increase
Hatchobaru	500 (800)	250 (300)	S + W	50	—
Onikobe	1000 (1350)	230 (288)	S + W	25	—
Onuma	800 (1700)	>200 (—)	S + W	10	—
Mexico					
Cerro Prieto	1300 (2630)	300 (370)	S + W	75	75
New Zealand					
Wairakei	800 (2300)	230 (260)	S + W	192	—
Kawerau	800 (1250)	250 (285)	S + W	10	increase
Broadlands	1100 (2420)	255 (300)	S + W	—	150
Philippines					
Tiwi	950 (2300)	250 (—)	S + W	110	increase
Turkey					
Kizildere	700 (1000)	190 (220)	S + W	0.5	10
United States					
The Geysers	1500 (3120)	250 (285)	S	600	300
Salton Sea	1500 (2470)	— (360)	S + W	—	50+
East Mesa	— (2448)	200 (—)	S + W	—	10?
Soviet Union					
Pauzhetsk	— (800)	185 (200)	S + W	5	7

[a] S, steam; W, water.

were made under United Nations sponsorship and deep wells were drilled in the Berlin and Ahuachapan thermal areas. At present 16 wells tap water at 225°-250° and discharge considerable quantities of steam and water. A 60-MW power station is operating, and increased capacity is planned. The waters are dilute, neutral-pH chloride waters but the rather high boron concentrations present an effluent disposal problem in this agricultural area. Water disposal is currently mainly by reinjection back into the field, but a concrete canal to the sea is being constructed as an alternative.

Ethiopia

Geological, geophysical, and geochemical surveys of some of the 600 geothermal areas in Ethiopia were made recently in a joint Ethiopian-United Nations project. The Afar–Danakil and Tendaho grabens appear to be promising localities and more detailed investigations aimed at power developments are proceeding there and in the southern lakes district.

France

At Melun in the Paris Basin, two 1800-m holes, one for production and the other for cool water reinjection, are the basis of a community heating service supplying 3000 flats. The water tapped is at 70 °C.

Greece

Several geothermal prospects are currently being surveyed. The island of Milos, 150 km south of Athens in a setting of faulted Mesozoic marbles and schists, is the most promising area. Four 1000-m wells have been proposed (*Geothermal Energy*, June 1974, p. 37). On the mainland, prospects are at Thermopylae, on the Methana Peninsula, the Sousaki region, and North Euboea.

Guadeloupe, West Indies

At La Bouillante, within a recent volcanic area, several wells drilled to a maximum depth of 800 m encountered water at 220°-240°; one well had an output of 95 tons of water and 45 tons of steam per hour at

5 bars pressure (Facca, 1970). Two more recent and deeper wells were not productive.

Guatemala

An area of 100 km^2 at Moyuta was covered by geophysical surveys and a deep well drilled (little production). At Zunil in the eastern highlands, surveys are in progress.

Hungary

Near the Yugoslavian border, over 80 hot water wells are in production, drilled to an average depth of 1800 m in sandstones of Upper Pliocene age. Their total outflow of 6000 ton/h is used to heat 1200 residences and glasshouses (Boldizsar, 1970).

Iceland

Icelandic records going back to 1300 note that hot water was led from a warm pool to washbasins in a farm building at Reykholt. Apparently it was also common for food to be cooked in either hot earth or hot springs. In the middle of the nineteenth century, salt was recovered by pumping seawater into shallow pans and evaporating it with the aid of heat from hot springs (Barth, 1950). Heating of greenhouses by geothermal energy began to expand in Iceland during the 1920s and a rapid growth in this field occurred after 1940. In the early 1960s, the total area covered by greenhouses amounted to some 100,000 m^2 (Lindal, 1961).

Drilling for hot water in Iceland commenced near Reykjavik in 1928, and the hot water produced was used to heat a small number of houses. Larger-scale drilling in the Reykir area in the 1930s and at Reykir, Reykjahlid, and Reykjavik in the 1940s and 1950s produced water flows of over 500 l/sec at temperatures between 86° and 128° (Einarsson, 1973). Most of the buildings in Reykjavik are now heated by hot water pumped and reticulated from over 100 wells in or near the city. At Reykir some of the recent deep wells (800–2045 m) are of large diameter (22–31 cm) and produce a total of over 1000 l/sec of 81.5° water (Thorsteinsson, 1975).

Geothermal systems in Iceland can be broadly divided into two categories. In some areas, of which Reykjavik is typical, hot water occurs at 1–2 km depth at temperatures in the vicinity of 100° (not over

150°). These are in areas of older basalts of Tertiary age. High-temperature geothermal areas associated with recent volcanic centers have water temperatures of 200°–300° at depths of 1–2 km. Thirteen of these areas exist, some covering up to 100 km², with natural heat outflows between 3×10^3 and 3×10^5 kcal/sec (Bodvarsson, 1961, 1970).

At Hengill, 30 km east of Reykjavik, eight wells (the maximum depth is 1200 m) tapped water at temperatures up to 230° and produced over 400 tons of steam and 1300 tons of hot water per hour at 5 bars pressure. Some wells are used for heating greenhouses and houses.

On the Reykjanes peninsula in the southwest of Iceland three high-temperature geothermal areas have been investigated by deep drilling. Wells drilled in the late 1960s to depths of 1750 m at Reykjanes on the tip of the peninsula encountered temperatures up to 286° (Lindal, 1970), and highly saline water consisting of seawater which had been modified in composition by the high temperature. Recovery of salts was proposed. At Svartsengi and at Krisuvik (which is slightly to the northeast), deep wells tapped less saline and rather lower-temperature waters (235° and 260°, respectively). Use of the Svartsengi well discharges for space heating is planned (Arnorsson and Sigurdsson, 1974).

At Namafjall in northern Iceland, shallow (60-m) wells drilled in the 1950s in an extensive fumarole area tapped wet steam high in hydrogen sulfide (up to 0.2% by weight). More recently deeper wells drilled to 700–1400 m encountered temperatures up to 280°. A flow of about 110 tons of steam and 600 tons of water per hour at 10 bars was produced from seven wells (Ragnars *et al.*, 1970). The steam is used for drying diatomite and for producing a minor amount of electricity (~3 MW). At Krafla to the north of the area a 55-MW power station is currently being constructed (1976).

India

The principal high-temperature geothermal prospects are in the northwest at the junction of the Indian and Asian crustal plates. In the Himachal Pradesh State at Manikaran 5×10^5 l/h of 90°C water flows from springs (Behl *et al.*, 1975). At Puga in the Ladakh district high-temperature springs also occur, and shallow wells (130 m) supply 130°C water.

Further geothermal provinces lie between the Western Ghats and

the western coastline (Konkan) in the Cambay graben where petro-
leum wells have encountered waters much above 100 °C, and in the
northeastern states of Bihar and Orissa. Their potential is being
investigated.

Indonesia

The Kawah Kamojang hot spring and fumarole area occurs in a
volcanic complex in west Java. In 1925–1928 five shallow wells were
drilled, one of which (66 m deep) still produces 8 tons/h of steam at
123 °. Following a recent resurvey of the field by geological, geochemical, and electrical resistivity methods, five wells have been drilled
(maximum depth 761 m). One well with a maximum well temperature
of 238 °C produced dry steam; another produced steam plus water.

Other areas being investigated include Dieng in central Java, Salak
south of Jakarta, Manuk southwest of Jakarta, and Cisolok-Cisukarame
on the southwest coast of Java. Prospecting surveys are also under way
on Bali, Sumatra, and North Sulawesi.

Italy

Larderello Region.   In the boraciferous region of southeast Tuscany
on the northern side of the Metalliferous Hills, the Larderello geothermal complex produces about 3100 tons of steam per hour. The total
installed electricity-generating capacity is 360 MW (Cataldi *et al.*, 1970;
Ceron *et al.*, 1975).

There are approximately 200 producing wells drilled in a sequence
of Triassic–Jurassic mudstones, anhydrite, limestone, and quartzite
(see also Chapter 2). The average well depth is about 650 m, although
more recent wells have averaged 1100 m. The wells extend over an area
of 180 km² and produce steam at pressures of 2–7 bars and temperatures of 140 °–220 °. The maximum temperature and pressure measured
downhole were about 240 ° and 31 bars. The steam has a gas concentration between 3% and 8% (mainly carbon dioxide), with minor
methane, hydrogen sulfide, boric acid, and ammonia. In the initial
stages of development some of the chemicals from the steam (boric
acid mainly) were recovered during power production but there is
now little interest in this operation.

A 691-m well drilled in late 1971 at Travale, east of Larderello,
encountered a temperature of 264° at depth and gave a very large
output of steam—over 300 tons/h at a wellhead pressure of 8 bars, but

with a high gas content of 9%. A further well (1370 m) drilled in 1974 produced 90 tons of steam per hour at 5 bars with a very high gas content (mainly CO_2). A 15-MW generator is operating.

Monte Amiata Region. In the Bagnore and Senna geothermal fields 80 km southeast of Larderello, some 60 wells have been drilled to an average depth of 780 m, with a maximum of 1500 m (Cataldi *et al.*, 1970). One 50-cm diameter well at Bagnore had a flow of 300 tons/h of steam exceptionally high in carbon dioxide (originally as high as 80% by weight, decreasing later to about 8%). Steam was tapped in the central part of the production area and hot water on the periphery. The maximum temperature and pressure at Bagnore were 150° and 21 bars (Cataldi, 1967), but recently encroachment of cold water has lowered the temperatures. At Piancastagnaio about 10 km to the east a similar area has been developed. Again the gas concentrations in the discharges were high (19% by weight), but the wells encountered higher temperatures (185°). The installed generating capacity in this region is 25 MW.

Naples Region. In the Campi Flegrei, in the Naples Province, several wells were drilled during the period 1939–1955. The area is one of Quaternary volcanic ash, trachytic tuffs, and lavas. Minucci (1961) described the drilling of the deepest well (Agnano 1, 1842 m) on the eastern slopes of the Solfatara area, which encountered temperatures up to 325°. Wells were also drilled on the island of Ischia, tapping hot water up to 175° (Penta, 1954). In both areas the waters produced had salinities ranging from 1% to 3%. A small 0.5-MW power plant was installed on Ischia but there were great difficulties with mineral deposition in the drill pipes. The geothermal potential of the Campi Flegrei and the area north of Naples is currently being reevaluated.

Other Regions. The various areas of development and investigation are shown in Fig. 1.2. Intensive geological, hydrological, geochemical,

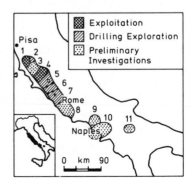

Fig. 1.2 Map of central Italy locating various areas of geothermal exploration and development (adapted from Ceron *et al.*, 1975).

and geophysical work has been carried out in several areas, and in some cases wells have been drilled (Ceron *et al.*, 1975).

At Monte Volsini, a high heat flow area, five wells were drilled at Torre Alfina, four being productive and giving a total of 950 tons/h of water, steam, and gas. Deep temperatures were 120°–140° and waters were of low salinity (6000 ppm). A 3.5-MW generator plant supplied by flashed steam was proposed.

Two 1000-m wells at Monte Cimini encountered temperatures of only 60–80°, but farther south in the Monte Sabatini region a well (Cesano 1) drilled to 1435 m produced 250 tons/h of highly saline water (25–35% T.D.S.) and steam from a base temperature of 250°. The salts present in the water were mainly sodium and potassium sulfates and chlorides, accompanied by 1.5% boric acid. Detailed investigations of this unique geothermal system are progressing.

Surface investigations are proceeding in the regions of Roccastrada, Colli Albani, Roccamonfina, and Vulture.

Japan

The hot springs of Japan have been used for many centuries for bathing and horticultural purposes. In the 1500 hot spring resort areas many wells have been drilled (average 500 m) to produce steam and hot water. The total rate of geothermal energy dissipation in the form of waste hot water is 5000 MW (*Geothermal Energy*, February 1975, p. 5).

The first attempts at producing electric power were in the 1920s when the Tokyo Electric Light Company made exploratory drillings in the Beppu, Kyushu, hot spring area and installed a 1-kW generator. In 1946 at Atagawa in the Izu Peninsula an experimental 38-kW generator was operated, and this was followed by similar experiments in other areas. A reevaluation of geothermal fields for power production commenced in 1947. Geological, geochemical, and surface temperature surveys were made, together with geophysical prospecting and investigational drilling to shallow depths in some areas (Saito, 1961).

At Otake, Kyushu, intensive investigations and well drilling began in the early 1950s, and at Matsukawa, Iwate Prefecture, in the early 1960s. The first major geothermal station (20 MW), at Matsukawa, was completed in 1966, and a 13-MW station at Otake was finished the following year (Fig. 1.3). Deep wells have now been drilled in many areas and several power stations have been built or are scheduled for construction.

Fig. 1.3 The major areas of geothermal power development in Japan.

Matsukawa. In an area of Pleistocene andesite, wells ranging in depth from 900 to 1500 m now produce sufficient steam to generate 22 MW of electricity (Hayakawa *et al.*, 1967; Nakamura and Sumi, 1967; Katagiri, 1975). Temperatures at Matsukawa range up to 280°. The steam utilization pressure is 3.5 bars and the gas content of the steam is 1.0% by weight. The wells first produced wet steam (75–100% steam) but the discharges later became almost completely dry. The waters produced from the wells were of low pH (about 5) and were dilute iron sulfate solutions, rather than the sodium chloride solutions found at depth in most hot water areas. Some problems have arisen from iron sulfate and silica scale formation in pipelines.

Otake. The 13-MW power station uses steam from five wells (350–550 m) drilled in an active hot spring and fumarole area in a zone of andesite volcanism (Noguchi, 1966). They tap dilute sodium chloride hot waters at temperatures of 230°–250° (see also Chapter 2). Wells produce up to 50 tons/h of separated steam (gas content 0.4%) at 2 bars pressure. The adjacent Hatchobaru area has been drilled to 800 m, the wells giving a good output of steam and water. Temperatures range up to 300°. A 50-MW power station is now being constructed at Hatchobaru.

Onikobe. Test wells to about 700 m in this hot spring and fumarole area produced hot water at about 190° from andesite and dacite tuff beds (Nakamura, 1959). Later drilling to 980 m encountered temperatures up to 288° (Facca, 1970) and tapped dilute neutral-pH sodium chloride waters. A 25-MW power station commenced operation in 1975.

Onuma. This field is located on the northwestern part of Hachimantai volcano. Exploration started in 1965 and there are currently six exploration wells and several production wells to a maximum depth of 1700 m in rock consisting of dacite and andesite welded tuffs overlying siliceous shale. The deep waters are weakly alkaline and slightly saline, and the temperatures exceed 200° (Sato, 1970). A 10-MW power station was completed in 1974.

Other Areas. At Takenoyu, 7 km from Otake, there has been exploratory drilling and a 50-MW power station is planned. Other current investigations include areas on Izo-Oshima Island; Ibusuki in Kyushu; Nasu and Oshirakawa in central Honshu; Takinokami in northern Honshu; and Kutcharo in eastern Hokkaido.

Kenya

The geothermal potential of hot spring and fumarole areas in the Rift Valley of Kenya, extending from Lake Hannington in the north to Lake Naivasha in the south, have been investigated by scientific surveys and pilot drilling. Three areas, Olkaria south of Lake Naivasha, Hannington, and Eburru, were found to be good prospects, and recently six wells were drilled at Olkaria (maximum depth 1650 m). The best two wells in this area produced approximately 30–40 tons/h of wet steam from a 245° steam zone at about 700–800 m. Others encountered impermeable conditions. Temperatures at 1650 m are approximately 300° (R. B. Glover, personal communication).

Mexico

Mexicali. The Cerro Prieto area in the rift valley system of the Gulf of California has surface thermal activity over about a 30-km² area. (Mercado, 1967). Chapter 2 discusses the field in greater detail. Geothermal developments began about 1960, and wells tapped high-temperature water in very permeable sediments. A total of 30 wells have now been drilled to depths of 2630 m, but the average well is about 1300 m deep (Guiza, 1972). Temperatures are very high (up to 370°), the water salinities are about 2%, and the production from wells is high (50–117 tons/h of separated steam). The field is currently being extended to double the size of the present 75-MW capacity power station.

Pathe. A pilot geothermal plant of 3.5-MW capacity was set up in the 1950s at Pathe, a hot spring and fumarole area in Hidalgo Province (De Anda *et al.*, 1961). In this volcanic region wells drilled down to 530 m produced water at temperatures up to 175°, but there were difficulties arising from mineral deposition in pipes.

At Ixtlan de los Hervores in Michoacan, an area of boiling pools and steaming vents, prospecting wells were drilled to 1000 m, tapping water at 160°. In the same state at Los Negritos, where there are mud volcanoes and fumaroles, resistivity surveys suggested a potential 100-km² production area and wells drilled to 700 m encountered water at 140°.

Other areas which have been investigated include Los Humeros, Puebla; Los Azufres, Michoacan; and La Primavera, Jalisco (Banwell and Gómez Valle, 1970).

New Zealand

In the North Island of New Zealand and particularly in the towns of Rotorua and Taupo, shallow wells have supplied private households and hotels with hot water for at least 30–40 years. In Rotorua over 1000 wells have been drilled (Kerr *et al.*, 1961). Bruce and Shorland (1932) suggested making a detailed compilation of data on New Zealand hot spring areas, to be followed by small-scale experiments along the lines of the Larderello geothermal scheme. Little was done, however, until the immediate postwar years, when an electric power shortage caused a renewed interest in geothermal power, and major scientific survey work commenced in the Taupo zone of rhyolitic and andesitic volcanism (Fig. 1.4). Geothermal developments began in the early 1950s with prospecting drilling at Wairakei, 8 km north of Taupo. This led to large-scale steam and hot water production, and since 1958 a power station has been in operation (Figs. 1.5 and 1.6). The total heat output of New Zealand thermal areas (thermal energy above 0°) is 17,000 MW (Bolton, 1975).

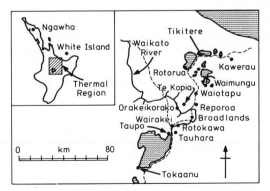

Fig. 1.4 The geothermal areas of the North Island of New Zealand that have been explored by drilling.

Fig. 1.5 View of part of the Wairakei geothermal field showing wellheads and pipelines. (Photo, A. L. Tilbury.)

Wairakei.   Wairakei is one of the largest geothermal areas (about 20 km^2) in the Taupo Volcanic Zone. Approximately 100 20-cm diameter wells have been drilled, mainly varying between 600 and 1200 m in depth (one well reached a depth of 2300 m) (Smith, 1958). They tap liquid water at 230°–260°C and discharge a mixture of steam and water. The steam, which has a low gas content, is separated from the water at the wellheads either by successive U-bend and cyclone separators or simply by cyclone separators, and is piped to the power station for direct use in turbines operating at 7.3, 3.4, and 0.07 bars pressure. From some wells additional steam is obtained from the residual hot water in "flash plants" operating at 4.8 or 1.7 bars. The operating pressures were influenced by an early proposal to run a heavy-water concentration plant in conjunction with power generation.

The Wairakei power station has a generating capacity of 192 MW, and wells withdraw energy from the field at a rate of 5 × 10^5 kcal/sec, compared with the original natural heat flow of about 1.6 × 10^5 kcal/ sec. There is no recovery of chemicals from the steam or from the slightly saline water, and mineral deposition in pipes is not a significant problem.

Kawerau.   The Onepu hot springs area at the northern end of the

Taupo Volcanic Zone was investigated in the late 1950s and subsequently developed to produce steam and electricity for a large wood-pulp and paper industry. Nineteen wells ranging in depth from 430 to 1250 m tap slightly saline water at 250°–280°. Rather low rock permeability causes steam to boil from the hot water in the country under flow conditions, and most wells produce a higher fraction of steam than would be expected from the underground water temperatures. This boiling and concentrating has led to calcite deposition in several well casings. Approximately 200 tons/h of clean process steam is produced in heat exchangers, and surplus geothermal steam is passed into a 10-MW generator (Smith, 1970). A recent well (No. 21) produced an exceptionally high output of steam (190 tons/h) and water (520 tons/h).

Broadlands.   In this area 22 km northeast of Wairakei in the Taupo–Reporoa basin, 31 wells were drilled to depths of between 750 and 1400 m and one well reached a depth of 2420 m. Temperatures ranged up to 300° in the deepest wells, but were usually about 250°–260° at the major production levels of about 750 m. A number of wells had a high discharge rate (over 250 tons) of steam and water per hour (Mahon and Finlayson, 1972). The gas content of the steam separated from the water at 10 bars pressure was moderately high (3–8%) and was mainly carbon dioxide. Further drilling is progressing, and a 150-MW power station is planned.

Fig. 1.6   The Wairakei geothermal power station, which produces 16% of the total electricity generated in the North Island of New Zealand. (Photo, A. L. Tilbury.)

Waiotapu. In the mid-1950s, seven wells were drilled to depths of between 450 and 900 m in this area, about 30 km northeast of Wairakei and in the Reporoa basin (D.S.I.R., 1963). The natural heat escape in the area was 1.3×10^5 kcal/sec. Although some wells tapped high-temperature water (maximum 295°), the area appeared to be underlain by thin aquifers connected by low-permeability paths. The overall low permeability and rather high discharge enthalpies led to problems with calcite deposition in drill pipes.

Other Areas. Several other New Zealand hot spring areas have been drilled to depths of about 1000 m but no immediate developments are proposed (Fig. 1.4). In the Taupo Volcanic Zone the areas include Orakeikorako, Tauhara, Reporoa, and Rotokaua (Smith, 1970). In each case the wells encountered hot water (200°–280°). A 580-m well drilled in sedimentary rocks at Ngawha hot springs, Northland, tapped hot water at about 240° with a high gas concentration (Ellis and Mahon, 1966).

Nicaragua

Two geothermal areas are currently being examined at the Volcano Momotombo and at San Jacinto. In the former area, following resistivity surveys, wells from 500–900 m in depth are being drilled, so far with promising results. Temperatures of the order of 230° were obtained.

Philippines

In southern Luzon, a geothermal power development of up to 200 MW is planned at the Tiwi field, where 11 wells are drilled to an average depth of 1200 m and tap high-temperature (about 250°) water in Quaternary andesites. Construction of the first 110-MW plant commenced in late 1974. In a pilot scheme, a 2.5-kW generator is supplied by steam from a 195-m well and excess steam is used for salt making.

At Mount Makiling in the province of Laguna four production wells have been drilled to an average depth of 1050 m and a power plant is planned.

The Tongonan and Burauen fields in northern Leyte have recently been systematically surveyed, and ten wells drilled in the former area to depths of 228–626 m. The waters tapped were at temperatures up to about 200°, but rock permeabilities were not high and there were problems with calcium carbonate deposition.

At Palimpinon-Dauin in Negros Oriental extensive and promising survey work is to be followed by drilling.

Taiwan

In the Tatun volcanic region of northern Taiwan, wells to depths of 500–1500 m penetrated andesite and underlying sandstone, and tapped waters at temperatures of up to 200° in the Tahuangtsui area and up to 293° at Matsao. In both areas acidic sulfate chloride waters were encountered at depths where sulfur-containing rocks occurred (as deep as 1100 m at Matsao). Most wells of reasonable output discharged acidic waters which rapidly corroded the drill pipes, preventing utilization.

To the southeast, at Tuchang in an area of metamorphic slate rocks, four wells have been drilled, obtaining approximately 180° high-bicarbonate water.

Turkey

At Kizildere, in the Denizli-Sarayköy area of the Menderes Valley, groups of boiling springs occur in an area of Pliocene-Miocene sandstones and limestones, and Paleozoic marbles (see Chapter 2). In a joint program with the United Nations, 14 wells were drilled to depths of up to about 1000 m, encountering waters at 190°–210°. Although some wells had high discharge rates, there were problems arising from calcite incrustations in pipes; also, the water temperatures were rather lower than is desirable for efficient power generation by steam turbines. A 0.5-MW portable generator is in operation in the meantime.

Drilling is now proceeding in other areas (Alpan, 1975). To provide hot water for district heating, wells are being drilled about Ankara and Afyon. In the latter area a 905-m well produced 20 l/sec of 100 °C water from Paleozoic schists and marbles. Hot spring areas at Agamemnon, Izmir, and at Tuzla, Canakkale, both have subsurface temperatures much in excess of 100°.

United States

The Geysers. At The Geysers in Sonoma County, California, the original thermal activity consisted of a few minor steam vents and pools of steam-heated waters (McNitt, 1963) in a northwest-trending graben zone of greywackes, shales, serpentinites, and basalts. Between the years 1921 and 1924, three experimental wells were drilled on the eastern bank of Geyser Creek. Two successful wells to 62 and 66 m produced dry steam at a pressure of 4.1 bars. In 1925–1926 five more wells were drilled, to between 127 and 198 m, in which static pressures of up to 19 bars were measured (Allen and Day, 1927). Shortly after, a small power plant was installed to provide lighting at The Geysers resort. It was not until much later, however, from 1955, that more and deeper wells were drilled, and the installation of the first stage of what is now a major power-producing complex was initiated in 1960.

The drilling operations of several private companies at The Geysers have now revealed a major steam-producing field along a zone about 11 km long covering about 50 km^2. Initial deep wells in the Big Geyser Field were near The Geysers resort, but as development proceeded, wells with a high production of dry steam were located at Sulphur Bank to the northwest and Castle Rock Springs to the southeast (McMillan, 1970).

Over 170 wells with diameters ranging from 20 to 40 cm have been drilled to a maximum depth of 3120 m (but depths are usually 1200–2000 m). Reservoir temperatures are usually in the range of 240°–245 °C, and shut-in pressures of deeper wells are of the order of 31–33 bars. Recent wells have averaged a 50 tons/h output of dry or slightly superheated steam containing about 0.5% gases (Koenig, 1970), with the largest well producing approximately 180 tons/h.

In 1976 there were eleven turbine plants supplied by 75 wells and producing 500 MW of electricity. A further unit will soon increase production to 600 MW and extensions to 900 MW are planned at an early date. Current drilling is constantly extending the size of the productive areas. The total power capacity of the field is rated at 5000 MW over 1000 km^2 (Reed and Campbell, 1975).

The Imperial Valley Region. In the Imperial Valley, a major rift valley system, several hot water geothermal areas have been discovered in deltaic sediments derived from the Colorado River. The thickness of sandstones and shales ranges from about 1500 m in the north to 6000 m at the Mexican border. Three major northwest-trending faults cross the valley, forming a block system. At the southeastern shore of the Salton Sea there was natural surface activity consisting of cool mud pots and a few gaseous springs covering about 30–50 km^2. In the 1930s there was commercial production of carbon dioxide from shallow (100 to 200-m) wells (Helgeson, 1968). Between 1957 and 1963, 13 wells drilled to 2470 m tapped highly saline sodium and calcium chloride brines ranging in temperature from 250° to 350°C. The salinity of waters increased with depth from a few tenths of a percent to over 25% at 1000 m, and the brines were remarkable for their high concentrations of metals such as manganese, zinc, lead, copper, and silver. About eight of the wells individually produced sufficient separated steam to generate between 5 and 10 MW of electricity. Power development and extraction of chemicals from the waters have been considered by several companies, but the difficulty of disposing of waste brine in this low-lying area and uncertain markets for the extracted chemicals curtailed exploitation. At present a local power utility company is assembling a test facility to investigate the scaling problem, heat transfer, and power generation. To the south of Salton Sea in the Brawley area further deep geothermal wells are being drilled.

The East Mesa and Heber geothermal areas at the Mexican border of the Imperial Valley are being developed. At East Mesa five deep wells (maximum 2448 m) tap low-salinity 200°C water, and the U.S. Bureau of Reclamation is investigating a joint water desalination–electricity generation scheme. Three 900-m wells were recently drilled at Heber, where water temperatures are lower (160°–180°) and the use of downhole pumps and a binary cycle generation system would be necessary for power production.

Other Areas. Wells have been drilled in numerous other hot spring and fumarole areas in the United States, mainly in California and Nevada (Anderson and Axtell, 1972). In California these include Casa Diablo in Mono County and Sulphur Bank in Lake County. In the former area, water at a temperature of 178° was found at 320 m (McNitt, 1963) but there were problems with scaling of pipes and with the disposal of waste fluids. Sulphur Bank is an area with a geological setting and thermal waters very similar to those of Ngawha, New Zealand (White and Roberson, 1962). Drilling to 1500 m located water at temperatures in excess of 180° (Koenig, 1970).

A 2150-m well was drilled recently in the Puna area on the island of Hawaii. This produced a high output of steam and water (generation potential 5–10 MW). The deep temperature was about 280° (Hawaii Tribune–Herald, July 23, 1976).

At Steamboat Springs, Washoe County, Nevada, over 30 wells were drilled to depths of up to 980 m (Anderson and Axtell, 1972). Waters at temperatures up to 187° were tapped, but trouble was experienced with calcite deposition in the pipes (White, 1968). Other deep wells in Nevada, at Beowawe and at Brady's Springs, tapped waters with temperatures in the vicinity of 200°.

In Yellowstone Park, Wyoming, during 1967–1968 the U.S. Geological Survey drilled 13 wells for scientific purposes. The measured natural heat flow from the area was approximately 10^6 kcal/sec (Fournier *et al.*, 1975). Temperatures of 200°–237°C were found at depths of up to 332 m, the highest temperature being in Norris Basin (White *et al.*, 1975).

In the early 1960s wells drilled to a maximum depth of 1700 m at Valles Caldera, New Mexico, encountered water at temperatures in excess of 200°.

In the Gulf of Mexico "geopressurized" waters at high temperature have been encountered by oil wells (Jones, 1970). The maximum temperature recorded was 273°C at 5860 m, and the waters contained a high content of dissolved methane. A current study is being made over a 100-mile front. Because of the high confining pressures, the waters also have a potential energy content similar in magnitude to their thermal energy.

In Utah, between the mineral range and Milford graben, the Roosevelt geothermal prospect is being investigated by surveys and drilling, revealing hot water (over 200°) at 1000 m in fractured crystalline rocks.

In many western states, notably California, Oregon, and Idaho, use is made of moderate-temperature waters for space heating and agricultural purposes.

In west-central Alaska, hot springs up to 77° were reported by Miller *et al.* (1975) in areas of granite plutons. High-temperature hot springs also exist throughout the Aleutian volcanic chain of islands.

Soviet Union

Kamchatka. At Pauzhetsk, a large hot spring and fumarole area (350 km²) occurs in an active volcanic region at the northwestern end of the Kambalny Mountains (Piip *et al.*, 1961). Wells drilled to depths of 800 m produced slightly saline waters at temperatures of about 170°–195°, the highest output (70 tons of water and steam per hour at 4 bars pressure) being from fissures in agglomerate andesitic tuff. A power station at Pauzhetsk began production in 1967, developing 5 MW. Expansion, first to 12 MW and then to 22 MW, was planned (Naymanov, 1970). An experimental 0.5-MW power unit was constructed in the late 1960s at Paratunsk, near Petropavlovsk-Kamchatskiy, utilizing hot water (80°–90°) from wells and using freon as a heat exchange fluid (Moskvicheva, 1970). Experimental drillings have also been carried out at Bolshe-Bannoe, where temperatures were about 170°.

Other Areas.   According to Tikhonov and Dvorov (1970), 50–60% of the territory of the Soviet Union contains thermal waters which could be commercially exploitable, and these form energy reserves which are comparable with those of the total coal, gas, and peat reserves. Twenty-eight hot water fields are in operation. In 1970 the value of geothermal energy being utilized (mainly for space heating) was about 10^6 (Makarenko *et al.*, 1970). The use of hot water to produce refrigeration by the lithium bromide absorption process is being investigated for use in areas with hot summers.

Saline waters (3–9 g/l) are associated with the oil-bearing region between the Caucasus Mountains and the Caspian Sea. During deep oil-drilling operations to over 4500 m in the Caucasus region, temperatures of up to 220° were recorded. At Makhachkala in Dagestan, wells drilled to 1200–1500 m produce water at 60°–70° which is used on a large scale for house and hothouse heating. One well produces 80 tons of water per hour. There are several other community heating projects in the region. To the northwest at Grosny, wells produce water with temperatures up to 96° at rates of up to 100 tons/h. The heat is used in a huge greenhouse project. At Prikumsk between Makhachkala and Stavropol, waters with temperatures of about 120° occur at a depth of about 3000 m. In Georgia, in a complex folded and block-faulted region, there are many thermal waters with temperatures of about 100°; for example, at Zugdidi, waters of 80°–100° are used for industrial and community projects.

Thermal waters with moderate temperatures are being used for community heating and assisting agriculture in the cold northeast of the Soviet Union. Hot waters occur at depth in Siberian regions from the north Urals to Magadan. For example, at Chukotsk there are springs issuing at 80°–90° (Tikhonov and Dvorov, 1970), and at Magadan a well produces 85° water which is used for hospital and house heating.

Vast storage of 150° hot water occurs at depths of 2500–3500 m near Panfilov in southeast Kazakhstan, and a 12-MW power station was proposed (*Times Rev. Indus. Technol.*, Feb. 1964, p. 83). At Chimkent hot waters (50°) are used for community heating.

There is also interest in the possibility of utilizing the energy of boiling springs at Khodzhi-Obi-Garm in Tadzhikhstan.

Other Countries

Other countries which are examining the possibilities of further utilization of geothermal energy include Algeria, Colombia, Czechoslovakia, Fiji, Israel, Panama, Peoples Republic of China, Peru, Poland, Tanzania, Thailand, Uganda, and Zambia.

1.4 SUMMARY

The examples given show the rapid upsurge in the use of geothermal energy during the last decade. Now that many of the prospecting and engineering problems of geothermal power development have been solved, geothermal electricity generation is particularly attractive to developing countries because of the favorable economics for small generating units. The costs of steam production are lower than in a conventional fuel-fired station, and capital, installation, and running costs are of a similar magnitude per kilowatt capacity. The economics of geothermal power were summarized by Armstead (1973). The costs of geothermal power are variously quoted to be from 0.3–0.8 U.S. cents per kilowatt-hour for the older and larger units to on the order of 1–1.5 cents per kilowatt-hour for the newer or smaller stations. In the United States these costs compare more than favorably with nuclear-based electric power (Meidav, 1975). In remote areas the alternative is often diesel-based power at about 7 cents/kWh, since transmission costs per kilowatt-hour become very high for small power use and long distances (Golding, 1961).

In spite of these favorable factors, geothermal energy utilization is necessarily limited to parts of the world with the appropriate geological conditions, and these do not always coincide with population centers.

The efficiency of converting geothermal energy into electrical energy is rather low because of the relatively low temperatures of geothermal steam. From thermodynamics, the limiting conversion efficiency is about 30%. For dry steam geothermal fields, the actual efficiency may be 15%; for hot water fields it may be 8–10%. Because of this, increasing attention is being given in many countries to the direct use of geothermal steam or hot water for industrial or domestic heating, in which a major proportion of the thermal energy is usefully employed (Howard, 1975). With modern pipeline techniques hot water can be transmitted long distances (up to 100 km) at an economic price for household heating (Haseler, 1975).

From time to time geothermal energy has been promoted as "pollution free." However, the accelerated production of natural fluids can cause local environmental problems; for example, undesirable elements such as arsenic and boron can be liberated into waterways, or sulfur gases into the atmosphere, and slight earth subsidence may occur. The situation was reviewed by Axtmann (1975). Fortunately, most of the problems are minor and can be solved with known technology.

With over 50 countries currently displaying interest in geothermal energy utilization, the proportion of the world's energy needs being met by geothermal heat resources will undoubtedly rise. In only a few countries will geothermal plants provide a major fraction of the total energy requirements, but they will continue to be one of the most economical means of production.

In spite of current research on the extraction of useful heat from sources such as hot dry rock or the molten lava of basaltic volcanoes, it is considered that in the next decade useful geothermal production will be predominantly from fields such as those at present utilized and through only minor modifications of present technology. With the accelerating pace of geothermal exploration, many new deposits of natural hot water will undoubtedly be utilized.

REFERENCES

Allen, E. T., and Day, A. L. (1927). Steam Wells and other thermal activity at The Geysers California. Carnegie Inst. Wash. Publ. No. 378.

Alpan, S. (1975). *Proc. U.N. Symp. Develop. Use Geothermal Resources, 2nd, San Francisco, California, May 1975* **1**, 25.

Anderson, D. N., and Axtell, L. H. (1972). Geothermal Overviews of the Western United States. Geothermal Resources Council, Davis, California.

Armstead, H. C. H. (1973). *In* "Geothermal Energy" (H. C. H. Armstead, ed.), pp. 161–174. UNESCO Earth Sci. Ser. No. 12, Paris.

Arnorsson, S., and Sigurdsson, S. (1974). *Geothermics* **3**, 127.

Axtmann, R. C. (1975). *Science* **187**, 795.

Banwell, C. J., and Gómez Valle, R. (1970). *Geothermics (Spec. Issue 2)* **2** (Pt 1), 27.

Barth, T. F. W. (1950). Volcanic Geology, Hot Springs and Geysers of Iceland. Carnegie Inst. Wash. Publ. 587.

Behl, S. C., Jagadeesan, K., and Reddy, D. S. (1975). *Proc. U.N. Symp. Develop. Use Geothermal Resources, 2nd, San Francisco, California, May 1975* **3**, 2083.

Bodvarsson, G. (1961). *Proc. U.N. Conf. New Sources Energy, Rome, 1961* **2**, 82.

Bodvarsson, G. (1970). *Geothermics (Spec. Issue 2)* **2** (Pt 2), 1289.

Boldizsar, T. (1970). *Geothermics (Spec. Issue 2)* **2** (Pt 1), 99.
Bolton, R. S. (1975). *Proc. U.N. Symp. Develop. Use Geothermal Resources, 2nd, San Francisco, California, May 1975* **1**, 37.
Bruce, J. A., and Shorland, F. B. (1932). *N.Z. J. Agr.* **45**, 272; **46**, 29.
Bullard, E. (1973). *In* "Geothermal Energy" (H. C. H. Armstead, ed.), pp. 19-29. UNESCO Earth Sci. Ser. No. 12, Paris.
Cataldi, R. (1967). *Bull. Volcanol.* **30**, 243.
Cataldi, R., Ceron, P., Di Mario, P., and Leardini, T. (1970). *Geothermics (Spec. Issue 2)* **2** (Pt 1), 77.
Ceron, P., Di Mario, P., and Leardini, T. (1975). *Proc. U.N. Symp. Develop. Use Geothermal Resources, 2nd, San Francisco, California, May 1975* **1**, 59.
De Anda, L. F., Septien, J. I., and Elisondo, J. R. (1961). *Proc. U.N. Conf. New Sources Energy, Rome, 1961* **2**, 149.
D.S.I.R. (1963). Waiotapu Geothermal Field, New Zealand. Dept. Sci. and Ind. Res. Bull. 155.
Einarsson, S. S. (1973). *In* "Geothermal Energy" (H. C. H. Armstead, ed.), pp. 123-134. UNESCO Earth Sci. Ser. No. 12, Paris.
Ellis, A. J., and Mahon, W. A. J. (1966). *N.Z. J. Sci.* **9**, 440.
Facca, G. (1970). *Geothermics (Spec. Issue 2)* **1**, 8.
Fournier, R. O., White, D. E., and Truesdell, A. H. (1975). *Proc. U.N. Symp. Develop. Use Geothermal Resources, 2nd, San Francisco, California, May 1975* **1**, 731.
Golding, E. W. (1961). *Proc. U.N. Conf. New Sources Energy, Rome* **1**, 149.
Guiza, J. (1972). *In Compendium First Day Papers, Conf., El Centro* (D. N. Anderson and L. H. Axtell, eds.), pp. 39-44. Geothermal Resources Council, Davis, California.
Haseler, A. E. (1975). District Heating: An Annotated Bibliography. Property Service Agency, Dept. of the Environment, London.
Hayakawa, M., Takaki, S., and Baba, K. (1967). *Bull. Geol. Surv. Jpn.* **18**, 147.
Helgeson, H. C. (1968). *Am. J. Sci.* **266**, 129.
Howard, J. H. (1975). *Proc. U.N. Symp. Develop. Use Geothermal Resources, 2nd, San Francisco, California, May 1975* **3**, 2127.
Jones, P. H. (1970). *Geothermics (Spec. Issue 2)* **2**(Pt 1), 14.
Katagiri, K. (1975). *U.N. Symp. Develop. Use Geothermal Resources, 2nd, San Francisco, California, May 1975*, Abstract IV-25.
Kerr, R. N., Bangma, R., Cook, W. L., Furness, F. G., and Vamos, G. (1961). *Proc. U.N. Conf. New Sources Energy, Rome*, **3**, 456.
Koenig, J. B. (1970). *Geothermics (Spec. Issue 2)* **2**(Pt 1), 1.
Lahsen, A., and Trujillo, P. (1975). *Proc. U.N. Symp. Develop. Use Geothermal Resources, 2nd, San Francisco, California, May 1975* **1**, 170.
Lindal, B. (1961). *Proc. U.N. Conf. New Sources Energy, Rome* **3**, 471.
Lindal, B. (1970). *Geothermics (Spec. Issue 2)* **2**(Pt 1), 910.
McMillan, D. A. (1970). *Geothermics (Spec. Issue 2)* **2**(Pt 2), 1705.
McNitt, J. R. (1963). Exploration and Development of Geothermal Power in California. Spec. Rep. California Div. Mines Geol., No. 75.
Mahon, W. A. J., and Finlayson, J. B. (1972). *Am. J. Sci.* **272**, 48.
Makarenko, F. A., Mavritsky, B. F., Lokshin, B. A., and Kononov, V. I. (1970). *Geothermics (Spec. Issue 2)* **2**(Pt 2), 1086.
Meidav, T. (1975). *U.N. Symp. Develop. Use Geothermal Resources, 2nd, San Francisco, California, May 1975*, Abstract X-10.
Mercado, S. (1967). Geoquimica hidrotermal en Cerro Prieto, B.C., Mexico. Comision Federal de Electricidad, Mexicali.

Miller, T. P., Barnes, I., and Patton, W. W. (1975). *J. Res. U.S. Geol. Surv.* **3**, 149.

Minucci, G. (1961). *Proc. U.N. Conf. New Sources Energy, Rome* **3**, 223.

Moskvicheva, V. N. (1970). *Geothermics (Spec. Issue 2)* **2** *(Pt 2)*, 1567.

Nakamura, H. (1959). *J. Jpn. Assoc. Miner. Petrol. Econ. Geol.* **43**, 158.

Nakamura, H. and Sumi, K. (1967). *Bull. Geol. Surv. Jpn.* **18**, 58.

Nasini, R. (1930). I Soffione e i Lagoni della Toscana e la Industria Boracifera. Tipografia Editrice Italia, Rome.

Naymanov, O. S. (1970). *Geothermics (Spec. Issue 2)* **2** *(Pt 2)*, 1560.

Noguchi, T. (1966). *Bull. Volcanol.* **29**, 529.

Penta, F. (1954). *Ann. Geofis.* **8**, 1.

Piip, B. I., Ivanov, V. V., and Averiev, V. V. (1961). *Proc. U.N. Conf. New Sources Energy, Rome* **2**, 339.

Ragnars, K. K., Saemundsson, K., Benediktsson, S., and Einarsson, S. S. (1970). *Geothermics (Spec. Issue 2)* **2** *(Pt 1)*, 925.

Reed, M. J., and Campbell, G. E. (1975). *Proc. U.N. Symp. Develop. Use Geothermal Resources, 2nd, San Francisco, California, May 1975* **2**, 1399.

Saito, M. (1961). *Proc. U.N. Conf. New Sources Energy, Rome* **2**, 367.

Sato, K. (1970). *Geothermics (Spec. Issue 2)* **2***(Pt 1)*, 155.

Shimozura, D. (1968). *Bull. Volcanol.* **32**, 383.

Smith, J. H. (1958). *N.Z. Eng.* **13**, 354.

Smith, J. H. (1970). *Geothermics (Spec. Issue 2)* **2***(Pt 1)*, 232.

Thorsteinsson, T. (1975). *Proc. U.N. Symp. Develop. Use Geothermal Resources, 2nd, San Francisco, California, May 1975* **3**, 2173.

Tikhonov, A. N., and Dvorov, I. M. (1970). *Geothermics (Spec. Issue 2)* **2***(Pt 2)*, 1072.

Villa, F. (1961). *Proc. U.N. Conf. New Sources Energy, Rome* **3**, 416.

White, D. E. (1968). Hydrology, Activity and Heat Flow of the Steamboat Springs Thermal System, Washoe County, Nevada. Prof. Pap. U.S. Geol. Surv. 458-C.

White, D. E., and Roberson, C. E. (1962). *In* "Petrological Studies: A Volume to Honor A. F. Buddington," pp. 397–428. Geol. Soc. Am.

White, D. E., Fournier, R. O., Muffler, L. J. P., and Truesdell, A. H. (1975). *U.N. Symp. Develop. Use Geothermal Resources, 2nd, San Francisco, California, May 1975,* Abstract II-56.

Chapter 2

NATURAL HYDROTHERMAL SYSTEMS

2.1 GENERIC TYPES OF WATER

Certain generic terms referring to various types of water are used in the discussion and should be explained at the outset. The definitions are adapted from White (1957a,b).

Juvenile water. "New" water derived from primary rock magma, and which has not previously been part of the hydrosphere.

Magmatic water. Water derived from magma, but not necessarily juvenile water, since magma may incorporate meteoric water of deep circulation or water from sedimentary material.

Meteoric water. Water recently involved in atmospheric circulation.

Connate water. "Fossil" water which has been out of contact with the atmosphere for geologically long periods. The water enclosed by deep rock formations. In sedimentary basins, connate waters such as oil-field brines may commonly have originated from ocean water, but are much altered by chemical and physical processes.

Metamorphic water. A special modification of connate water, derived from hydrous minerals during their recrystallization to less hydrous minerals during metamorphic processes.

2.2 DISTRIBUTION OF HYDROTHERMAL AREAS

Most countries have springs with water at temperatures appreciably above the average atmospheric temperature. Water at temperatures approaching 100°C occurs within 1–2 km of the surface in many regions, but areas with water at temperatures in excess of about 150°C at these depths occur only in special geological situations. Regions of high heat flow (several times the average 1.4×10^{-6} cal cm^{-2} sec^{-1}), in which the major high-temperature geothermal systems occur, are often associated with zones of volcanism and mountain building at the edge of crustal plates. Molten mantle material is added to the crust in zones of plate divergence, as in the mid-Atlantic, where geothermal systems occur both on land (Iceland) and on the sea floor (Fyfe, 1977). Further examples are the rift zones of Southern California/Mexico and of East Africa, where many geothermal areas occur in Uganda, Kenya, and Ethiopia. Where plates meet, crustal material may be forced down to great depths beneath a continental mass and become melted. The geothermal areas associated with the andesitic volcanism of the Circum-Pacific belt occur in this type of situation.

High-temperature hydrothermal systems also occur in a zone stretching from Italy, Greece, and Turkey through the Caucasus to the Himalayas and southwest China, associated with major tectonic activity and mountain building at the junctions of the African, European, Asian, and Indian crustal plates.

Tamrazyan (1970) noted that the separation of the Siberian and the European plates created a wide rift zone which was subsequently filled with thick sedimentary beds. An unusually high heat flow insulated by sedimentary structures created many of the important hot water resources of the Soviet Union. The hot water resources within sediments of much of the Texas and Louisiana areas of the Gulf of Mexico basin and in the Cambay basin of India may prove to be of equal significance.

2.3 PERSISTENCE OF HYDROTHERMAL SYSTEMS

Many natural hydrothermal systems have been in operation for at least 10^4–10^5 years. Grindley (1965) showed that there was hydrothermal activity at Wairakei, New Zealand, 500,000 years ago, while Barth (1950) considered that the age of the Great Geysir, Iceland, was at least 10,000 years. The Steamboat Springs, Nevada, system was given an age of the order of 10^6 years by White (1974). However, direct quantitative information on the trends with time in the compositions and temperatures of hot springs is available only over the last hundred or so years. Before this it is necessary to rely on the qualitative observations in historical records, or on indirect geological evidence such as fluid inclusions within hydrothermal minerals.

As reviewed by Waring *et al.* (1965), hot springs at Bone in Algeria supplied baths in ancient times and probably have not changed by more than about 4°C in 2000 years. The hot springs of Tiberius near the Sea of Galilee were presumably also known in biblical times, since the town name Hammath means warm springs, and the name Emmaus means hot springs. Near Reykjavik, Iceland, the Thvottalaugar hot springs were in existence in the ninth century when a colonizing expedition arrived. Piip (1937) gave records of the temperatures of springs in Kamchatka from the early 1700s, showing less than 1°–2°C change since those times. More exact temperature and chemical records from the United States, New Zealand, and Iceland are shown in Table 2.1. Within the accuracy of the measuring techniques, the changes are negligible over a period of about 100 years.

Considerable rates of water flow exist in many areas; for example, the outflow of Iceland hot springs totals 1000 l/sec; Yellowstone Park springs 3000 l/sec; and Wairakei, New Zealand, 400 l/sec. One spring at Deildartunga, Borgarfjardarsysla, Iceland, has an output of 250 l/sec of near-boiling water (Barth, 1950). From many geothermal areas the accumulated outflow of water must be estimated in terms of cubic kilometers. This indicates the size and capacity of natural hydrothermal systems. The large outflow of dissolved materials from hot spring systems is also evident when it is considered that a cubic kilometer of average hot spring water may contain several million tons of dissolved salts.

2.4 CHARACTERISTICS OF GEOTHERMAL FIELDS

Geothermal fields occur in a wide variety of geological environments and rock types. The hot water geothermal fields about the Pacific basin, such as those in Japan, New Zealand, Indonesia, and Chile, are in zones of predominantly rhyolitic or andesitic volcanism. In contrast, the widespread hydrothermal activity in Iceland occurs in extensively fractured and predominantly basaltic rocks, although rhyolitic rocks occur in several of the areas. In other Cenozoic tectonic regions, geothermal fields occur in many types of metamorphic and sedimentary rocks. The Larderello steam fields in Italy are in a region of metamorphic rocks, dolomite, limestone, marble, and shale, whereas The Geysers steam field in California is largely in fractured greywacke. Both the Cerro Prieto field in Mexico and the geothermal areas of the Imperial Valley in Southern California occur in deltaic river sediments associated with nearby rhyolite volcanic centers.

TABLE 2.1

Comparison of Solute Concentrations (in ppm) in Spring Waters at Different Dates[a]

Spring areas	Date	Na^+	K^+	Mg^{2+}	Ca^{2+}	Cl^-	SO_4^{2-}	SiO_2
Iceland								
Hveragerdi	1847	230	12	trace	0	172	—	320
	1948	—	—	0.8	2.5	186	—	386
Geysir	1847	254	8	2	0	144	108	519
	1960	252	22	—	0.2	131	100	525
Reykjavik	1847	62	—	trace	0	30	15	135
	1958	70	—	—	3	30	17	123
New Zealand								
Champagne Pool,	1884	1120	120	18	68	1940	31	340
Wairakei	1959	1170	153	3	32	1999	146	—
Kuirau, Rotorua	1878	317	16	4	2	410	99	287
	1962	330	32	<0.5	<0.5	326	45	318
Yellowstone Park								
Old Faithful Geyser	1884	367	27	0.6	1.5	439	18	383
	1935	372	31	0	4	435	21	357
	1961	326	22	<0.1	0.7	434	45	390
Excelsior Geyser	1884	419	33	2	2	279	18	221
	1935	410	14	0	3	271	21	237
	1960	385	15	<0.1	<0.1	270	34	303

[a] Analyses: New Zealand, Chemistry Division, D.S.I.R.. Iceland, Barth (1950); Chemistry Division; and Dr. G. Bodvarsson (personal communication). Yellowstone, Gooch and Whitfield (1888): Allen and Day (1935); and U.S. Geological Survey.

Similar types of rock environments may produce different types of geothermal fields; for example, at Larderello, Italy, wells produce mainly dry steam, whereas at Kizildere, Turkey, in a zone of similar rocks, deep wells tap high-temperature water. The processes which determine whether a geothermal field produces steam or hot water are examined later in this chapter (Section 2.6).

It is difficult to define a typical geothermal field, since each has its own characteristics. Nevertheless, areas have points in common. As pointed out by Banwell (1970), the known high-temperature geothermal fields are often associated with Quaternary or Recent volcanic activity, with faulting, graben formation, tilting, and commonly, nearby intrusions of rhyolitic rocks.

The way in which geothermal systems operate is now examined in more detail, noting both the observable features and the processes operating in various types of systems. Specific fields are described as examples.

The difficulty in categorizing geothermal systems arises from the early stage of knowledge of the processes operating. A useful classification should relate to a field situation. McNitt (1970), after reviewing various attempts at classification, proposed one based on geological processes and the positions of fields with respect to orogenic and volcanic belts.

For convenience in discussion the various hydrothermal systems are conveniently (if somewhat loosely) classified into two major types, cyclic and storage systems.

Cyclic systems. The hot water is meteoric water which has passed through a cycle of deep descent, heating, and rising. Convective forces are important in these systems. Isotopic analysis of hydrogen and oxygen in the hot waters or steam (Craig, 1963) demonstrates their predominant local surface water origin, but minor additions of other waters are not excluded. The cycle may not be completed continuously, and temporary storage of hot water or steam may occur in suitably sealed permeable horizons.

Storage systems. The water is stored in the rocks for geologically long periods and heated *in situ*, either as a fluid within the formation, or as water of hydration in minerals.

Since an aquifer in a sedimentary basin may be part of a regional artesian system, there is clearly a transition between the storage and cyclic types of systems. The storage type stresses the small inflow or outflow of the system, and the probability that the waters may contain an appreciable proportion of connate water from sediments.

Geothermal systems are referred to as *open* systems when little or no sealing rock formation exists at present near the surface, and *closed* systems when a low-permeability cap-rock structure prevents the ready outflow of water or steam. In different geothermal fields various types of cap rock include igneous rock flows, fine-grained sedimentary rocks, and rocks whose permeabilities have been decreased by mineral deposition from hot waters. The presence of one or more cap rocks often causes lateral spread and accumulation of hot water in permeable formations, and may create large hot water aquifers like those found at Wairakei and Broadlands in New Zealand. In contrast, at Orakeikorako, where there is no cap-rock structure and the permeable aquifers outcrop at the surface, temperatures were lower than in other New Zealand geothermal areas and the outputs of wells did not encourage development.

Cyclic Systems

The formation of this type of hydrothermal system requires (1) suitable rock formations which allow water to circulate to deep levels; (2) a source of heat; (3) an adequate availability of water; (4) sufficient time and surface area of heat exchange to enable the water to be heated; and (5) a return path toward the surface.

Cyclic hydrothermal systems reflect the general tendency for precipitation in every catchment to circulate down and through the rocks to depths controlled by the local structure. Warm springs occur in most countries where water reaches the surface again after circulating to deep levels and receiving heat from the rocks, but it is rare for the spring water to attain boiling temperatures. The latter situation is limited mainly to zones of recent volcanic or tectonic activity.

Cyclic systems can be subdivided as follows: (a) high-temperature systems associated with recent volcanism; (b) high-temperature systems in nonvolcanic zones of Cenozoic tectonic activity; (c) warm water systems in near-normal heat flow zones.

(a)   High-Temperature Systems Associated with Recent Volcanism

These systems occur in a variety of situations. They are more often associated with areas of andesitic, dacitic, and rhyolitic rocks than with basaltic eruption centers (McNitt, 1970). Many geothermal fields have structures produced by tectonic activity, such as block faulting, graben formation, or rift valleys, but have no obvious relationship to a particular volcanic center. Locations which are particularly favorable are at the intersection of faults bordering major structural blocks. The majority of the New Zealand geothermal fields are situated in major graben structures, as are the fields at the Salton Sea, California, and Cerro Prieto, Mexico. Several fields are associated with volcanic caldera structures (e.g., Matsukawa, Pauzhetsk, Valles Caldera, and Ahuachapan), whereas others are related to specific volcanoes (e.g., Momotombo, Nicaragua; Kawah Kamojang, Indonesia).

The general operation of hydrothermal systems of this type is given in Fig. 2.1. Waters are derived from local meteoric waters which may circulate to considerable depths (many kilometers) through fault–fissure systems, become heated, and rise again through convective

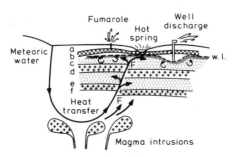

Fig. 2.1 Schematic diagram of the operation of a typical cyclic hydrothermal system. a–b is a fine-grained sedimentary cap rock; b–c and d–e, porous volcanic breccia or ash; c–d and e–f, impermeable volcanic rock, lava flows, or ignimbrite. F–F is a major fault zone, and w.l. is the steam–water interface level.

forces. Major upflow routes are commonly through faults and fissured zones caused by magma intrusions. In porous horizons hot water may spread out for considerable distances. At shallow levels in the system there may be convective recirculation of waters which have been cooled by boiling under the lower-pressure conditions, while mixing with water outside the thermal system occurs on the margins at all levels. Frequently a cap rock may limit the output of fluid and heat, but leakages of water give rise to hot springs at lower ground levels, and steam leakages create fumaroles at higher levels or shallow steam-heated groundwaters. Under natural conditions there may be a sub-surface steam–water interface which is depressed locally when a well provides a high-permeability route to the surface. The proportions of steam and water discharged by the well vary according to the local interface level.

The heating of water at depth is usually attributed to magma intrusion, or intrusions, with heat being conducted through a zone of solidified rock of unknown thickness about the magma. Heat transfer could be aided by cracking of the solidified crust by thermal stresses. Elder (1965) considered that to provide the sustained local heat flows observed in high-temperature geothermal areas would require convection within a magmatic body. Alternatively, a sequence of magma intrusions could maintain heat flows.

High-temperature fluids may also be expelled from a magma body. Mahon and McDowell (1977) considered that the expansion of high-temperature and high-pressure magmatic fluid would create a dense highly saline solution at subcritical-point temperatures, plus a wet steam phase which in most systems would rise to merge into circulating meteoric waters. A. McNabb (personal communication) suggested that a high-density salt brine of this type would form a good heat-conducting medium between the magma and water.

Some of the difficulties of a continuous heat flow model requiring constant heat flow from low thermal conductivity rock are avoided if it is considered that the water turnover time in geothermal systems is very long (of the order of 10^4–10^5 years), and the outflow intermittent. Systems could then have long periods of conductive heating of water with little outflow, followed by comparatively brief periods (of order of 10^3–10^4 years) of high flow when new channels to the surface are formed. This may be triggered by tectonic activity, or by hydrothermal explosions should the temperature cause steam pressures to exceed the lithostatic pressure. After a period of flow, the channelways may become sealed with deposits of silica and calcite.

There is some field evidence for this proposition. Extensive silica terraces in some areas give evidence of larger water flows in the past. Experience in New Zealand and overseas has shown that there is little relationship between the natural rate of water outflow from springs or fumaroles and the size of a geothermal system as revealed by drilling.

In high-temperature geothermal fields in high-porosity rocks (e.g., Wairakei), it has been suggested that hot water may rise from considerable depths (several kilometers) as a narrow plume surrounded by comparatively cold water, the hot water spreading out laterally in shallow permeable strata (Elder, 1965). The mushroom-shaped body of high-temperature water (Fig. 2.2) would mean that deep geothermal production wells are restricted to a comparatively

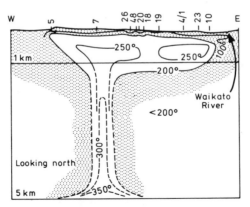

Fig. 2.2 Mushroom-shaped profile of high-temperature water beneath the Wairakei field, according to Elder (1965). Temperature contours in solid lines are as measured; broken lines, hypothetical. Numbers show well positions.

small area, whereas the positioning of shallower wells is less critical. Horizontal, near-surface flows of hot water occur over considerable distances in several geothermal fields (e.g., El Tatio, Chile). Donaldson (1970) gave a mathematical analysis of a simple convective geothermal model with variations in permeability at different points in the cycle.

Chapter 9 summarizes the characteristics of some of the geothermal areas in the New Zealand volcanic region. Brief descriptions are given of the Otake field of Kyushu, Japan, and the Mexicali field of Mexico, as further examples of high-temperature geothermal fields in recent volcanic areas.

The Otake Geothermal Field. The Otake geothermal field is 6 km northwest of Mount Kujyu-zan about halfway between Mount Aso Caldera and Beppu. It is about 900–1100 m above sea level and is dissected by the Kusu River. Geophysical surveys suggest the area to be at least 1 km wide from east to west and about 3–4 km from north to south. Natural activity at Otake consists of hot springs and fumaroles, while to the south at Hatchobaru there are steam fumaroles. The integrated natural heat output of the Otake–Hatchobaru–Komatsu areas was about 600 kcal/sec (Fukuda *et al.*, 1970).

The regional structure has the form of a caldera and the most prominent northwest fault across the area cutting the caldera flank is the location of several springs and fumaroles, and is intersected by a well. Much of the natural activity is closely related to many northwest-southeast faults.

Figure 2.3 shows a cross section of the area (Yamasaki *et al.*, 1970). The area is one of andesitic volcanism, and the Quaternary activity is divided into the middle Pleistocene Kujyu complex, with many lava dome volcanoes of viscous hornblende andesites, and the lower Pleistocene Hohi complex, about 1000 m thick, consisting of pyroxene

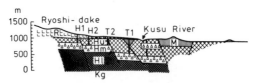

Fig. 2.3 Geological cross section of the Otake geothermal field. Rocks of the Kujyu volcanic complex are R, Ryoshi-dake lava, and M, Misokobushiyama lava. The Hohi volcanic complex is represented by the Hu (upper), Hm (Middle), and Hl (Lower) formations, and the Kusu group of sediments by Kg. The positions of some of the wells are shown. From Yamasaki *et al.* (1970).

andesites and lava and tuff breccias. Underlying the Hohi complex is the Miocene Kusu group of sediments, with alternating tuffs, sandstone pebbles, and mudstone, which is considered to form a deep permeable aquifer. However, tuff breccias in the middle formation of the Hohi complex are the main reservoir for production wells supplying the power station. Production is often good at depths of 200–400 m at the top of the permeable tuff breccia, and through fissures and cracks in overlying compact andesite lava, which forms a cap rock. The wells produce steam and water, the water being of neutral pH and moderate salinity (Table 2.2).

The maximum temperatures encountered by drilling in the field increased from north to south. In the northern Otake production area maximum water temperatures in wells ranged from 215° to 250°. Three kilometers south at Hatchobaru, the temperatures of water tapped by wells ranged up to 290°.

Strongly hydrothermally altered rocks are associated with faults, fissures, and joints, which are considered to form the main passages for geothermal fluids.

The hydrothermal alteration of the andesites has several zones with increasing depth. At the surface, zones of kaolin and alunite occur where there have been acid hydrothermal solutions. In the southern Komatsu–Hatchobaru part of the field, acid alteration persists, with kaolin and pyrophyllite to considerable depths of 600–700 m, indicating acid conditions. In agreement with this, Well H-2 at Hatchobaru discharged an acid sulfate–chloride water. In the northern Otake production area a zone of montmorillonite–calcite–gypsum–quartz–mica alteration underlies the acidic zone, and at deeper levels there is a zeolite zone, reflecting neutral-pH conditions. The latter has the mineral assemblage wairakite–laumontite–montmorillonite–sericite–calcite–quartz–pyrite, and has probably formed from highly porous tuff breccias. It is recognized in all the production wells in the northern Otake area. At deep levels in compact andesite the alteration mineral assemblage is chlorite–calcite–quartz–mica–epidote–pyrite.

The Mexicali Geothermal Field. The Cerro Prieto area near Mexicali was described by Alonso (1966) and by Banwell and Gomez Valle (1970). It is situated on the landward extension of the Gulf of California rift zone, which also continues farther northward into the Imperial Valley. Its structure is shown in Fig. 2.4, from Mercado (1967).

A thick deposit of deltaic sediments with successive horizons of sand, mud, and clays occurs within the Mexicali valley. To the west

TABLE 2.2

Water Analyses for Various Types of Fields[a]

Component[b]	(a)	(b)	(c)	(d)	(e)	(f)	(g)	(h)
Temp.	248°	350°	—	200°	73°	~100°	85°	69.5°
pH	8.15	6.40	—	9.0	7.65	5.70	—	6.8
Li	5.2	15.5	—	4.5	3.3	15	—	4.4
Na	936	6429	82	1280	1718	13600	[c]	1190
K	131	1176	35	135	104	404	[c]	23
Rb	2.4	—	—	0.0	—	0.6	—	—
Cs	0.84	—	—	0.33	—	<0.5	—	—
NH$_4$	0.06	—	—	2.6	0.1	134	—	464
Ca	12.3	347	372	2.5	102.5	12200	6.5	20
Mg	0.19	18.6	305	0.2	46.5	275	2.2	55
Fe	0.03	—	23	0.09	0.1	0.1	—	0.0
Mn	0.01	—	—	0.0	—	7.0	—	0.1
F	4.65	—	—	23.7	2.4	1.6	—	1.0
Cl	1474	11735	7.5	117	617	44000	23	644
Br	3.4	—	—	1.2	1.4	238	—	1.6
I	0.26	—	—	—	—	56	—	3.2
SO$_4$	136	15	2058	770	1662	16	23	598
HCO$_3$	46	303	—	1860	2100	80	1575	3290
B	19.7	11.6	780	26.2	1.0	50	—	620
SiO$_2$	665	1133	109	325	68	63	—	42
Reference	Koga (1970)	Molina and Banwell (1970)	Nasini (1930)	Chemistry Division	Ovchinnikov (1955)	White (1965)	Boldizsar (1970)	White et al. (1973)

[a] Key: (a) well 9 (550 m), Otake, Japan; (b) well 8 (1300 m), Mexicali; (c) Larderello steam-heated pool; (d) well 1A (430 m), Kizildere, Turkey; (e) main spring, Carlsbad, Czechoslovakia; (f) oil well (3650 m), Leda formation, Kings County, Calif.; (g) well, Szentes, Hungary; (h) Geyser Spring, Sulphur Bank, California.

[b] Concentrations are in ppm (mg/kg) in waters collected at atmospheric pressure, and pHs are as measured in cooled waters.

[c] Combined Na and K concentration; 595 ppm.

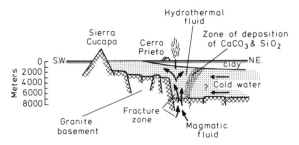

Fig. 2.4 Schematic cross section of the Mexicali geothermal field, from Mercado (1967).

this sandwich-type sequence is terminated by an outcrop of the granitic basement at Sierra Cucapa, whereas the sediments stretch east to the Colorado River. The Cerro Prieto volcano within the field is of basalt and pyroclastics, with associated fumarolic activity to the southwest. The regional structure corresponds to a tectonic trench which rises in a series of stepped blocks toward Sierra Cucapa to the west and descends abruptly to the east of Cerro Prieto where the basement crystalline rocks are displaced downward by about 3000 m. The basement is broken by a series of northwest-southeast faults which provide upflow channels for high-temperature water.

The geothermal system is capped by approximately 700 m of plastic clay which acts as a barrier, forcing the hot fluids to flow horizontally away from the fractured upflow zones. Hot water flows mainly toward the west due to flat stratification and good permeability in the sandstone. To the east the sedimentary strata are more compacted, and the lower permeability is considered to be due to precipitation of calcite and silica in a zone of interaction between the hot fluids and a cold water front. This type of self-sealing may be common in high-temperature geothermal systems.

An inflow of colder water at depths of 1000–2000 m causes cooler temperatures at deeper levels in wells on the eastern and western edges of the field.

Figure 2.5, from Mercado (1970), shows an interpretation of water flow patterns and temperatures in the central part of the production field. Temperature differences are shown in terms of the Na/K ratio in the hot waters, which shows an inverse relationship to temperature (Chapter 4). The highest temperatures in the central upflow zone are approximately 375°, while temperatures fall rapidly to the northeast to 150°, and less rapidly to the southwest where high-temperature water sweeps through the sandstone layer.

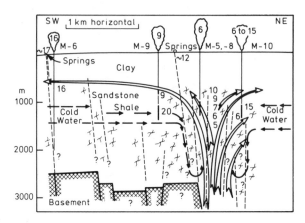

Fig. 2.5 An interpretation of water flows in the Mexicali field (from Mercado, 1970). Small heavy arrows show cold water inflows; large open arrows show the directions of hot water flow. Wells are designated by numbers such as M-9, and the atomic ratios of Na to K in hot waters by simple numbers such as 7, 9, 11 (see Chapter 4).

Because of the exceptionally high permeability in the area, wells have a very high output. The waters tapped are highly saline (15,000–17,000 ppm total solids; example (b), Table 2.2) but this is an order of magnitude less than the concentrations in the waters of the Salton Sea field in the Imperial Valley to the north. The high salinities may arise from the solution of evaporite sequences in the sediments filling the graben structure. The source water for the geothermal system is considered to be to the northeast in the direction of the Colorado River (Mercado, 1970).

(b) High-Temperature Systems in Nonvolcanic Zones of Cenozoic Tectonic Activity

Examples of this type of geothermal field have been examined by deep drilling for power development projects. The Larderello region of Tuscany, Italy, is the best known, and its geological structure is briefly summarized here. The structure of the Kizildere field in the Menderes Valley of Turkey is also outlined.

The Larderello Region, Italy. The Larderello region has many hills with peaks reaching altitudes of 1000 m and forming part of a 50-km-long range called the Metalliferous Hills. The region is characterized by the occurrence of small islands of Paleozoic schists and quartzites and Mesozoic limestones, outcropping through a sedimentary cover of clays, shales, and sandstones.

The structure of the region results from compression which formed elongated folds parallel to the Apennine Range (Burgassi, 1964) and faults converging at depth and parallel to the axis of the folds. Considerable masses of rock were also forced downward to depths of sufficiently high temperature to form a granitic magma. At a later stage of crustal relaxation the granite magma rose near, or to, the surface. Schistose clays were formed and spread by overthrusting along the flanks of the folds during upheaval. These clays partially covered and submerged the folds, especially after a phase of subsidence. During a marine transgression in the Upper Miocene and Pliocene, a series of sands, clays, limestones, and conglomerates were formed. Post-Pliocene emergence was accompanied by further faulting, and the last movements are of Quaternary age (Burgassi, 1964).

Marinelli (1969) considered that Larderello constitutes a graben formed at the crest of a dome, caused by the intrusion of granite beneath the area. The vast gravity low could be attributed to the presence of a magmatic body at a depth of 6–8 km. The presence of a rising intrusive body could create fault and fracture systems which would enable the transfer upward of considerable heat over long periods of time through convective water movements (Marinelli, 1969).

Figure 2.6 shows a schematic cross section of Larderello (ENEL, 1970). In summary, the stratigraphic sequence consists of faulted and folded schists and quartzites which form the ancient crystalline basement of the region. These formations are overlain by the main pervious complex, which at Larderello is often thinned to about 50–100 m. It contains Triassic anhydrite-bearing evaporite formations which have been partly transformed by the preferential solution of calcium sulfate to an open structure cemented by calcite, and with veins of celestite. In places such as this, which frequently correspond to minor anticlines, the evaporite series often is covered only by

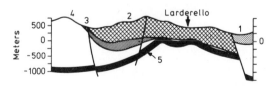

Fig. 2.6 A schematic cross section of the Larderello region (from ENEL, 1970). Rocks designated 1 are Pliocene–Upper Miocene clays, sandstones, conglomerates; 2, "Argille scagliose" (shales, marbles, limestones of Eocene–Jurassic age); 3, Oligocene–Cretaceous shales; 4, a Jurassic–Upper Triassic evaporite series; and 5, quartzites and phyllites of Upper Triassic age.

schistose clays (and sometimes Neogene formations). It is in these anticlinal structures that geothermal exploration has attained greatest success.

The upper clay complex constitutes an effective impermeable cap rock to the system. The natural thermal activity in the area, consisting of steam vents and steam-heated waters, occurred in places where the clay cover was thin and lay directly over the anhydrite series. An example of the composition of a steam-heated pool (*lagoni*) is given in Table 2.2. The depth of wells for steam production varies with the thickness of the impermeable clay cover and with the depth of the three main steam-bearing strata (either the evaporite series, an upper horizon of Jurassic jasper formation, or quartzites of the basement). The depth of drilling averages 600 m, with a maximum depth of 1600 m.

The wells produce steam, sometimes after a brief period of emitting a steam–water mixture in newly developed parts of the field. With time the steam discharge becomes increasingly superheated, with decreasing pressures and slightly increasing temperatures. The temperature of the steam tapped by wells varies considerably over the field, from a minimum of 150° to a maximum of 260°, and the maximum steam pressure registered in the region is 39 bars absolute. Figure 2.7 (ENEL, 1970) shows the distribution of pressure at the top of the reservoir in the Larderello geothermal field in 1969.

From petrological evidence, Marinelli (1969) concluded that in the

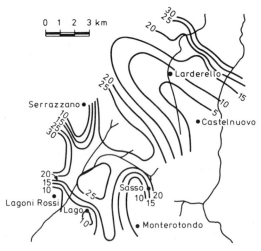

Fig. 2.7 Contours showing the pressures (in bars absolute) at the top of the Larderello reservoir in 1969. (Adapted from ENEL, 1970).

natural condition convective circulation of hot liquid water occurred in
the rocks. At the base of a discharging steam well the steam supply is
probably formed by the evaporation of water held within rock pores
(Truesdell and White, 1973).

Typical hydrothermal alteration minerals in the basement quartzite
and slates include adularia, zeolites, chlorites, calcite, quartz, anhydrite, and pistacite (Marinelli, 1969). Marinelli concluded that calcium
minerals had formed at these levels through recirculation of calcium
salts from the higher carbonate strata. Hydrothermal alteration was
not prominent in the cover beds due to their low permeability, but
wairakite was found in a siltstone in contact with the production
horizon.

The Kizildere Field, Turkey. The Kizildere geothermal field is in the
Denizli Province of western Turkey, lying north of the Menderes River
between Cubukdag and Sarayköy (Fig. 2.8, from Dominco and Samilgil, 1970). The area is dominated by block faulting. The Büyük-
Menderes Valley, which runs from east to west for over 100 km, is a
graben, or a fault-angle depression, bounded to the north by a fault
along the southern margin of the Menderes Massif. The surface
geology indicates a series of roughly east–west and northeast–southwest striking fault zones. At Kizildere, a major fault across the

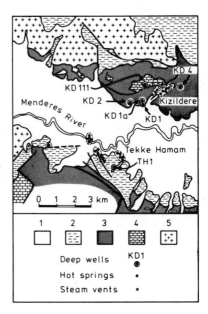

Fig. 2.8 The Kizildere geothermal field,
showing surface rocks and faults, hot springs
and fumaroles, and well positions. The rocks
are 1, alluvium; 2, Pliocene; 3, Upper Miocene; 4, Lower Miocene; 5, metamorphic
basement (adapted from Dominco and Samilgil, 1970).

northern part of the field separates the crystalline basement to the north from Miocene–Pliocene rocks to the south. The dips of the faults in the rather plastic Miocene–Pliocene series are erratic, and may be steep (Ten Dam and Erentöz, 1970). In addition, there is gravity sliding and slumping. The Miocene strata are broken into a series of stepped blocks (see Fig. 2.9), and several faults and deep folds divide the large stepped blocks into alternate horsts and depressions. The largest of the faults may have a throw of as much as 600 m. An important fault striking northeast–southwest across the field has boiling springs and steam vents aligned along it at Kizildere and at Tekke Hamam. The Kizildere field is situated in one of the most active seismic zones of western Anatolia, and fractures in the hot aquifer may be kept open by continuous tectonic movements (Ten Dam and Erentöz, 1970).

The oldest rocks in the area are the gneisses of the Menderes Massif; these are overlain by mica and quartz schists with occasional marble bands. Some of the deeper geothermal wells intersect the schists, which form a good production aquifer because of their highly fractured nature (Ten Dam and Erentöz, 1970). A rather impervious Miocene sedimentary series begins with approximately 250 m of argillaceous sandstones with occasional limestones and anhydrite. This is overlain by a similar thickness of hard, fractured, and cavernous limestones which formed a production aquifer for some of the earlier shallow wells. The Upper and Middle Miocene is represented

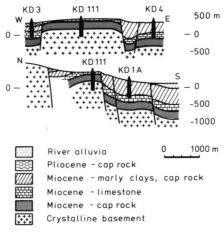

Fig. 2.9 Cross sections of the Kizildere field showing the positions of wells. (Adapted from Ten Dam and Erentöz, 1970).

by 400–600 m of argillaceous sandstones and marls, and at the top by a more argillaceous series with plastic clays. The Pliocene consists of marls, argillaceous sandstones, and anhydrite. The Upper Miocene and Pliocene sediments form an impermeable cap-rock series over the hydrothermal system (Fig. 2.9).

The similarity of the structure of the Kizildere field to that of the Larderello field is noteworthy. However, there has been no appreciable production of dry steam from any well in the Kizildere field, even in the most favorable structural position. Instead, water at moderate temperatures (usually 175°–200°) was tapped. The highest temperature (approximately 200°) was measured in wells drilled to deep levels into the basement rocks. Hot water appears to rise through fractures in the basement, and then into the fractured limestone aquifer where in places it mixes with cold waters of a shallow origin.

The water from springs and wells is a sodium bicarbonate type, high in sulfate, low in chloride, and containing relatively high fluoride and boron (example (d) in Table 2.2). Chemistry indicates a common origin for the hot waters (Dominco and Samilgil, 1970). The high carbon dioxide content of the underground waters (up to 1.6% by weight of total well discharges) and the high carbonate content of the waters have given rise to strong calcium carbonate encrustation in pipes (calcite above about 11 bars and aragonite below 9 bars). The rocks around the steam vents and springs are intensely altered, with deep argillization and with deposition of pyrite, sulfur, sulfates, and carbonates.

(c) Deep Circulation Warm Water Systems in Near-Normal Heat Flow Zones

In many areas of the world recurrent faulting and fracturing of igneous and metamorphic rocks allows water to circulate to great depths. This gives rise to warm springs with temperatures usually below 100°C. The warm waters frequently are dilute chloride–bicarbonate–sulfate solutions containing dissolved nitrogen, methane, and carbon dioxide. The total salt concentrations usually are less than about 1 g/l. The discharges of the springs are usually small and the storage of water within the systems may not be large.

Wells are drilled to intersect fissured zones in the solid rock beneath the levels where the thermal waters become dispersed in surface alluvium.

Examples of this type of water occur at Carlsbad in Czechoslovakia; in the Greater Caucasus, the Pamirs, and Tien-Shan in Russia; in the

Vosges mountain area of France; in northern India; and in the South Island of New Zealand.

Carlsbad, Czechoslovakia.   The Carlsbad spring system of Czechoslovakia is given as an example. This area, which has been known for more than 600 years because of the therapeutic use of its springs, is situated in Bohemia in the western end of the Krüsné Hory graben. It is within an area of ancient granites, where Tertiary age sediments fill the graben (Klir, 1970).

The central part of the spring system is located on the intersection of two major faults in the granite massif. Drilling into other faults, running east–west, has shown the rocks to be impermeable. Klir considered that meteoric water infiltrates the granite massif south of Carlsbad, and suggested slow water movement through the system to a circulation depth of about 2 km.

The temperature of the Carlsbad springs reaches 73°C and there is a total outflow of 33 l/sec of a sodium sulfate–bicarbonate–chloride type of water having a total dissolved solids content of about 6 g/kg. An analysis is given in Table 2.2. The water has a high concentration of carbon dioxide.

Storage Systems

(a)   Sedimentary Basin System

Marine sediments contain up to 60% water. As they are buried and compacted, the water content decreases and chemical, mineralogical, and bacterial reactions produce water of diverse chemical types. Hot waters which occur in this type of geological situation are of current interest for geothermal development only in zones of unusually high geothermal gradient and where there are storage horizons of high permeability in the rock strata.

The total salinity of hot waters within sedimentary rocks varies widely and is generally higher than in waters of volcanic rock areas. Differences in salinity cause a resultant difference in the relative levels of various elements in the waters (e.g., in hot brines there are higher calcium, magnesium, iron, and manganese concentrations relative to sodium than in dilute waters of the same temperature). With increasing total salinity, chloride usually becomes dominant over other anions. The waters of sedimentary basins usually contain higher carbon dioxide and hydrocarbon concentrations than volcanic area waters.

Drilling in oil fields has shown a trend with depth in the chemistry of the waters intercepted. Chebotarev (1955) noted that near-surface waters (at depths of the order of 500 m) are commonly high in sulfate; waters at intermediate depths may be dominated by bicarbonate; whereas at depths exceeding about 700 m, the dominant anion is generally chloride.

Many authors have considered the processes which cause changes in water solute compositions and concentrations in the sedimentary environment, although in a particular situation the predominant mechanism is not always clear. The processes include (a) precipitation of minerals (e.g., calcite, pyrite, anhydrite); (b) recrystallization of rock to new mineral assemblages; (c) hydration or dehydration of minerals; (d) bacterial action; (e) changing redox or pH conditions; (f) dilution, or mixing with other waters; (g) solution of elements from material within sediments; (h) ultrafiltration or reverse osmosis of waters by enclosing clay or shale beds.

The operation of the last process may need clarification. White (1965) suggested that compacted negatively charged clay minerals could form a membrane which under a pressure gradient would be capable of passing neutral molecules, such as water, carbon dioxide, boric acid, and ammonia, but would have a restricted permeability to ions. Sodium, bicarbonate, and iodide were considered to be more mobile than chloride ions, magnesium less mobile, and calcium and sulfate ions considerably less mobile than chloride ions. A calcium-rich brine was expected to form on the high-pressure side of the clay membrane, whereas on the low-pressure side a sodium bicarbonate water would be produced containing a higher proportion of boron and ammonia, and with a lower chloride concentration than the original water. Example (f), Table 2.2, is possibly a brine formed by membrane filtration (White, 1969).

Sourirajan (1970) showed that the operation of reverse osmosis membranes could be explained on a simple model with membrane channels of a specific size (to be correlated with the degree of compaction of clays). With increasing channel size, the selective passage of solutes is possible according to size of the solute entity. Uncharged molecules and univalent ions pass through relatively small openings in the membrane, but divalent ions with their large hydration shells require larger channels. A selectivity in solute transfer as well as an ability to produce a concentrated residual solution has been demonstrated in laboratory experiments with clay membranes (e.g., McKelvie and Milne, 1962; Hanshaw and Coplen, 1973). The process is particularly effective for solutions of low salt concentration.

Highly saline waters may, however, be produced by less complex processes, such as the solution of evaporite sequences in sedimentary basins, which can provide an explanation for some of the highly concentrated warm brines of the Siberian platform. Thermal waters of sedimentary basins include brines with up to saturation concentrations of chlorides, sulfates, and carbonates.

Bacterial action may have a considerable effect on the waters and the coexisting minerals, particularly in the earlier stages of sediment compaction and alteration. For example, bacterial reactions may form hydrogen sulfide and carbon dioxide from sulfate and organic material; iron sulfides and calcium carbonate consequently may be precipitated (White, 1965).

At an advanced stage of sediment alterations, metamorphic reactions will produce waters which are derived from the decomposition of crystalline minerals and from the associated organic matter. In particular, metamorphic waters can be expected to contain unusually high concentrations of carbon dioxide, boron, ammonia, iodide, and mercury. A gradual transition could be expected between waters filtered from compacting fine-grained sediments and waters resulting from the metamorphism of the sediments at higher temperatures and pressures.

At any stage during compaction, reaction, solution, membrane filtration, or metamorphism in a sedimentary sequence, a water phase could pass into a storage horizon or to the surface through fissures or joints in the formations. An immense range of water compositions may result.

The Hungarian Basin.   As an example, the Hungarian Tertiary sedimentary basin is considered. Here, temperature gradients of between 50° and 70°C per kilometer occur, and heat flows between 2 and 3.4 × 10^{-6} cal cm^{-2} sec^{-1} have been measured (Boldizsar, 1970; Korim, 1972). The highest values occur where the basement rock is elevated. The volume of porous rocks beneath 1000 m in depth in the Hungarian basin is over 4000 km^3, and their average porosity is about 20%.

The most extensive and important aquifer, the Pannonian formation of Pliocene age, is more than 3000 m thick in the central part of the basin (Fig. 2.10, from Korim, 1972), and it stores more than 50% of the geothermal energy of Hungary. Over 95% of the drilling activity has been concentrated in the Upper Pannonian, 800–2400 m from the surface, where permeable sands and sandstones are interbedded with low-permeability siltstones. Horizontal permeability is high but vertical permeability is low. Water temperatures in wells range up to 100°.

Fig. 2.10 Map of Hungary showing the positions of wells producing hot water from Upper Pannonian aquifers. Contours show the depth in meters of the bottom of the aquifer sequence. Typical water temperatures are given for various areas. Adapted from Korim (1972).

The Upper Pannonian has alkali bicarbonate hot waters containing about 1800–2500 ppm soluble material (example (g), Table 2.2), while methane and carbon dioxide are present at a total gas/water ratio between 0.2 and 1.2 m^3/m^3.

Miocene limestone and dolomite have also been exploited for hot water, where they occur deeper than 1000 m. Fractured and interconnected Paleozoic and Mesozoic limestones at the base of the Tertiary sequence contain some of the highest-temperature waters. However, drilling is expensive and is limited to tectonic fracture zones.

The fluid pressure in the aquifers remains approximately hydrostatic down to about 3.5 km. Boldiszar (1970) considered that any water recharge would take 10^3–10^6 years, and was probably negligible.

(b) Metamorphic Systems

There are no proved examples of geothermal systems based on waters expelled during regional metamorphism. However, White *et al.* (1973) and Barnes (1970) suggested a metamorphic origin for hot springs at several localities in the northern Californian Coast Ranges. Mercury deposits are commonly associated with the areas. The springs produce dilute sodium bicarbonate waters with appreciable ammonia and boron, and the oxygen and hydrogen isotope compositions of the waters show them to be mainly of nonmeteoric origin. The springs issue from a wide variety of surface rock types, and it was suggested that the waters arise from fluids yielded from the current metamorphism of marine sedimentary rocks beneath the area. An analysis of a

hot spring at the Sulphur Bank mercury mine is given in Table 2.2.
Drilling in this area revealed a temperature of 186°C at 430 m.

2.5 GEOTHERMAL RESOURCES OF THE SOVIET UNION: AS EXAMPLES

Makarenko *et al.* (1970) outlined the occurrences of natural hot
waters in the Soviet Union in a wide variety of geological situations.
This provides a convenient review of this chapter. The vast extent of
hot water resources in the Soviet Union is evident from Fig. 2.11,
showing the potential uses related to the water temperatures.

Makarenko divided the country into regions of similar geological
associations. The highest-temperature waters are in regions of recent
volcanic activity within Cenozoic geosynclines in Kamchatka and the
Kuril Islands. Boiling springs and fumaroles occur, temperatures at

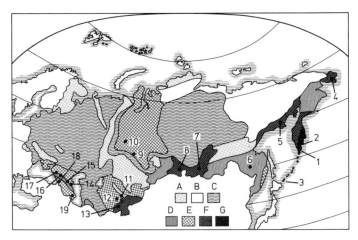

Fig. 2.11 Map of Soviet Union showing the thermal water resources of various
regions. A, areas with no thermal water resources; B, areas not yet surveyed; C–G, areas
of increasingly higher water temperatures; C–E, areas of balneological interest; E–G,
areas where thermal waters are used for district heating; G, areas where thermal waters
are used for electricity production. Numbers denote areas using geothermal water: 1,
Pauzhetsk; 2, Paratunsk; 3, Goryachy Plyazh; 4, Magadan; 5, Talaya; 6, Kuldur; 7,
Iljinka; 8, Ush-Beldyr; 9, Omsk; 10, Tobolsk; 11, Chimkent; 12, Tashkent; 13, Khodzhi-
Obi-Garm; 14, Makhachkala; 15, Grosny; 16, Cherkessk; 17, Maikop; 18, Zugdidi; 19,
Astara. (Adapted from Makarenko *et al.*, 1970).

depth are well in excess of 100°C, and waters are often of the dilute neutral-pH, sodium chloride type.

In zones of Cenozoic orogenesis, as in the Lesser Caucasus and northern Kamchatka, dilute sodium chloride–sulfate waters occur, circulating in zones of deep faulting. Water temperatures are usually less than 100°.

In mountain structures of Cenozoic geosynclines and recent tectonic activity, in conditions of higher heat flow (average 1.75×10^{-6} cal cm^{-2} sec^{-1}) waters flow in tectonic crush zones and reach the surface at temperatures approaching the boiling point. Examples occur in the Greater Caucasus, Pamirs, and Tien-Shan. Some springs have a high water output (2–50 l/sec) and salinity is usually less than 5 g/l. The highest temperature recorded is the Khodzhi-Obi-Garm springs in Tien-Shan, where waters emerge at 98°.

Cenozoic foredeeps, inner troughs, and tectonic depressions contain sedimentary basins with thermal waters which have temperature and compositional zoning with depth. Heat flows are about 0.95×10^{-6} cal cm^{-2} sec^{-1}, and water temperatures may reach 100° at depths of 4 km. Upper parts of the basins often contain dilute chloride–bicarbonate–sulfate waters, but at deeper levels chloride brines predominate with high bromide and iodide contents. Examples are the depressions of Predverkhoyanie, Ferghana, Rioni, Kura, and West Turkmenia.

High geothermal gradients occur over vast areas of the West Siberian, Turanian, and Skiphian plateaus. These areas occur between the edges of the Siberian suture (Tamrazyan, 1970), where the east European and Siberian platforms moved apart, resulting in large tectonic depressions which were later filled with thick beds of terrigenous sediments of low thermal conductivity. Temperatures reach 125–150° at depths of 3–5 km in the lower sedimentary horizons, and artesian basins of stratified thermal waters exist over vast areas. The largest area in the West Siberian basin covers about 3×10^6 km^2 where the mean heat flow is 1.25×10^{-6} cal cm^{-2} sec^{-1} (Makarenko et al., 1970). Most of the important hot water resources of the Soviet Union, including the West Siberian lowland, the Turana lowland, the pre-Caucasus, the Crimean Steppe, and the southeastern Caspian lowland, occur within the Siberian suture region.

The Mesozoic folded mountain structures in the vast northeastern territories of the Soviet Union, the Paleozoic mountain structures, the pre-Cambrian shields and platform regions do not appear to have waters of sufficiently high temperature accessible for utilization. Heat flows and geothermal gradients range from average to below average.

171032

2.6 HOT WATER OR STEAM SYSTEMS

Wells drilled into most high-temperature hydrothermal systems tap hot liquid water from fissures and fractures in competent rocks or from porous rock strata. The wells initially discharge a mixture of steam and water in a proportion which depends on the original deep-water temperature and gas content. After a period of steam and water discharge, the well characteristics may gradually change and a rather higher proportion of steam may appear in the discharges than would be expected from the original underground water temperature. At a later stage the steam and water ratio may decrease again. This occurred for some wells at Broadlands and Wairakei, but was particularly marked at Broadlands, where the rock permeabilities were not as high as at Wairakei.

In a few high-temperature systems, such as Larderello, The Geysers, and Matsukawa, most wells discharge steam unaccompanied by water, sometimes after a very brief period of mixed steam and water flow. The amounts of water discharged are related to the operating pressures of the wells, and increasing the wellhead pressure tends to promote wetter discharges.

Whether wells tap steam or liquid water in a geothermal field at levels of a few hundred meters to at least 2.0 km is a matter of balance between heat flow and water availability, and between the permeabilities of the system at particular levels. Several authors have produced theories to explain the differences between water- and steam-producing fields: Facca and Tonani (1964); James (1968); White *et al.* (1971); Truesdell and White (1973). Only a brief discussion is given here, but the following model utilizing their ideas outlines some of the features which determine whether a geothermal field is dominated by steam or hot water discharges.

Consider Fig. 2.12a, a geothermal system in which the liquid water level reaches the surface. High-temperature water rises from considerable depths (many kilometers) through a flow route of permeability Π_1 and passes to the surface through a channel system of permeability Π_2. For this system Π_1 is greater than Π_2. From point A at the surface to point B in the water column, pressures rise from 1 atm to a minimum of the combined steam and gas pressure at the temperature of the water. As an approximation (neglecting gas pressures), temperatures and pressures are those for a column of water at boiling point throughout its length (boiling point–depth curve). Point B has been found to have temperatures ranging up to at least 350° in various hot

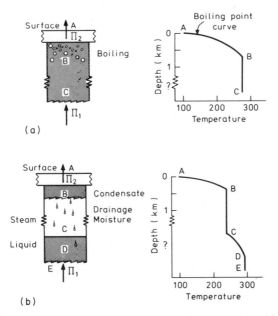

Fig. 2.12 Schematic cross sections of geothermal systems dominated by hot water (a) and steam (b) in near-surface zones. The temperature versus depth graph for each system is also given.

water geothermal systems. Wells placed in the reservoir ideally discharge steam and water in a ratio determined by the original water temperature.

The presence of high gas concentrations in a water system has a marked effect on temperature and pressure gradients. Whereas in a stream reservoir a considerable proportion of gas must be present to contribute appreciably to the total vapor pressure, in a water system the partial pressure of dissolved gas can make up a high proportion of the total pressure. For example, at Broadlands most production wells had a CO_2 concentration of approximately 0.6% by weight in the discharges. The partial pressure of this gas dissolved in 250°–260° water was between 8 and 15 bars. At greater depths in the field there is evidence that CO_2 concentrations approach 6% by weight, and at temperatures of 300°C the partial pressure of the carbon dioxide would exceed the water vapor pressure. The main effect of carbon dioxide pressures is to depress isotherms below the normal boiling point for depth relationship.

Pressure reduction in a gas-rich hot water system produces a gas-rich steam phase and it is likely that two phases, water and steam plus gas, are always present above a certain depth. Broadlands, for example, is characterized at the surface by extensive gas emissions. An irreversible decompression of these systems with exsolution of gases is likely to occur very rapidly as a result of exploitation.

Figure 2.12b shows a system of steam production where Π_2 is greater than Π_1. In general, this type of system is found in tight and incompetent rocks (White *et al.*, 1971). Possibly the near-surface system was filled with water originally but later the inflow permeability became limited by mineral deposition. A restricted water inflow and a continuous outflow would cause the water level to fall and a steam phase to form. Over a long period as steam was discharged the water level would fall further, leaving a considerable reservoir dominated by steam. The pores of the rocks in this steam zone would retain a proportion of liquid phase from the original solution and from the downward percolation of condensate which forms at the cooler near-surface levels.

The condensation of high-temperature steam in the cooler near-surface levels would also tend to clog outflow channels with condensate and to reduce permeability through the creation of clay minerals by rock alteration by the acidic oxidized condensate. In this situation a dynamic balance is set up, with a steam chamber existing in moist rock beneath a layer of rock containing condensate and forming a cap to the system. Steam-heated surface waters may occur in the immediate subsurface and at the surface.

In Fig. 2.12b, temperatures in the condensate zone A–B may lie close to the boiling point curve; within the steam zone B–C temperatures are almost constant, at least in the central part of the field; and in the deep water the temperatures may increase further along a boiling point curve C–D to a maximum temperature which is retained to very great depths E. These ideas are still in part hypothetical and no evidence has yet been obtained for the lower portion of the curve.

Neglecting gas pressures, the temperatures in the undisturbed steam zone would reflect the temperatures of the deep-water surface up to a maximum value which is imposed because of the thermodynamic characteristics of steam. Saturated steam has a maximum enthalpy at 236° (about 31 bars pressure). Saturated steam at higher temperatures, if expanded at constant enthalpy, condenses out liquid water until the condition 236°C, 31 bars is reached. Saturated steam can only expand at constant enthalpy to become a dry steam flow at temperatures below 236°.

If the temperature of the water reservoir at depth C was originally above 236°, the temperature would soon adjust to about this value through the draining back of liquid water formed during the expansion of the high-temperature steam. Steam reservoirs would therefore be expected to have a temperature of about 230°–240° or less. It is significant that this is about the temperature of the steam flows in the Larderello region, The Geysers, and Matsukawa. It should be noted, however, that in some of these areas temperatures higher than this have been measured in wells before discharge.

Lower temperatures may occur where there are high gas concentrations in a portion of a steam reservoir, since a high partial pressure of the gases requires lower saturated steam pressures to allow a pressure balance with the remainder of the steam reservoir system; for example, when the vapor phase contains 50% carbon dioxide, the temperature of the vapor would be about 200°.

Wells drilled into a steam zone of a reservoir increase the outflow above the natural rate and lower the pressures. The effects are complex. There is an evaporation of liquid water held in the rock pores around the well intake zone. Part of the heat of evaporation is extracted from the rock and part obtained by cooling of the steam plus water phases. If the rock is highly porous and contains considerable liquid, the well could produce a wet discharge, at least for a period. However, where the porosity and the water content of the rock are low, the steam plus water system in the rock may cool below 236° during evaporation of the liquid phase. The steam may then become superheated by extracting heat from the rock as it flows down the pressure gradient toward the well. For example, the early Larderello steam discharges were at about 190°–220° (Truesdell and White, 1973). An expanding zone of dried rock surrounds the well and this retains a major proportion of the heat storage in the system.

For a given rock temperature and production pressure (WHP) there is a maximum rock porosity below which the well should produce steam only. For a system in thermal equilibrium at 250°–260° the maximum porosity for dry discharges is of the order of 8–10% for a well production pressure of approximately 5 bars.

Following extensive production of fluid from a reservoir there may be a tendency for steam to be drawn at least partly from the surface of the deep-water body (C) and for the moist rock reservoir to become dried out and dep'eted. Steam temperatures will rise to values of about 230°–240°, but eventually could decrease gradually again as the deep-water temperatures are lowered by accelerated boiling.

In the natural state of the Broadlands field, the permeability sequence maintained an all-liquid system, but the additional outflow from wells caused boiling to occur within the rock, and a retreating water-steam interface formed about some wells. This progressed to the stage of forming a dry steam discharge from one well and very high steam-to-water proportions in other wells. However, the high discharge enthalpies were not sustained and decreased with time toward original or lower values.

There are many variations to these simplified models and obvious transitions from one type of system to the other. Steam systems are frequently imposed on deeper chloride water systems. The depth of the steam portion of such systems may be considerable and of the order of 500 m or more, and wells drilled within these depths in a number of areas have produced steam discharges. Deep drilling into the Hakone Volcano, Japan (Oki and Hirano, 1970), has shown the transitional features likely to occur in many systems.

The artificial formation of a localized steam zone in a water system may not necessarily be a favorable situation for continuing well production. The very rapid nature of the changes brought about in the rock-water-steam system may create difficulties, such as initiating deposition of minerals (quartz or calcite) by the intensive boiling of waters.

REFERENCES

Allen, E. T., and Day, A. L. (1935). Hot Springs of the Yellowstone National Park. Carnegie Inst. Wash. Publ. 466.
Alonso, H. (1966). *Bol. Soc. Geol. Mexicana* **29,** 17.
Banwell, C. J. (1970). *Geothermics (Spec. Issue 2)* **1,** 32.
Banwell, C. J., and Gomez-Valle, R. (1970). *Geothermics (Spec. Issue 2)* 2(Pt 1), 27.
Barnes, I. (1970). *Science* **168,** 973.
Barth, T. F. W. (1950). Volcanic Geology Hot Springs and Geysers of Iceland. Carnegie Inst. Wash. Publ. 587.
Boldizsar, T. (1970). *Geothermics (Spec. Issue 2)* **2** *(Pt 1),* 99.
Bullard, E. (1973). *In* "Geothermal Energy" (H. C. H. Armstead, ed.), pp. 19–29. UNESCO Earth Sci. Ser. No. 12, Paris.
Burgassi, R. (1964). *Proc. U.N. Conf. New Sources Energy, Rome 1961* **2,** 99.
Chebotarev, I. I. (1955). *Geochim. Cosmochim. Acta* **8,** 22.
Craig, H. (1963). *In* "Nuclear Geology on Geothermal Areas (E. Tongiorgi, ed.), pp. 17–33. Consiglio Nazionale della Richerche, Laboratorio di Geologia Nucleare, Pisa.
Dominco, E., and Samilgil, E. (1970). *Geothermics (Spec. Issue 2)* **2** *(Pt 1),* 553.
Donaldson, I. G. (1970). *Geothermics (Spec. Issue 2)* **2** *(Pt 1),* 649.
Elder, J. W. (1965). *In* Terrestrial Heat Flow, pp. 211–239. Am. Geophys. Un., Geophys. Monograph No. 8.
ENEL (1970). "Larderello and Monte Amiata: Electric Power by Endogenous Steam," pp. 1–42. Ente Nazionale per l'energia Elettrica, Rome.
Facca, G., and Tonani, F. (1964). *Bull. Volcanol.* **27,** 1.

Fukuda, M., Ushijma, K., Aosaki, K., and Yamamuro, N. (1970). *Geothermics (Spec. Issue 2)* **2** *(Pt 2)*, 1448.

Fyfe, W. S. (1977). *In* "Geochemistry 1977" (A. J. Ellis, ed.). New Zealand Dept. Sci. and Ind. Res., Bull. 218, Wellington (in press).

Gooch, F. A. and Whitfield, J. E. (1888). Analyses of Waters of the Yellowstone National Park. Bull. U.S. Geol. Surv. No. 47, Washington, D.C.

Grindley, G. W. (1965). The Geology Structure and Exploitation of the Wairakei Geothermal Field Taupo, New Zealand, pp. 131. Bull. Geol. Surv. New Zealand, No. 75, Wellington.

Hanshaw, B. B., and Coplen, T. B. (1973). *Geochim. Cosmochim. Acta* **37,** 2311.

James, C. R. (1968). *N.Z. J. Sci.* **11,** 706.

Klir, S. (1970). *Geothermics (Spec. Issue 2)* **2** *(Pt 2)*, 1055.

Koga, A. (1970). *Geothermics (Spec. Issue 2)* **2** *(Pt 2)*, 1422.

Korim, K. (1972). *Geothermics* **1,** 96.

McKelvie, J. G., and Milne, I. H. (1962). *In* "Clays and Clay Minerals" (*Conf. Clays Clay Miner., 9th, Lafayette, Indiana*), (A. Swineford, ed.) pp. 248-259.

McNitt, J. (1970). *Geothermics (Spec. Issue 2)* **1,** 24.

Mahon, W. A. J. and McDowell, G. D. (1977). *In* "Geochemistry 1977" (A. J. Ellis, ed.). New Zealand Dept. Sci. and Ind. Res. Bull. 218, Wellington. (in press).

Makarenko, F. A., Mavritsky, B. F., Lokshin, B. A., and Kononov, V. I. (1970). *Geothermics (Spec. Issue 2)* **2** *(Pt 2)*, 1086.

Marinelli, G. (1969). *Bull. Volcanol.* **33,** 1.

Mercado, S. (1967). Geoquimica Hidrotermal en Cerro Prieto B.C. Mexico. Comision Federal de Electricidad, Mexicali.

Mercado, S. (1970). *Geothermics (Spec. Issue 2)* **2** *(Pt 2)*, 1367.

Molina, R., and Banwell, C. J. (1970). *Geothermics (Spec. Issue 2)* **2** (Pt 2), 1377.

Nasini, R. (1930). I Soffione e i Lagoni della Toscana e la Industria Boracifera. Tipografia Editrice Italia, Rome.

Oki, Y., and Hirano, T. (1970). *Geothermics (Spec. Issue 2)* **2** (Pt 2), 1157.

Ovchinnikov, A. M. (1955). *Vopr. Kurortol. Fiz. Lecheb.* **3,** 66 (*Chem. Abstr.* **52,** 2311).

Piip, B. I. (1937). "The Thermal Springs of Kamchatka," pp. 268. Akad. Nauk C.C.C.P., Moscow.

Souririjan, S. (1970). "Reverse Osmosis." Academic Press, New York.

Tamrazyan, G. P. (1970). *Geothermics (Spec. Issue 2)* **2** (Pt 2), 1212.

Ten Dam, A., and Erentoz, C. (1970). *Geothermics (Spec. Issue 2)* **2** (Pt 1), 124.

Truesdell, A. H., and White, D. E. (1973). *Geothermics* **2,** 154.

Waring, G. A., Blankenship, R. R., and Bentall, R. (1965). Thermal Springs of the United States and other Countries of the World. U.S. Geol. Survey. Prof. Paper 492, Washington, D.C.

White, D. E. (1957a). *Geol. Soc. Am. Bull.* **68,** 1637.

White, D. E. (1957b). *Geol. Soc. Am. Bull.* **68,** 1959.

White, D. E. (1965). *In* Fluids in Subsurface Environments—A Symposium, pp. 342-366. Am. Soc. Petrol. Geologists Memoir No. 4.

White, D. E. (1969). *Proc. Int. Geol. Congr., 23rd* **19,** 269.

White, D. E. (1974). *Econ. Geol.* **66,** 75.

White, D. E., Barnes, I., and O'Neil, J. R. (1973). *Geol. Soc. Am. Bull.* **84,** 547.

White, D. E., Muffler, L. J. P., and Truesdell, A. H. (1971). *Econ. Geol.* **66,** 75.

Yamasaki, T., Matsumoto, Y., and Hayashi, M. (1970). *Geothermics (Spec. Issue 2)* **2** (Pt 1), 197.

Chapter 3

THE CHEMICAL NATURE OF GEOTHERMAL SYSTEMS

The first two chapters emphasized the wide variety of natural hydrothermal systems in many different geological situations. The remainder of the book concentrates principally on high-temperature geothermal systems—their physical and chemical characteristics, and the chemical problems associated with their utilization. The authors have been associated mainly with these systems and with projects directed toward utilizing geothermal fluids for the development of electric power. To date, appreciable quantities of electricity have been produced only in areas where there is water or steam at temperatures in excess of about 180°C within 1–2 km of the surface. This selected coverage is not meant to imply that high-temperature geothermal systems and electric power generation are the only areas of significance. Indeed, because of their wider occurrence, warm water systems may prove in the long term to be of greater economic importance in connection with community heating projects.

3.1 WATERS

Classification

Most chemical types of water found in high-temperature geothermal areas can be classified under the general headings that follow (White, 1957; Ellis and Mahon, 1964). The classification can be applied to high-temperature spring and well waters in nonvolcanic as well as volcanic areas, and is more for convenience in discussion than for defining origins of waters. Examples of the compositions of various types of hot water are given in Table 3.1.

58

TABLE 3.1

Analyses Typical of Each Water Classification Group[a]

	Source	pH	Li	Na	K	Rb	Cs	NH₄	Ca	Mg	Fe	Mn	F	Cl	SO₄	HCO₃	B	SiO₂	Ref.[b]
a.	Geyser 238, El Tatio, Chile	7.32	45	4340	520	6.7	12.6	3.8	272	0.5	0.1	0.4	3.1	7922	30	46	178	260	1
b.	Taumatapuhi Geyser, Tokaanu. N.Z.	7.8	22.6	1710	168	0.9	3.3	1.3	32	0.2	0.01	—	1.7	3021	63	2	90	270	1
c.	Green-black hot pool, explosion crater, Waiotapu, N.Z.	2.8	—	43	11	—	—	6.2	27	3.5	8.2	—	—	32	347	0	2.5	280	2
d.	Yellow hot pool, explosion crater, Waiotapu, N.Z.	2.8	—	405	74	—	—	65.5	40	7.5	5.0	—	—	612	666	0	10.1	370	2
e.	Well 205, Matsao, Taiwan, 1500 m deep	2.4	26	5490	900	12	9.6	38	1470	131	220	42	7.0	13400	350	0	106	639	1
f.	Crater Lake, Ruapehu, N.Z.	1.20	1.6	740	79	0.4	0.1	11	1200	1030	900	34	260	9450	10950	0	13.8	852	3
g.	Well 5, Wairakei, 471 m deep	8.6	1.2	230	17	—	—	0.2	12	1.7	—	—	3.7	2.7	11	680	0.5	191	1

[a] Concentrations in mg/kg (ppm) in waters reaching the surface; pHs measured in cooled waters.
[b] References: 1, Chemistry Division, DSIR; 2, Wilson (1963a); 3, Giggenbach (1974).

59

Alkali Chloride Waters (A)

The dissolved salts in these waters are mainly sodium and potassium chlorides, although in the more concentrated waters appreciable calcium concentrations may occur. The waters also contain high concentrations of silica, and usually significant concentrations of sulfate, bicarbonate, fluoride, ammonia, arsenic, lithium, rubidium, cesium, and boric acid. The chloride/sulfate ratio is usually high, and the pH ranges from slightly acid to slightly alkaline (pH 5–9). The main dissolved gases are carbon dioxide and hydrogen sulfide. The waters often occur in areas with boiling springs and geyser activity and are common to many developed geothermal areas in both volcanic and sedimentary rocks (e.g., Wairakei, Otake, Mexicali, El Tatio, Salton Sea; examples a and b in Table 3.1).

Acid Sulfate Waters (B)

Acid waters, low in chloride content, may be formed in volcanic geothermal areas when steam below about 400° condenses into surface waters. Hydrogen sulfide from the steam is subsequently oxidized to sulfate. Acid sulfate waters are found in areas where steam arises from underground high-temperature water and in volcanic areas where in the cooling stages of volcanism only carbon dioxide and sulfur gases remain in the vapors rising through the rocks. The constituents present in the waters are mainly leached from the rocks surrounding the pools (example c, Table 3.1). Because of their generally superficial nature, their geochemical significance is usually minor in survey work.

Acid Sulfate–Chloride Waters (C)

Hot spring waters containing chloride and sulfate in comparable concentrations are found in many areas. These waters are usually acidic (pH 2–5), and may originate in several ways:

1. Mixing of water types A and B; example d, Table 3.1.
2. The sulfide in waters of type A may become oxidized at depth to bisulfate ions, perhaps through an association with oxidized lavas. The water may have a near-neutral pH at depth due to the neutralizing and buffering action of the confining rocks. However, since the dissociation constant for bisulfate ion increases markedly with a lowering of temperature, a high-bisulfate water of close to neutral pH underground may become acid on rising to cooler conditions at the surface.

3. This type of water could also form when high-temperature chloride water comes into contact at depth with sulfur-containing rocks. The hydrolysis of sulfur to hydrogen sulfide and sulfuric acid produces an acidic solution (example e, Table 3.1).

4. In active volcanic areas, high-temperature steam may rise from molten rock at shallow depth to condense into surface or near-surface waters. The resultant thermal waters often contain high fluoride, chloride, and sulfate concentrations derived from the volcanic steam. With decreasing steam temperature, the fluoride, chloride, and sulfur gases, in order, decrease in abundance, so that acid sulfate–chloride–fluoride waters merge into acid sulfate–chloride water, then into acid sulfate waters. Many of the constituents present in these acid waters are derived by superficial surface leaching of rocks by hydrochloric and sulfuric acids (example f, Table 3.1).

Bicarbonate Waters (D)

Low-chloride hot waters containing high bicarbonate and variable sulfate concentrations may occur near the surface in volcanic geothermal areas where steam containing carbon dioxide and hydrogen sulfide condenses into an aquifer. Under stagnant conditions, reaction with rock produces neutral-pH bicarbonate or bicarbonate–sulfate solutions (example g, Table 3.1). Sodium is often the main cation in the waters, since calcium carbonate is not very soluble at high temperature and potassium and magnesium are fixed in clays. At high temperature, sulfate concentrations are limited by $CaSO_4$ solubility. Bicarbonate waters of greater complexity are also common at deep levels in geothermal systems within metamorphic or sedimentary rocks (e.g., analyses *d, e*, and *h* of Table 2.2).

Compositions

There is a wide range of information on the chemical composition of hot spring waters (Waring *et al.*, 1965), but analytical data on high-temperature waters from deep levels are more restricted. Table 3.2 gives representative chemical analyses of waters from various areas where there are geothermal developments. The composition of water from a major hot spring in the area is compared with water produced from a representative deep well. The concentrations are in milligrams per kilogram (parts per million) for waters separated at atmospheric pressure from the well discharges (with two exceptions). These concentrations (for gaseous solutes in particular) differ from the concentrations in the deep waters due to steam formation. For example, 250°

TABL

Source and reference[a]	Approx. temp. (°C)	Depth (m)	pH (20°)	Li	Na	K	
				Concentrations (mg/kg) for water collecte[d] atmospheric pressure from discharge			
Spring, Reykjavik, Iceland (a)	88	0	9.4	0.00	159	1.4	
Drillhole, Reykjavik, Iceland (b)	100	600	8.6	<0.1	95	1.5	<(
Spring, Hveragerdi, Iceland (b)	100	0	9.3	—	—	—	
Hole G-3, Hveragerdi, Iceland (b)	216	650	9.6	0.3	212	27	
Spring, Pauzhetsk, Kamchatka, USSR (c)	100	0	8.4	—	1010	88	
Pauzhetsk drill holes, Kamchatka, USSR (d)	170–195	300–400	8.9	—	Na + K 940		
Jubilee Bath, Ngawha, N.Z. (e)	50	0	6.5	11	870	79	
Drillhole 1, Ngawha, N.Z. (e)	230	585	7.4	12.2	950	80	
Ohaki Pool, Broadlands, N.Z. (e)	95	0	7.05	7.4	860	82	
Drill hole 2, Broadlands, N.Z. (e)	260	1030	8.3	11.7	1050	210	
Champagne Pool, Wairakei, N.Z. (e)	99	0	8.0	10.8	1070	102	
Drill hole 24, Wairakei, N.Z. (e)	250	830	8.3	13.2	1250	210	
Spring 6, Rotokaua, N.Z. (e)	65	0	2.5	7.8	990	102	
Drill hole 2, Rotokaua, N.Z. (e)	220	880	7.8	10.2	1525	176	
Spring, Tahuangtsui, Tatun, Taiwan (f)	91	0	—	3.0	611	25.9	
Hole E103, Tahuangtsui, Tatun, Taiwan (f)	200	1000	3.2	—	282	54	
Shofuso Hotel, Matsukawa, Japan (g)	78.5	0	3.1	—	40	9.6	
Hole 1, Matsukawa, Japan (g)	~300	945	4.9	—	264	144	
Spring 22, Mexicali, Mexico (h)	—	0	—	11.6	4250	535	
Hole 5, Mexicali, Mexico (h)[b]	340	1285	—	19	5820	1570	
Wister mudpots, Salton Sea, Calif. (i)	21	0	7.1	9.6	6470	466	
Hole 1, IID, Salton Sea, Calif. (i)[b]	340	1600	4.7	215	50,400	17,500	13
Spring, Makhachkala, Dagestan, USSR	63	0	8.6	0.2	1137	5.4	

[a] *References:*   (a) Bodvarsson (1961).
     (b) Chemistry Division, DSIR, N.Z. and pers. comm. from Dr. G. Bodvarsson, and Dr. G. E. Sigvaldason
     (c) Ivanov (1958).   (d) Averyev *et al.* (1961).   (e) Chemistry Division,
     (f) Y.-C. Liu (personal communication).   (g) Nakamura and Suumi (1967).   (h) Mercado (1966).
     (i) Muffler and White (1969).   (j) Ivanov and Nevraev (1964).

mposition of Thermal Waters from Springs and Drill Holes

Concentrations (mg/kg) for water collected at atmospheric pressure from discharge

Cs	Mg	Ca	Mn	Fe	F	Cl	Br	I
—	0.3	1.9	0.00	0.00	1.0	30	0.9	0.1
0.02	—	0.5	—	—	—	31	—	—
—	0.8	2.5	—	—	1.1	186	—	—
<0.02	0.0	1.5	0.00	0.1	1.9	197	0.45	0.0
—	10	64	—	0.0	0.8	1684	3.2	0.0
—	7	119	—	—	—	1470	—	—
0.5	2.5	8	—	—	0.3	1336	—	—
0.4	Ca + Mg 28		0.02	0.1	0.8	1625	—	1.3
1.2	Ca + Mg 2.6		—	—	5.2	1060	3.0	0.6
1.7	0.1	2.2	0.009	≤0.01	7.3	1743	5.7	0.8
2.7	0.4	26	—	—	6.6	1770	4.0	0.7
2.5	0.04	12	0.015	≤0.01	8.4	2210	5.5	0.3
—	11.3	11	—	—	<1	1433	—	—
—	—	50	—	—	6.6	2675	—	0.2
—	2.8	3.5	2.5	1.1	7.3	387	—	—
—	89	0	—	1368	—	1223	—	—
—	8.2	31.8	—	8.6	—	3.0	—	—
—	8.7	22.9	—	508	—	12.4	—	—
—	24	340	—	—	—	8600	12.3	—
—	8	280	—	0.2	—	10,420	14.1	3.1
—	325	79	0.9	0.8	14	8480	—	—
14	54	28,000	1400	2290	15	155,000	120	18
—	2.4	16.4	—	—	0.4	335	1.6	—

[b] Concentrations and pH in deep aquifer.
[c] Total CO_2, SiO_2, etc., is the total $CO_2 + HCO_3^- + CO_3^{2-}$ expressed as CO_2, silica + silicate as SiO_2, etc.

Table 3.2

Source and reference[a]	SO₄	As	Total[c] CO₂	Total SiO₂	Total B	Total NH₃	Total H₂S
Spring, Reykjavik, Iceland (a)	17	0.00	32	126	0.06	0.00	0.2
Drillhole, Reykjavik, Iceland (b)	16	—	58	155	0.03	0.1	—
Spring, Hveragerdi, Iceland (b)	65	—	—	386	—	—	—
Hole G-3, Hveragerdi, Iceland (b)	61	—	55	480	0.6	0.1	7.3
Spring, Pauzhetsk, Kamchatka, USSR (c)	83	1.0	32	160	39	—	—
Pauzhetsk drill holes, Kamchatka, USSR (d)	164	—	61	170	31.3	0.7	—
Jubilee Bath, Ngawha, N.Z. (e)	500	—	240	186	1020	80	6
Drillhole 1, Ngawha, N.Z. (e)	17	—	61	460	1200	46	<1
Ohaki Pool, Broadlands, N.Z. (e)	100	1.0	490	338	32	3.8	1
Drill hole 2, Broadlands, N.Z. (e)	8	8.1	128	805	48.2	2.1	<1
Champagne Pool, Wairakei, N.Z. (e)	26	2.5	55	—	21.9	0.7	1.8
Drill hole 24, Wairakei, N.Z. (e)	28	4.5	17	670	28.8	0.2	1
Spring 6, Rotokaua, N.Z. (e)	520	—	144	340	45.0	1.6	0.2
Drill hole 2, Rotokaua, N.Z. (e)	120	—	55	430	102	3.2	—
Spring, Tahuangtsui, Tatun, Taiwan (f)	456	—	71	167	13.0	9.5	369
Hole E103, Tahuangtsui, Tatun, Taiwan (f)	1462	—	—	170	—	—	1.7
Shofuso Hotel, Matsukawa, Japan (g)	315	—	—	68	3.3	2.4	5.0
Hole 1, Matsukawa, Japan (g)	1780	—	26	635	61.1	—	tr
Spring 22, Mexicali, Mexico (h)	20.0	—	23	121	—	—	12
Hole 5, Mexicali, Mexico (h)[b]	0	—	1653	740	12.4	—	700
Wister mudpots, Salton Sea, Calif. (i)	900	—	3130	59	54	32	—
Hole 1, IID, Salton Sea, Calif. (i)[b]	5	12	7100	400	390	386	16
Spring, Makhachkala, Dagestan, USSR	1299	—	560	—	5.2	—	—

Concentrations (mg/kg) for water collected at atmospheric pressure from discharge

water expands into a mixture of approximately 70.3% water and 29.7% steam at 1 atm pressure; hence, the solute concentrations in the water at the base of well 24, Wairakei, are approximately 0.7 times the concentrations given in Table 3.2. The Mexicali and the Salton Sea area water analyses are for the original deep water before steam formation.

ontinued)

			Molecular ratios				
Cl/SO$_4$	Cl/B	Cl/F	Cl/Br	Cl/As	Na/Li	Na/K	Na/Ca
4.8	150	16	75	—	>5000	190	145
5.2	400	—	—	—	>300	110	330
7.7	—	90	—	—	—	—	—
8.7	100	55	800	—	200	13	250
55	13.2	1100	1200	3600	—	19.5	27
24	14.3	—	—	—	—	—	~13
7.2	0.40	2400	—	—	23.9	18.7	190
260	0.41	1100	—	—	23.5	20	59
29	10.1	109	800	2300	35	17.8	575
590	11.0	128	690	450	27	8.5	830
185	24.5	145	1000	1500	30	17.8	72
210	23.5	140	910	1040	28.5	10.1	180
7.5	9.7	>800	—	—	39	16.5	155
60	8.0	2.7	—	—	45	14.7	53
2.30	7.3	28	—	—	61	40	300
2.26	—	—	—	—	—	8.9	>500
0.026	0.22	—	—	—	—	7.1	2.2
0.019	0.062	—	—	—	—	3.1	20
1150	—	—	1580	—	111	13.5	21.8
—	256	—	1670	—	92	6.3	36
2.55	39	325	—	—	203	23.5	14.2
80,000	121	5500	2900	27,000	705	4.9	3.15
0.70	19.6	450	470	—	1700	360	121

The loss of CO$_2$ and H$_2$S from the water with steam formation and the consequent readjustment of the acid–base equilibria in the water is discussed later in detail (Chapter 7). The waters collected at the surface are of a higher pH than the deep high-temperature water.

The situation is complex for spring waters. Although some concentration by steam loss may occur as waters rise in natural upflow channels, cooling by conduction and dilution with other waters can also occur. In Chapter 6 the use of mixing models to relate hot spring compositions to underground water supplies is discussed. Steam heating of a pool causes evaporation and often increases sulfate concentrations.

The most common type of water at deep levels in geothermal systems is an alkali chloride solution at a pH within 1–2 units of the neutral-pH water at the temperature. The more concentrated waters also contain appreciable calcium. Silica concentrations are very high, and solutes such as boron, fluoride, arsenic, ammonia, and hydrogen sulfide are usually present at much higher concentration levels than are common in cold waters. Fluoride concentrations in thermal waters are commonly in the range of 1–10 ppm. Boron concentrations may be extraordinarily high in hot waters passing through organic-rich sedimentary rocks (e.g., at Ngawha, New Zealand, and Sulphur Bank, California). Arsenic concentrations up to 48 ppm have been found in hot spring waters of the El Tatio geothermal area in Chile.

Appreciable concentrations of lithium, rubidium, and cesium occur in many high-temperature waters (not in the Iceland basaltic areas). In some thermal waters the molal concentration of lithium approaches that for potassium, while cesium concentrations as high as 13–14 ppm are found in hot spring waters at El Tatio. Values of the ionic Na/K ratio are lowest in the waters of highest underground temperature.

Common rock-forming elements such as magnesium, aluminum, iron, and manganese are at very low concentration levels, except in waters of very high salinity or acidity. (Before the advent of atomic absorption analysis, magnesium concentrations reported in thermal waters from titration methods were frequently too high). Aluminum concentrations are usually below 1 ppm.

The Cl/SO$_4$ ratio is usually higher in the deep water than in the local springs, due to surface oxidation of sulfide and to the low solubility of sulfate minerals at high temperature. Chloride waters may become modified to acid sulfate–chloride solutions when they come into contact with rocks containing sulfur, as in the Tatun volcanic zone in Taiwan and at Rotokaua, New Zealand. In the Rotokaua area the contact with sulfur occurs at shallow levels, and deep geothermal wells penetrated to neutral-pH alkali chloride water beneath. In contrast, in the Tatun zone sulfur-containing rock and acidic water occur to depths of at least 800 m.

At Matsukawa, Japan, wells at first tapped hot dilute iron sulfate solutions, low in chloride. The wells produced only small quantities of water and later changed to dry steam discharges. This may be an aquifer system heated by high-temperature steam containing sulfur gases, or alternatively, a steam-heated aquifer in contact with sulfur-bearing rocks.

In most explored geothermal areas the ratios of the most soluble constituents, such as Cl/B, Cl/Br, Na/Li, and Na/Cs, usually differ little between waters from springs of good flow and waters from deep wells. Marker ratios of this type are very useful in examining the areal extent of an aquifer.

Wells to depths of 3000–4500 m near Makhachkala, Dagestan, USSR, tapped water of a similar general character to the spring analysis shown in Table 3.2, with a total mineralization of 3–9 g/kg and a temperature of 130°–150° (Dr. V. Zyka, personal communication).

The most important point to be made at this stage is that high-temperature springs with a good outflow of chloride-containing water usually can be interpreted to give an accurate indication of the deep-water composition in an area. Small springs or water seepages which are much diluted, oxidized, and cooled (e.g., in the Salton Sea area, Table 3.2) lose much of their original chemical identity through near-surface reactions.

Table 3.3 gives a summary of some trace metal concentrations in a series of geothermal waters, calculated for the water at depth in the field. There is a close relationship between the concentration of many minor metals and the salinity of the water. The examples have been chosen to cover a wide range of chloride concentrations. In the low-salinity, high-temperature waters which are most frequently encountered, trace metals seldom exceed the order of tens of parts per billion (10^9). However, for highly saline geothermal waters, even those of moderate temperatures, heavy metal concentrations become appreciable. There is an approximate relationship between many heavy metal concentrations and the square of the water salinity (approximately m^2_{Na}), and the reasons for this are discussed in Section 3.5. The unusually low pH (~3) of the Matsao water in comparison with the other waters is to a large extent responsible for its high trace metal concentrations. The concentrations are not typical of the more usual near-neutral-pH waters.

Examples of concentrations of other heavy metals in Broadlands waters are (concentrations in parts per billion) W, 59; Tl, 0.47; V, 1.5; Cd, 0.01; Ge, 3; Sn, 1.5; Be, 0.1; Au, 0.03 (analysts R. L. Goguel and J. A. Ritchie, Chemistry Division, DSIR). The partitioning of trace metals between water and precipitates, which form from cooling waters, is considered in Section 3.6.

TABLE 3.3

Some Minor Element Concentrations in Geothermal Waters[a]

Source and temperature	Cl (ppm)	Mg (ppm)	Mn (ppm)	Fe (ppm)	Ni	Cu	Pb	Zn	Sb	Ag	Ref.[b]
Mean, 1 and 2 IID wells, Salton Sea (300°–350°)	155,000	10–54	1400	2150	—	5500	91,000	520,000	400	1400	(a)
Well, Cheleken Pen., Caspian Sea (80°)	157,000	3080	46.5	14	330	1410	9200	3060	—	—	(b)
Aquifer, Cerro Prieto, Mexicali (340°)	7420–11,750	6–33	0.64	0.2	2.2	5	4.6	6	400	4	(c)
Well E205, Matsao, Taiwan (245°)	9000	88	28	148	—	35	500	8800	—	—	(d)
Average Wairakei wells (250°)	1500	<0.01	0.001	0.008	0.7	1.3	1	1.5	70	—	(d)
Well 2, Broadlands, N.Z. (260°)	1180	<0.01	0.001	0.01	0.1	0.6	0.8	0.6	130	0.5	(d)

[a] In parts per 10^9 in the underground water unless otherwise noted. Chloride concentrations give a measure of salinity.
[b] References: (a) White (1968), (b) Lebedev (1972), (c) Mercado (1967), (d) Chemistry Division.

3.2 STEAM

Although the term fumarole is often used for all natural steam outlets, Ivanov (1958) noted a useful classification which would require a more restricted use of the term. *Fumarolic* steam can be used as a term for high-temperature volcanic steam which arises directly from a magmatic origin and which has not passed through a hot water body. It contains gases such as HCl, HF, CO_2, H_2S, and SO_2. *Solfataric* steam is a useful term for the steam boiling from an underground hot water phase, as in the area of Solfatara, near Pozzuoli in Italy. Both types of steam may condense in surface waters, when the steam origin may be distinguished through the relative solution concentrations of constituents such as Cl, F, B, SO_4, and CO_2. The Larderello and The Geysers steam flows are considered to be special cases of solfataric activity. Typical analyses for examples of each type—a fumarole on White Island volcano, New Zealand, and Karapiti Blowhole, a steam vent at Wairakei, New Zealand—are given in Table 3.4.

Steam and Gas Compositions

The concentration of gases in the steam is an important factor in planning geothermal power generation plants and in assessing the effect of exploitation on underground conditions in a system. In steam-producing systems, such as Larderello and The Geysers, the gas concentration in the steam samples does not vary with the collection pressure. However, in high-temperature water areas, boiling occurs either within the well or in natural flow channels as waters rise toward the surface. After the formation of only a few percent of steam, the

TABLE 3.4

Comparison of Compositions of Steam from a Fumarole on an Active Volcano (White Island, Bay of Plenty, N.Z.) and from Underground Hot Water (Karapiti Blowhole, Wairakei, N.Z.) [a]

	H_2O	CO_2	SO_2	H_2S	HCl	HF	H_2	CH_4	N_2	NH_3
White Island fumarole (650°)	79.6	13.9	4.8	1.5	0.17	0.2	0.16	0.0003	0.018	1×10^{-4}
Karapiti Blowhole (115°)	99.8	0.16	0	0.004	0	10^{-6}	0.002	0.001	0.0002	4×10^{-4}

[a] Compositions in mole percent.

steam phase contains a large proportion of the gases originally dissolved in the water. The concentrations of such gases as CO_2, H_2S, CH_4, H_2, and N_2 in the steam are inversely related to the percentage of steam flashed from the water, and therefore to the separation pressure at which the steam is collected from the discharge.

In most areas where comparison was possible, natural steam from major fumaroles gave a good indication of the quality of steam produced later by wells. However, some hydrogen sulfide may be lost from steam rising in natural channels, and near-surface reaction of

TABLE 3.5

Source	Source depth (m)	Temp. (°C)	Pressure for composition (bars abs.)	Steam fraction in discharge at pressure	Total gas in steam ((mole %)
Fumaroles, Larderello	0	100	1[c]	1.00[c]	3
Average well, Larderello	500	200	1	1.00[c]	2.0
Average well, The Geysers	200–2000	230	1	1.00[c]	0.59
Well MR-2, Matsukawa	1080	230	1	High	0.22
Wells, Reykjavik	500	100	————————Water only———————		
Well G-3, Hveragerdi	400	216	1	0.20	0.015
Karapiti fumarole, Wairakei, N.Z.	0	115	1	1.00	0.17
Average well, Wairakei, N.Z.	650	260	1	0.32	0.063
Well 11, Broadlands, N.Z.	760	260	1	0.355	0.61
Well 1, Ngawha, N.Z.	585	228	1	0.33	20
Fumarole, Matsao, Tatun	0	100	1	1.00	3.2
Well E205, Matsao, Tatun	1500	245	1	~0.3	0.15
Well 5, Mexicali	1285	340	1	0.44	0.54
Fumarole, El Tatio, Chile	0	86.6	0.6	1.00	0.10
Well 2, El Tatio, Chile	650	220	0.9	0.24	0.11
Wells, Salton Sea	1500–1800	300–350	16	0.18	0.1–1.0
Well 1, Ahuachapan	1195	230	1	0.23	0.007
Well 7, Otake, Japan	350	230	3.0	0.215	0.10

[a] $\overline{HC}$ means total hydrocarbon gases.
[b] References:
  (a) Nasini (1930).
  (b) ENEL (1970).
  (c) Kruger and Otte (1973).
  (e) Dr. G. E. Sigvaldason (personal communication).
  (g) Y-C Liu (personal communication).
  (i) Helgeson (1968).
[c] Area of steam discharges only.

  (d) Nakamura and Sumi (1967).
  (f) Chemistry Div., DSIR.
  (h) Alonso (1966).
  (j) Noguchi (1966).

steam with organic material may create additional methane, ammonia, and carbon dioxide in steam from minor steam vents.

Table 3.5 gives the composition of steam from geothermal wells and natural steam vents in several areas. For some hot water areas the analysis of steam from wells is given for atmospheric pressure separation, but for utilization studies the approximate gas concentrations at higher separation pressures can be calculated if the steam fraction of the well discharge is known.

Although gas concentrations and compositions cover a wide range,

Composition of Steam from Fumaroles and Wells[a]

			Gas composition (mole %)					
CO_2	H_2S	$\overline{HC}$	H_2	$N_2 + A$	O_2	NH_3	H_3BO_3	Ref.[b]
92.3	2.06	1.4	2.6	1.07	0.05	—	0.5	(a)
94.1	1.6	1.2	2.3	0.8	—	0.8	0.33	(b)
55	4.8	9.5	15	3	—	12.5	0.25	(c)
81.8	14.1			Remainder 4.1%				(d)
—	—	—	—	100	—	—	—	(e)
73.7	7.3	0.4	5.7	12.9	—	—	—	(e)
94.6	2.3	0.74	1.0	1.1	—	0.26	—	(f)
91.7	4.4	0.9	0.8	1.5	—	0.6	0.05	(f)
94.8	2.1	1.2	0.2	1.5	—	0.2	—	(f)
93.9	0.7	4.1	0.5	0.8	—	0.04	—	(f)
55.0	37.2			Remainder 7.8%				(g)
92	5	0.7	0.8	1.5	—	—	—	(f)
81.4	3.6	7.0	0.5	7.0	0.4	—	—	(h)
75	0.6	0.1	0.0	19.4	4.9	—	—	(f)
99	0.7	0.01	0.03	0.2	—	—	—	(f)
90			Remainder mainly H_2S + minor $\overline{HC}$ and H_2					(i)
50–80	4	—	10–40	2–10	—	—	—	(e)
96.7	0.65			Remainder 2.7%				(j)

the predominant gases associated with geothermal steam from both hot water and steam-producing areas are carbon dioxide and hydrogen sulfide. In lower-temperature waters of Tertiary basalt areas of Iceland, the gas is mainly nitrogen derived from air originally dissolved in the circulating meteoric water. The oxygen is presumably lost in the rock through oxidation processes.

Hydrogen sulfide concentrations vary widely. Ahuachapan, El Salvador, and El Tatio, Chile well discharges have very low concentrations of hydrogen sulfide. At the opposite extreme, wells at Matsukawa and in the Tatun zone of Taiwan produce steam which contains very high hydrogen sulfide concentrations. A further example is Namafjall in northern Iceland, where sulfur deposits occur, and where the gas in steam flows contains about 20% H_2S. Areas which have major natural steam flows with gases containing more than about 10% hydrogen sulfide may be suspected of containing deep high-temperature acidic waters associated with sulfur deposits or of an association with active volcanism.

The Larderello and The Geysers areas have appreciable concentrations of boric acid in the steam, while the gas at The Geysers has a high proportion of hydrogen, methane, and ammonia. A high proportion of hydrogen also occurs in the gas in steam from some high-temperature water wells in Iceland (Namafjall) and at Ahuachapan, El Salvador.

Where high-temperature waters come into contact with organic-rich sedimentary rocks, as at Ngawha, New Zealand, the gas concentrations in the steam may be high (up to 50%). The gas at Ngawha was mainly carbon dioxide and methane. In the Broadlands geothermal system, where the hot waters at depth contact greywacke and argillite, much higher carbon dioxide and methane concentrations occurred in well discharges than in the Wairakei system, where the aquifer rocks were predominately volcanic.

The oxygen concentration in steam and hot water from deep levels in geothermal systems is below detection levels for uncontaminated samples. At Wairakei and Broadlands, for example, the calculated partial pressure of oxygen in the underground systems is of the order of 10^{-40} bars. (See Chapter 4.6.)

In warm waters of nonvolcanic systems the gas compositions are variable. For example, in the regions of Alpine folding (Carpathians; the Caucasus) a high proportion of nitrogen is found in the gases associated with the hot waters, while in thermal waters of the provinces of the Russian platform, nitrogen and methane are the most common gases (Makarenko and Mavritsky, 1965).

Where a geothermal well draws its output from a single deep hot liquid phase it is possible, by combining the steam and water analyses in the correct proportions, to calculate the concentrations of various gases in the water that existed before a steam plus gas phase was formed. The partial pressures of gases in the original water can then be estimated and correlated with mineral–solution equilibria (Chapter 7). However, inflow to a well sometimes occurs over a range of depths, and may include steam which has separated in the country rock. In this case, the integrated steam and water analyses for the well cannot be correlated with a deep-water composition. This situation, however, can usually be identified by the physical and chemical characteristics of the discharge.

Table 3.6 gives the partial pressures of gaseous solutes in the deep hot waters of several geothermal areas; it shows that values for most gases vary widely between areas, but the partial pressure of hydrogen usually is of the order of 0.1 bar. The pressure is much lower for El Tatio but there is other evidence that loss of steam and gas has occurred from the deep water tapped by the well. The table also gives the approximate partial pressures of gases in the underground steam phase at Larderello and The Geysers.

A limited number of analyses are available for the concentrations of

TABLE 3.6

Partial Pressure of Gases (P_g, in Bars) in Deep Hot Waters

Area	Temp. (°C)	P_{CO_2}	P_{H_2S}	P_{CH_4}	P_{H_2}	P_{N_2}	P_{NH_3}
Wairakei (average)	260	1.0	0.02	0.04	0.04	0.1	0.0003
Broadlands (well 11)	260	11	0.11	0.5	0.11	1.0	0.001
Hveragerdi (well G-3)	216	0.14	0.005	0.004	0.07	0.1	—
Mexicali (well 5)	340	7	0.2	1.3	0.11	1.7	—
Ahuachapan (well 1)	230	0.06	2.5×10^{-3}	—	0.1	0.05	—
El Tatio (well 2)	220	1.6	2×10^{-3}	8×10^{-4}	2×10^{-3}	2×10^{-2}	—
Larderello[a]	220	0.4	0.007	0.006	0.01	0.004	0.004
The Geysers[a]	230	0.09	0.01	0.015	0.025	0.005	0.02

[a] In steam phase.

helium, neon, argon, krypton, and xenon in gases from steam in
geothermal areas. Hulston and McCabe (1962a) gave results from mass
spectrometric determinations of gases from several New Zealand
areas, including the concentrations and isotopic compositions of
argon. For Wairakei and for Larderello wells, Wasserburg *et al.* (1963)
presented concentration and isotopic data for carbon gases, hydrogen,
nitrogen, helium, and argon. Mazor and Wasserburg (1965) reported
concentrations of helium, neon, argon, krypton, and xenon in gases
from hot springs at the Yellowstone and Lassen National Parks in the
United States, and later Mazor and Fournier (1973) gave analyses for
noble gases in steam from Yellowstone Park wells. A review of the
application of inert gas analyses to geothermal prospecting was given
by Mazor (1975). Table 3.7 contains some of the analytical information.

3.3 ORIGIN AND AGE OF THE WATER

Source of the Water

The source of the water in a geothermal area and whether the water
is static or part of a deep circulating system are important questions
when considering the long-term output potential of the area.

TABLE 3.7

Concentration of Some Minor Gases in Various Geothermal Discharges[a]

	N_2	He	Ne	Ar	Kr	Xe	Ref.[b]
Larderello wells	5500	12.6	—	8.3	—	—	(a)
Average, Wairakei wells	15,000	1.5	—	250	—	—	(b)
Wairakei well 18	15,000	12	—	242	—	—	(a)
Average, Yellowstone and Lassen hot spring gases	43,000	12.9	1.0	820	0.15	0.033	(c)
Yellowstone wells, gas phase	—	127[c]	0.3	300	0.07	0.01	(d)

[a] In parts per million by volume, cm^3/m^3, of total gas.
[b] *References:* (a) Wasserburg *et al.* (1963).   (b) Hulston and McCabe (1962a).
                    (c) Mazor and Wasserburg (1965).   (d) Mazor and Fournier (1973).
[c] One result only.

Fortunately, natural waters have built-in identification tags. The principal stable molecular species in natural waters are $H_2^{16}O$, $H_2^{18}O$, $H_2^{17}O$, and HDO. In seawater the proportions of these species are 10^6 : 2000 : 420 : 316 (Craig, 1963) and this composition is referred to as SMOW (standard mean ocean water). Atmospheric water derived from the ocean is depleted in ^{18}O and in deuterium. The isotopic composition of precipitation from the atmosphere depends on the fraction of water remaining in the air mass from which the rain or snowfall is derived, the first precipitation being richest in heavy isotopes. As reviewed by Craig (1963), a worldwide study of freshwater samples showed that the isotopic reactions in meteoric waters could be expressed by the equation* $\delta D = 8 \, \delta^{18}O + 10$. The percentage depletions of heavy isotopes in precipitation in general can be correlated with latitude, altitude, and distance from the sea. Surface waters affected by extensive nonequilibrium evaporation, as in inland basins, lie off this line, but at ordinary air temperatures they are connected approximately to the original precipitation composition $\delta^{18}O_0, \delta D_0$ by a line expressing the equation $\delta D = 5 \, (\delta^{18}O - \delta^{18}O_0) + \delta D_0$. The line slope is less at higher temperatures, for both equilibrium evaporation and nonequilibrium evaporation. For example, rapid surface evaporation of waters at 70°–90° gives water compositions approximately on a line $\delta D = 3(\delta^{18}O - \delta^{18}O_0) + \delta D_0$.

Igneous rocks have $\delta^{18}O$ values in the range of +6 to +12, and water in contact with magma will approach a similar oxygen isotopic composition range. Although the δD value of juvenile water is not known, it is likely to be below that of seawater by several tens of per mil units, since there has been a preferential loss of hydrogen gas from the atmosphere to space; for example, a value of $\delta D = -55.3‰$ was obtained for water from Surtsey volcano, Iceland (Arnason and Sigurgeirsson, 1968). The δD values for magmatic water will cover a rather wider range than for juvenile water due to recycling of hydrous crustal rocks. The range suggested by White (1974) is shown in Fig. 3.1.

Craig (1963) gave isotopic data from several hydrothermal areas of the world with high-temperature chloride water or steam flows. The deuterium content of thermal water from most areas is approximately constant, and equal to that of the local meteoric waters. However, there is usually enrichment in ^{18}O ("oxygen shift") to a varying extent in thermal waters of each area, ranging from a very large effect in the Salton Sea area to a small effect at Wairakei. Figure 3.1 summarizes

---

* $R/R_0 = 1 + \delta$. R is the isotope ratio in a sample and R_0 the ratio in a standard material. δ values are usually expressed as parts per thousand (‰).

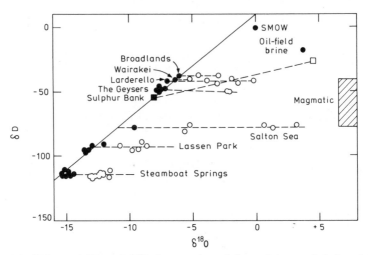

Fig. 3.1 Values of δD and δ¹⁸O for samples of thermal (open circles) and local meteoric waters (solid circles) in several areas.

data for several areas. If the thermal waters were formed at deep levels from a mixture of local meteoric waters with magmatic water of a hypothetical composition shown in Fig. 3.1, a horizontal series of points should not be expected in individual areas over a wide range of δD values. The simplest general model is that essentially all the hot water is local meteoric water which has been heated and enriched in ¹⁸O by exchange with silicates at about the observed water temperature (oxygen shift). The deuterium concentration is almost unchanged because of the generally lower abundance of hydrogen in the rocks. However, some exchange of deuterium/hydrogen with hydrous minerals does occur, and the possibility of a slight "deuterium shift" should not be ruled out in systems where a large proportion of clay and micaceous minerals occurs in the rocks (e.g., chlorite has an H/O ratio of about 4/9).

In the general model, areas showing little oxygen shift could be old systems of extensive water throughput in which the isotopic composition of rock has adjusted to equilibrium with the recharge water. A large oxygen shift could be expected in systems of small throughput, as well as in young hydrothermal systems where waters come in contact with relatively fresh rock.

The alternative model of mixing juvenile or magmatic water with local meteoric water would fit the isotope results for some areas where δD values of the local waters coincide with the magmatic range. However, in this case general correlations between temperature, salinity, and ^{18}O shift would be expected. The difficulty, if not the impossibility, of distinguishing magmatic water from meteoric water that has circulated to great depths and near-magmatic temperatures should also be kept in mind.

The model of circulating meteoric water does not fit the isotopic results for all geothermal systems. White *et al.* (1973) showed that for the hot spring waters of the Coast Ranges of California (of possible metamorphic origin) both δD and $\delta^{18}O$ values were considerably higher than those for local meteoric water (see Fig. 3.1, Sulphur Bank). Matsubaya *et al.* (1973) suggested from the correlation of increasing salinity of Arima hot spring waters with increasing δD and $\delta^{18}O$ values that the waters resulted from the dilution of a deep hot fossil brine by meteoric waters.

Steam at Larderello and The Geysers could be expected to give isotope results similar to those for the hot water areas, because at about 220°-240° (approximately the steam temperatures) the fractionation factor for deuterium/hydrogen distribution between steam and liquid water passes through unity (Combs and Smith, 1957; Craig, 1969).

Waters of stagnant acid sulfate pools and in mud pots which result from the steam heating and evaporation of surface waters show a different $\delta D/\delta^{18}O$ trend from that of waters and steam rising from deep levels. The waters fall on lines which approximately intersect the local meteoric water composition and have slopes commonly of the value of 2–3, depending on the temperature and mode of evaporation. This different trend helps to separate two possible origins for low-chloride acid sulfate waters: (1) steam-heated and evaporated surface waters, or (2) waters rising from deeper levels.

Excluding deuterium shifts form the discussion, we see in Fig. 3.2 a diagram of some variations in δD and $\delta^{18}O$ values which could be encountered in high-temperature geothermal areas. Cold meteoric water A is heated at depth, producing a high-temperature water B (after oxygen shift). Near-surface mixing of A and B may produce waters falling on a line AB. Water A may be evaporated at the surface by conductive heating, giving waters with compositions along lines with slopes which depend as shown on the temperature and the rate of evaporation.

If the deep water B boils off steam, the composition of the latter will be C_1, C_2, C_3, C_4, etc., in decreasing order of steam separation temperature (C_1, about 220°; C_2, 200°; C_3, 150°; C_4, 100°). The water composition would move slightly in the opposite direction.

If, for example, steam of composition C_3 evaporates surface water A

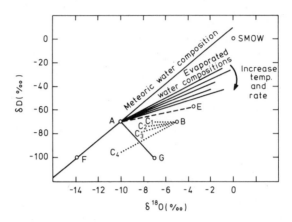

Fig. 3.2 The values of δD and δ^{18}O that may be found in geothermal areas, arising from various evaporation, mixing, and mineral–water exchange processes.

at 100°, the evaporating water compositions will lie along the dashed line AE. Alternatively, steam C_1, C_2, etc., could rise to the surface unchanged and discharge from fumaroles.

A further possibility is that the geothermal system is based on meteoric water F from outside the local water catchment area, and that taking into account an oxygen shift the deep water produced has a composition G. Mixing of G with local water A would result in a series of spring waters along the composition line AG (this situation was found at El Tatio, Chile).

The isotopic compositions of waters therefore give valuable information on hydrological processes and the directions of water movement within a geothermal area. For example, Arnason and Sigurgeirsson (1966) were able to draw conclusions about the recharge areas for geothermal areas in southwest Iceland from a knowledge of the regional distribution of deuterium in the precipitation.

Relatively little work has been done on hydrothermal systems in areas of marine sedimentary rocks. White (1965, 1974) gave some values for saline waters from this type of environment, showing δD values of the order of −10 to −20‰ and δ^{18}O values of +3 to +5‰ with respect to SMOW composition (a slight oxygen shift). Slightly negative δD values could be expected from the mixing of hydrogen from parent terrestial clays with ocean water retained in marine sediments after compaction. A typical point for an oil-field brine is shown in Fig. 3.1.

Age of the Water

Local meteoric waters appear to be the most probable ultimate source of the waters within most high-temperature geothermal systems. However, the situation is not simple, since with increasing depth in the upflow zones there may be several successive stages of mixing of cold surface waters with hot waters rising from considerable depths. The times taken for water to circulate through the system at particular levels are of importance in developing hydrodynamic models.

Tritium, a radioactive isotope of hydrogen, of half-life 12.5 years, occurs in all meteoric waters through nuclear reactions induced by cosmic-ray bombardment of the stratosphere. The built-in tritium "clock" can be used to measure the time that waters have been away from the surface. Large temporary increases in the tritium concentration in atmospheric waters have been caused in recent years by thermonuclear explosions, the effects varying between the northern and southern hemispheres.

Begemann (1963) discussed various circulation and mixing models by which tritium concentrations in waters can be examined. The tritium concentrations of waters from deep levels at Wairakei, Steamboat Springs, and Larderello are of the order of 0.1 tritium unit (T/H = 0.1×10^{-18}) or less. Deep-water circulation times well in excess of 50 years are indicated. Higher tritium concentrations may occur in spring and shallow well waters but are likely to be due to the admixture of a small proportion of relatively young waters; for example, 10% of "young" water in "old" water gives an apparent age for the mixture of about 40 years. Wilson (1963b) suggested the possibility that minor concentrations of tritium in hot waters could be due to a local accumulation of radioactive elements through hydrothermal solution processes.

Several attempts have been made to use carbon 14 (half-life, 5570 years) as an indicator of deep-water circulation times. However, there are difficulties in knowing exactly what proportion of the carbon in a hydrothermal system is derived from the atmosphere or the soil, and from rock carbonates or carbon (and possibly magmatic carbon). At various stages in the lifetime of a geothermal system, a large proportion of carbon dioxide (and ^{14}C activity) could be lost through calcite precipitation or by boiling of waters. Carbon-14 activity also could be affected by CO_2 exchange with old-rock carbonate. Some geothermal areas (The Geysers, Steamboat Springs, Salton Sea) have no measurable ^{14}C activity (Craig, 1963); that is, the carbon activity is less than 0.5% of modern carbon. Similar results were obtained for a deep Wairakei well (Fergusson and Knox, 1959).

It is doubtful whether ^{14}C work could give more than an order of magnitude estimate for water circulation times in deep convective systems due to the uncertainties over carbon loss and exchange. However, for a simplified model Craig (1963) suggested that at Steamboat Springs waters spend at least 30,000 years, and possibly much longer, underground. These long circulation times are in agreement with those estimated by Elder (1965) for a circulation model of Wairakei.

3.4 ORIGIN OF CHEMICALS IN GEOTHERMAL FLUIDS

General

The problem of the origin of the chemical solutes in natural hot waters and steam is a controversial one of long standing. Papers giving summarizing discussions include White (1965) for waters of sedimentary rocks; Ellis and Mahon (1964, 1967) and Truesdell (1975) for high-temperature geothermal waters; and Facca and Tonani (1964) for natural steam areas.

Many of the early discussions on the source of the solutes in waters were intuitive rather than based on quantitative evidence. The presence of lithium, rubidium, cesium, boron, fluoride, arsenic, carbon dioxide, and hydrogen sulfide in high-temperature waters was usually considered to be evidence of the presence of magmatic water. There are known geological processes, such as pegmatite formation and volcanic gas evolution, in which concentrations of at least some of these constituents occur. However, accumulation of experimental data on the reaction of hot water with rocks of various types, and a study of the isotopic composition of elements in thermal waters, led to a reexamination of the problem.

As already noted, the isotopic compositions of most high-temperature waters suggest that they are derived almost entirely from local meteoric waters that have become heated and charged with chemicals. The two simple models which are frequently discussed are shown in Fig. 3.3 (from Ellis, 1966). In model A, both heat and dissolved chemicals are derived from the addition of magmatic solutions to deep circulating water. This includes the possibility that the water may

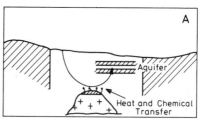

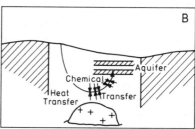

Fig. 3.3 Two simple alternative models for the origin of the heat and chemicals in geothermal waters.

diffuse into and through the magma. Later reactions in the outflow arm of the system are largely depositional processes associated with cooling. Steam and other volatiles may boil off from the geothermal waters near the surface. In model B, the waters achieve their chemical nature mainly by interaction with rock at all stages during the cycle, but particularly in the hottest zone. In this model, the unusual chemical nature of thermal waters is due to their high temperature and the local rock compositions. The heating need not be associated with a magmatic intrusion, and the model applies to any hot rock–water system. For similar temperatures, differences in the chemical composition of waters are due mainly to differences in local rock compositions rather than to the origin of the heat.

The mechanisms are not mutually exclusive. In many areas it is more relevant to pose the question which model predominates, rather than which is the model.

Solutes in natural waters can be divided approximately into two categories. The common rock-forming constituents, such as silica, aluminum, sodium, potassium, calcium, magnesium, iron, and manganese, are available in abundance. Their solubilities are limited by particular mineral–water equilibria, which in all cases favor the retention of the elements in the mineral phases. A second class of "soluble" elements includes constituents such as chloride, bromide, boric acid, arsenic, and cesium, which favor the liquid water phase.

An example of the type of investigation that is required to understand element origins in individual thermal areas is given with reference to the Taupo Volcanic Zone of New Zealand. Analytical information is available on minor elements in the range of volcanic rocks in this zone, and elements such as chlorine, fluorine, boron, and nitrogen occur in appreciable concentrations. Table 3.8 gives the annual natural output of particular elements from the Wairakei geothermal system, the typical concentration range of elements in the rhyolites and andesites of the zone, and the volume of these rocks that would be required to supply the annual output. For an assumption that the heat output of Wairakei is supplied by crystallizing magma beneath the area, the approximate magma volume required annually is also given.

The volume of rock required to supply the elements is of the same order as that of magma required to supply the heat, even for constituents such as lithium, cesium, and arsenic. However, the elements are available in the volcanic rocks throughout the system, whereas a magmatic heat source would be more localized. The availability and transfer of heat is as much a problem as the source of chemicals in proposing a model for volcanic geothermal systems.

TABLE 3.8

Concentration Range of Some Minor Elements in Taupo Volcanic Zone Rhyolites and Andesites, and Volume of Rock Required to Supply the Chemical Output of the Wairakei Geothermal Area[a]

Component	Wairakei output (10^4 kg/yr)	Conc. range in volcanic rock (ppm)	Rock required (km^3/yr)
Cl	2000	200–1000	0.01–0.05
B	100	10–25	0.02–0.05
F	8	100–400	0.0001–0.0004
Li	15	20–80	0.001–0.004
Cs	2.5	0.5–2	0.005–0.02
As	6	2–5	0.007–0.02
		Heat available molten rock	
Heat output 5×10^{15} cal/yr.		200 cal/g	0.01

[a] The volume of rock magma required to supply the heat output is given for comparison (Ellis, 1966).

Reaction of Hot Water and Rock

Experiments by Ellis and Mahon (1964, 1967) and by Mahon (1967) tested the availability to hot water solutions of various elements in volcanic and sedimentary rocks of the Taupo zone. Table 3.9 gives the partial compositions of solutions after reacting volcanic rocks with waters in equal proportions at various temperatures and pressures. Below 300° reaction rates were slow under the static conditions of the experiments, and in the experimental times of two weeks only partial reaction occurred. However, a high proportion (often 50–80%) of the chloride and boron in the rocks was dissolved, even before there was observable hydrothermal alteration. In particular, basalts and andesites liberated chloride and boron very readily. Fluoride was also dissolved readily from volcanic rocks, often producing metastable high concentrations in solution which later decreased to stable levels. Arsenic at the level of about 1 ppm was also found in rhyolite solutions (Ellis, 1967).

For constituents such as chloride and boron, where simple extraction from rock surfaces and fractures seems to occur, the concentration of the elements in the solutions could be expected to be in proportion to the rock/water ratio. In natural hydrothermal areas, a rock/water ratio of 10–20 is a realistic value, and chloride concentrations of the order of a few thousand parts per million and boron concentrations of the order of 20–30 ppm could readily be produced by hot water reacting with the local volcanic rocks. In the New Zealand hydrothermal areas hydrothermally altered rocks are depleted in chloride and boron.

At temperatures and times sufficient to produce hydrothermal alteration of the rocks, but well below magmatic temperatures, elements such as rubidium and cesium were liberated into solution, the latter almost quantitatively in some cases. Chelishev (1967) showed that the equilibrium distribution of rubidium and cesium between mica or feldspar and water was increasingly in favor of the solution phase as temperatures were lowered from 600° to 250°. Appreciable concentrations of lithium were not found in the experiments of Ellis and Mahon, but there was the suggestion that lower temperatures favored concentration in solution. However, recent experiments at 300° by F. W. Dickson (personal communication) using the same rhyolite but in a rotating vessel where rock particles were kept in suspension produced solutions containing 22 ppm Li, 6.6 ppm Rb, and 0.2 ppm Cs after two days. Reaction rates were much greater with

TABLE 3.9

Compositions of Solutions Resulting from Reaction of Volcanic Rocks with Water at 1500 bars Pressure for Two Weeks

Rock	Temp. (°C)	Rock/water ratio	Concentration in solution (ppm)								
			Cl	F	B	SO$_4$	NH$_3$	Li	Rb	Cs	Mg
Basalt	250	2^a	250	6	1.4	35	4	0.3	—	—	—
	350	2^a	390	3	2.5	10	4	0.5	<0.2	<0.2	0.2
	500	1	305	1	1.6	25	3	0.6	0.3	<0.2	—
	600	1	260	3	1.7	—	—	0.4	<0.2	<0.2	—
Andesite	250	2^a	52	10	2.5	—	3	0.5	—	—	—
	350	2^a	270	8	8	—	7	1.5	—	—	—
	400	1	125	14	5	73	1	1.8	1.4	0.3	0.4
	500	1	140	5	11	55	2	1.7	0.8	0.2	0.5
Dacite	250	2^a	40	15	1.9	42	3	0.7	—	—	—
	400	1	60	11	2.7	46	2	0.7	0.3	0.5	0.9
	500	1	110	9	7	61	2	0.2	1.2	0.6	1.8
	600	1	110	7	5	—	—	0.1	—	—	—
Rhyolite pumice	200	1^a	10	20	0.3	30	1	1.0	—	—	—
	300	1^a	150	50	0.6	35	1	—	—	—	—
	400	0.25	157	33	0.7	32	1	1.8	0.6	0.3	0.4
	500	0.25	168	34	1.6	34	2	0.8	0.9	0.7	0.8
Rhyolite	350	2^a	110	19	0.4	8	4	1.2	—	—	—
	400	1	105	24	1.1	175	8	1.4	0.3	<0.2	0.7
	500	1	270	35	5	125	7	0.2	1.3	0.8	0.4
	600	1	290	29	7	—	—	0.3	2.0	1.4	—

^{a} Runs at 500 bars pressure.

this technique. In a reaction of fresh pumice breccia with a Wairakei well water for a year at 215°, the lithium in solution increased from 10.5 to 20.5 ppm (Dr. R. B. Glover, personal communication).

Major hydrothermal alteration of rocks appears to be necessary to create appreciable concentrations of lithium, rubidium, and cesium in waters, and moderate temperatures (200°–300°C) are more favorable than very high temperatures (500°–600°).

F. W. Dickson (personal communication) also completed a series of experiments interacting basalts and seawater at temperatures of 200°–300°C. These experiments produced solutions with compositions very similar to those of waters of the Reykjanes geothermal area of the southwest coast of Iceland.

The rock–water experiments produce concentrations of solutes such as potassium, calcium, magnesium, silica, fluoride, sulfate, and ammonia at levels which correspond to those in high-temperature geothermal waters of similar salinity. Concentrations of these elements are controlled by temperature-dependent mineral–solution equilibria. These are discussed quantitatively in later sections.

In general, the experimental work has shown that in natural high-temperature geothermal waters, chlorine, bromine, boron, and cesium are among the few elements not involved to an appreciable extent in temperature- and pressure-dependent chemical equilibria between rock minerals and high-temperature water. Evidence from natural systems suggests that at deep levels, lithium also often acts as a rather unreactive solute of this type.

The interrelationships between this group of soluble elements can be used to identify particular waters in a geothermal system and to predict the type of rocks with which the waters have been in contact. For example, Table 3.2 shows the low lithium and cesium concentrations in waters in basalt areas, and the low Cl/B ratio in areas of sedimentary rocks, as at Ngawha, New Zealand. In the latter area, the lithium concentration is likely to have been derived from the local sediments and not the basalts. Ratios of Cl/B in the Taupo Volcanic Zone waters can be related to the ratio in the volcanic rocks.

Mahon (1967) gave an experimental background for New Zealand hydrothermal systems in greywacke, shales, and mudstones. He showed that solutions from reactions at temperatures of 200°–300° had chemical characteristics (Cl/B and Cl/NH_3 ratios and relative chloride concentrations) which were in line with the composition of hot waters found in areas with these rocks. Kissen and Pakhomov (1967) reported a similar study up to 200° with claystones and siltstones of the Caucasian mineral water region.

Facca and Tonani (1964) and Panichi (1975), in discussing the origin of chemical constituents in natural steam areas, pointed out that there was no evidence for a magmatic origin for the carbon dioxide, hydrogen sulfide, hydrocarbons, and ammonia contained in the steam. All of these constituents could be derived from simple hydrolysis and decomposition reactions of organic material and carbonates in hot sedimentary rocks.

The only high-temperature waters in Table 3.2 which contain appreciable concentrations of iron, manganese, and magnesium are those of high salinity. Highly saline waters such as those in the Salton Sea system may also contain high concentrations of other metals, such as lead, zinc, copper, and silver (Skinner *et al.*, 1967). Laboratory studies reacting salt solutions with rocks at high temperatures show a good correlation between salinity and the concentrations of metals taken into the aqueous phase during hydrothermal alteration (Ellis, 1968). For a sulfide-free system, reaction of an andesite with water and with sodium chloride solutions produced the results shown in Table 3.10. A rock of average composition can, when heated with salt solutions, produce waters which contain appreciable concentrations of "minor metals." In a system containing sulfide, the distribution is less in favor of the solution, although metal concentrations would still be a function of salinity. A complex interplay between metal sulfide solubilities, water pH, and solution complexes of the metals is involved.

TABLE 3.10

Concentration of Elements in Solutions after Reaction of an Andesite with Water and NaCl Solutions[a]

Reactant	Fe	Mn	Pb	Cu	B
Water	20	0.5	<0.5	<0.5	6
2 M NaCl	40	10	2	0.5	10
4 M NaCl	200	60	3	3	10
Conc. in andesite					
Original	—	950	10	70	12
After water reaction	—	950	8	30	<5
After 2 M NaCl reaction	—	950	2	5	5
After 4 M NaCl reaction	—	850	2	5	5

[a] Concentration in parts per million; reaction at 400°; 1500 bars pressure; rock/water = 1; 4–5 weeks. Spectrographic analyses are given of the andesite before and after reaction (Ellis, 1968).

Evidence from Isotopes and Inert Gases

The origin of gaseous constituents in natural hot waters may be sought through an examination of the ratios of gas concentrations and in the isotopic compositions of elements. For example, Ne, ^{36}Ar, and Kr act as tracers for atmospheric gases and are not produced from rocks in significant quantities. On the other hand, He, ^{136}Xe, ^{222}Rn, and ^{40}Ar are formed continuously from the radioactive decay of the elements U. Th, and ^{40}K (Wasserburg and Mazor, 1965). Whereas Xe, Kr, and Ar have solubilities in water which vary widely with ambient temperatures, Ne and He solubilities have little temperature dependence.

The isotopic composition of carbon gases depends on their origin (e.g., from marine limestone, decomposition of organic material in the rocks and soils, or atmospheric carbon dioxide). Similarly, the origin of sulfur constituents may be examined.

Inert gas analyses were reported for several areas in Table 3.7. The average N_2/Ar ratio in the Yellowstone and Lassen steam samples of about 50 (Mazor and Wasserburg, 1965) was higher than would arise from the degassing of a saturated solution of air in water formed at ambient temperatures (about 37). However, in further work at Yellowstone, Gunter (1973) analyzed gases in both the steam and water phases of springs, and the total nitrogen and total argon outflows correlated well with the proportions expected from circulating meteoric water. Hulston and McCabe (1962a) correlated the excess nitrogen in the New Zealand samples with the hydrogen concentration in the gases, suggesting that at least part of both may originate from the decomposition of organic material. Very high N_2/Ar ratios, as at Larderello (660), are a strong indication that an appreciable amount of the nitrogen (and probably of H_2, CH_4, and NH_3) is derived from organic material in the rocks.

Shukolyukov and Tolstikhin (1965) showed that in warm springs of Transbaikaliya the atmosphere was the only source of argon and xenon present in the nitrogen-rich gases associated with the waters.

Mazor and Wasserburg (1965) suggested that the isotopic composition of argon and the ratios of concentrations of the rare gases (except helium) in Yellowstone and Lassen samples were as expected from a solution of air in cold water. However, in comparison with an Ar/He ratio of about 7000, which would be expected for cold air-saturated water, the ratio was about 60, indicating a considerable contribution of radiogenic helium. Argon/helium ratios for Wairakei and Larderello wells were even lower (about 20 and 0.65). For the Yellowstone wells minor radiogenic ^{40}Ar was detected in the steam phase.

The ratio of radiogenic helium to radiogenic argon (^{3}He/^{40}Ar) in Larderello steam is about 10 and in the proportion expected from radioactive decay in normal rocks (Wasserburg *et al.*, 1963). Ratios of similar magnitude were found at Yellowstone Park (Mazor and Fournier, 1973). The proportion of excess radiogenic argon, ^{40}Ar, in total argon of Larderello steam discharges decreased with time from 18% in 1951 to 14% in 1963, and this type of information may offer the possibility of calculating reservoir volumes (Ferrara *et al.*, 1963a). At Monte Amiata, Italy, there was no excess radiogenic argon in well discharges. This was also the case at Wairakei (Hulston and McCabe, 1962a), and it was suggested that the low ratio (Ar/H$_2$O) in the total well discharges arose from a major loss of gas from the deep hot water during the history of the system.

Craig (1975) showed that in volcanic gas discharges and in high-temperature geothermal fluids there is an excess of ^{3}He over the normal atmospheric ratio ^{3}He/^{4}He of about 1.4×10^{-6}. For Salton Sea and Lassen Park geothermal gases the ^{3}He/^{4}He ratio was between 3 and 11 times the atmospheric ratio. Even higher ratios for Iceland, Kamchatka, and Kuril Islands geothermal areas were reported by Polak *et al.* (1975) and Gutsalo (1975). The suggestion is that the excess ^{3}He in geothermal gases is derived from the mantle through the medium of magmatic fluids. However, the alternatives of independent upward diffusion of ^{3}He through solid rock and of leaching of volcanic rocks also must be considered.

The isotope ratios for total carbon in the Wairakei well discharges range from δ^{13}C values of -3.5‰ to -4.4‰ with respect to the standard PDB, and about 0 to -7‰ for most steam and water samples in other New Zealand thermal areas (Hulston and McCabe, 1962b). Craig (1963) reported similar values (-1 to -5‰) for Yellowstone Park hot pool gases, and for Larderello steam the carbon dioxide values of -2 to -6‰ are also similar (Ferrara *et al.*, 1963b). Carbon dioxide at Lassen Park and Steamboat Springs has a δ^{13}C value of -8 to -9‰, while at The Geysers rather low δ^{13}C values of about -11‰ occur (Craig, 1963). In comparison, marine limestone δ^{13}C values range from about $+3$ to -3‰, ocean and atmospheric CO$_2$ is about -4‰, while organic carbon is appreciably negative, ranging down to about -30‰ for woods, shales, and graphites. In a steady-state crustal model, juvenile carbon would have a δ^{13}C value of about -8 to -11‰ (Craig, 1963).

In the gases from hydrothermal areas, methane has a δ^{13}C value more negative than for carbon dioxide, and is often in the range of −15 to −30‰ (see Chapter 4).

It is possible to correlate the overall carbon isotope composition of the gases in geothermal fluids from mixed, carbonate, organic carbon sources in the rocks which contact the hot waters, but an addition of juvenile carbon to the systems cannot be ruled out on present evidence. Atmospheric carbon dioxide in the cold recharge water is too low in concentration to affect the isotope ratios.

Steiner and Rafter (1966) suggested that the narrow range of $\delta^{34}S$ values for sulfide in the fluids and for iron sulfide minerals in New Zealand geothermal areas (+3 to +5.4‰ with respect to meteoritic sulfur) indicated that the sulfur was derived from an upper crust magma. However, from the results of Giggenbach (1977) it may be inferred that sulfur of an isotopic composition range similar to that at Wairakei and Broadlands could be extracted from the argillite and greywacke rocks which underlie the Taupo Volcanic Zone.

Evidence of the origin of other elements in geothermal waters has also been obtained by a comparison of the isotopic ratio with that in the rocks in the area. As an example, from isotopic compositions in the Salton Sea geothermal brines Doe *et al.* (1966) showed that much of the lead and strontium was derived by leaching of the aquifer sediments rather than from local volcanic rocks.

Summary

Ellis (1965) pointed out that the heat liberated through the hydrothermal alteration of glassy volcanic rocks could play an important part in heating up a stationary deep-water reservoir. The heat liberated during devitrification of glassy volcanic rocks ranges from 40 to 80 cal/ g, which would cause the temperature of water-filled glassy rocks of 10% porosity to rise to 200°–250°. In certain areas the need for external heat may even be questioned.

The presence of elements such as boron, fluorine, rubidium, and cesium in thermal waters is not proof of the presence of magmatic water in the system. Concentration of these elements into the water phase can occur during hydrothermal alteration of rocks at moderate temperatures. Whereas in areas of recent porous volcanic rocks, as in New Zealand , Iceland, and Japan, derivation of the solutes in hot waters by simple rock reaction may be reasonable, it can be argued in the manner of White (1957) that in hydrothermal areas of older dense rock, sufficient contact between rock and water could not occur to maintain the output of chemicals over the long lifetime of the systems.

The Wairakei system natural annual water output of 10^{10} kg could be held in about 0.1 km^3 of rock with 10% porosity. Grindley (1965) estimated the age of the Wairakei hydrothermal system to be about 500,000 years, so that for a maximum likely volume of the hydrothermal system of 500 km^3 (10 × 10 × 5 km), for constant water flow, the water volume of the system would have to be replaced 100 times. However, it is by no means certain that the water flow has been continuous, or the water composition constant. It is possible that over long periods (10^5 years) water is heated at deep levels under nearly stationary conditions and that appreciable convective inflow and outflow occurs only during comparatively short periods of the order of 10^3 years, when tectonic activity creates new flow channels. Flow may cease when these become sealed again with minerals such as quartz and calcite (Ellis, 1970).

With a magmatic origin for the dissolved chemicals in geothermal waters there are difficulties in explaining a continuous transfer of water and chemicals from a single igneous intrusion over a period of the order of 10^5 years. However, successive intrusions could occur, and with this explanation successive waxing and waning of surface outflows could be expected. Mahon and McDowell (1977) have suggested that the adiabatic expansion of a very high-temperature and high-pressure magmatic fluid (e.g., 800°C, 2000 bars) to lower temperatures and pressures (e.g., 300°C, 100 bars) would lead to the formation of a concentrated chloride-rich brine plus steam. Because of the high density of the brine it would remain at the base of the hydrothermal system, becoming a heat transfer fluid about the magma (A. McNabb, personal communication). Deep circulating waters could pick up heat and chemicals by contacting and mixing with the brine on its upper surface.

The origin of chemicals in geothermal systems is therefore still open to debate. From the geothermal development point of view, the origin of chemical constituents in water and steam is perhaps not so important as the realization that local rock–water interactions within a field at shallow levels can greatly influence the composition of the water and steam flowing from hot springs, fumaroles, and wells.

3.5 HYDROTHERMAL ALTERATION

General

Within high-temperature geothermal areas the reaction of the original rocks with hot water or steam results in a complex series of devitrification, recrystallization, solution, and deposition reactions which are referred to as hydrothermal rock alteration.

The order of decreasing susceptibility of rock-forming minerals to hydrothermal alteration is often as follows: volcanic glass, magnetite, hypersthene, hornblende, biotite = plagioclase (Steiner, 1968; Browne and Ellis, 1970). Quartz comes to saturation equilibrium with the hot solutions but primary crystals often remain in the altered rock.

The end products of hydrothermal alteration depend on many factors, the major ones being temperature and pressure of alteration, water compositions, original rock compositions, time of reaction, rate of water or steam flow, permeability of the rocks, and whether the permeability is of the fissure or bulk porosity type. The concentrations of carbon dioxide and hydrogen sulfide in the waters have an important control on the type of secondary mineralogy.

In most geothermal areas hydrothermal alteration of rocks shows a zoning with increasing depth, temperature, porosity, and changing chemical conditions. Usually there is a surface zone of argillitic alteration where acidic waters have been formed by the oxidation of hydrogen sulfide to sulfuric acid. Kaolin, alunite, sulfur, and gypsum are common in this surface alteration. In some fields (e.g., Matsukawa, Japan, and in the Tatun zone, Taiwan) argillitic alteration persists to deeper levels.

The type of alteration at deep levels in high-temperature water areas is commonly one of propylitic or potassium silicate mineral assemblages in the classification of Meyer and Hemley (1967). In cooler shallower zones, zeolites and montmorillonite clays are also common alteration products.

In areas of volcanic rocks the process of hydrothermal alteration is generally one of hydration, carbonation, and sulfide formation. In sedimentary rock areas, however, the highest-temperature alteration may even result in loss of water and carbon dioxide from the rock (e.g., at deep levels in the Salton Sea area).

Given stable chemical conditions and temperature, it could be expected that the rock–water system would adjust to a new set of equilibrium compositions and minerals. Temporarily ignoring iron and sulfur, we find that the rock–water system usually has the following main components: Na_2O, K_2O, CaO, MgO, Al_2O_3, SiO_2, H_2O, and CO_2. At constant temperature and pressure, at least seven mineral phases can be expected to coexist with water in the altered rocks at equilibrium. For example, in the Salton Sea geothermal field (Muffler and White, 1969) the deep, 300° mineral assemblage of quartz, epidote, chlorite, K-feldspar, albite, K-mica, and calcite was formed by recrystallization of detrital quartz, feldspar, clays, and carbonates in the original sediments.

A theoretical consideration of the kinetics of reactions between hot water and rocks in hydrothermal systems was given by Helgeson (1970), while Helgeson *et al.* (1969) provided a more detailed consideration of irreversible reactions in hydrothermal processes, including the prediction of mass transfer.

It can be expected from experimental knowledge of silicate reaction rates that in zones of slow water movement at temperatures over about 200°, an equilibrium alteration mineral assemblage is likely to form. However, in other cases equilibrium between rock and water may not be attained for reasons such as low rock permeability, rapid steam or water flow, boiling of water at particular depths, or condensation of steam into a water phase. Permeability is an important factor, since many mineralogical changes are not isochemical, and although some water is available in pore spaces, rocks must be open for the addition and removal of constituents. The water may impose a new composition on the rock if a temperature or compositional gradient exists along the direction of water flow.

In the New Zealand geothermal fields it was found that dense impermeable rocks, such as rhyolite and ignimbrite, or fine-grained sediments were little altered by near-neutral-pH solutions even at high temperatures. However, breccias or pumice of a similar composition were altered to an equilibrium mineral assemblage.

For the Salton Sea system, Clayton *et al.* (1968) showed that there was extensive oxygen isotope exchange between hot waters and rock, with carbonates being in isotopic equilibrium with the waters at temperatures as low as 100°, and most fine-grained silicates in the permeable media being close to equilibrium with the water at temperatures above 150°. Clayton and Steiner (1975) found for the Wairakei system there was oxygen isotopic equilibrium between water and hydrothermally grown quartz and calcite at the highest temperatures (250°C). At Broadlands there was similar equilibrium for fissure-grown quartz, calcite, and adularia at temperatures of 250°–290° (Blattner, 1975), as well as for secondary quartz and illite at temperatures of 160°–270° (Eslinger and Savin, 1973). In all the fields there was little tendency for primary quartz to exchange oxygen with the waters, even at 340° in the Salton Sea system.

Examples of Field Situations

Wairakei, New Zealand

The hydrothermal alteration of rocks in the Wairakei system was reviewed by Steiner (1968) and Clayton and Steiner (1975). The rocks were predominantly glassy volcanics consisting of pumiceous tuffs and breccias, ignimbrites, and rhyolite flows. Andesite occurred within the depth of drilling in the western field. At deep levels in the hottest central part of the field, upward flow was mainly through fissures in ignimbrite, but the water spread out laterally in permeable breccia formations closer to the surface. Rock alteration patterns showed a relationship to temperature, grading out from the most intense alteration close to the flow fissures (250°-260°). A general chemical trend was for potassium and silica to be added to the altered rocks and for sodium and calcium to be removed. The deep waters in the area had a pH of about 6.5 and a carbon dioxide concentration m_{CO_2} of 0.01.

Figure 3.4, from Steiner (1968), summarizes the rock formations, alteration, and temperatures for well 71, Wairakei. Beneath an acid-leached zone at the surface, ptilolite and calcium montmorillonite occurred at temperatures of 100°-120° in altered rhyolitic tuffs. With increasing temperature and depth, interstratified montmorillonite-illite appeared. The basal spacings of this mineral became smaller at increasing temperatures until at the deepest levels (650 m) illite (9.9-10.1 Å) appeared at temperatures of about 240°. Iron-rich chlorite and pyrite or pyrrhotite accompanied the micaceous phases at all depths, the chlorite forming from the groundmass of the original glass.

Steiner proposed the following genetic sequence of hydrothermal alteration of silicic volcanic glass to illite:

Silicic volcanic glass
↓
Ca-montmorillonite
↓
Mixed-layer illite-montmorillonite
rich in interstratified montmorillonite
10.40-12.45 Å
↓
Mixed-layer illite-montmorillonite
deficient in interstratified montmorillonite
10.16-10.28 Å
↓
Illite   9.93-10.04 Å

The general stability of calcium-rich phases at Wairakei with increasing temperature was in the sequence, ptilolite (about 100°) to laumontite (150°-200°) to wairakite and epidote, in agreement with the experimental studies of Coombs *et al.* (1959).

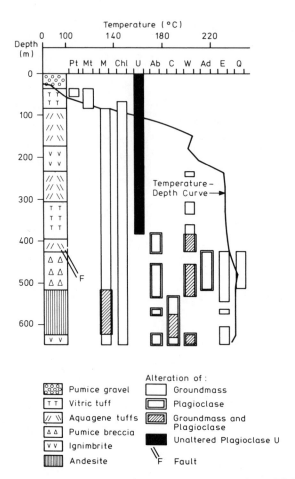

Fig. 3.4 Hydrothermal alteration of rocks in the Wairakei geothermal field (well 71). The stratigraphic sequence and faulting (F) of rocks is shown, along with temperatures at various depths. U denotes unaltered plagioclase. The minerals are K, kaolinite; Mt, montmorillonite; M, micaceous clay; Chl, chlorite clay; Pt, ptilolite; Lm, laumontite; W, wairakite; C, calcite; Ab, albite; Ad, adularia; Q, quartz; E, epidote (from Steiner, 1968).

The sequence of zeolite occurrences with increasing temperature is related to a series of dehydration reactions such as the following. For conditions close to the boiling point with depth curve, high temperature favors the less hydrous zeolite phases.

$$\underset{\text{mordenite}}{Na_2CaAl_4Si_{20}O_{48}\cdot14H_2O} = \underset{\text{laumontite}}{CaAl_2Si_4O_{12}\cdot4H_2O} + \underset{\text{albite}}{2NaAlSi_3O_8} + \underset{\text{quartz}}{10SiO_2} + 10H_2O \qquad (3.1)$$

$$\underset{\text{laumontite}}{CaAl_2Si_4O_{12}\cdot4H_2O} = \underset{\text{wairakite}}{CaAl_2Si_4O_{12}\cdot2H_2O} + 2H_2O \qquad (3.2)$$

Quartz, K-feldspar, chlorite, K-mica, albite, epidote, and wairakite were typical of the highest temperatures (over about 230°). The last three minerals persisted into slightly cooler zones than did illite and K-feldspar. In flow fissures at points where boiling of water occurred, quartz, potassium feldspar, and wairakite were major minerals.

Clayton and Steiner (1975) showed that on average there had been a $\delta^{18}O$ oxygen isotope change of -4‰ in the altered rocks at Wairakei, in comparison with a small oxygen shift in the geothermal water (about 1.5‰; J. R. Hulston, personal communication). The mass of water causing the alteration was therefore in excess of the mass of rock, and in the present flow routes a close to steady state has been achieved.

Broadlands, New Zealand

Hydrothermal alteration in this field was discussed by Browne and Ellis (1970), and Fig. 3.5 gives the rock alteration found in cores from well 9.

The rock types and general sequence of alteration were similar to those at Wairakei but there were also important differences, which are discussed on p. 106. Calcite was abundant in drill cores at Broadlands but not at Wairakei, whereas wairakite and iron-rich epidote were rare in comparison with Wairakei. In general, zeolites were not so common at Broadlands, and neither heulandite nor laumontite were positively identified, although both were present at Wairakei.

At Broadlands the pH of the underground water was about 6.0–6.5 and m_{CO_2} was about 0.12.

Matsukawa, Japan

The original rocks were a series of andesitic and dacitic welded tuffs and pyroxene andesites.

Rock alteration zones were vertically elongated and their distribution (Fig. 3.6 from Nakamura *et al.*, 1970) suggested that the alteration was caused by the lateral diffusion of acidic sulfate solutions from vertical channels at the center of the zones.

Near-surface rock alteration included saponite, montmorillonite, and kaolin. At depth there were radially distributed zones of alteration consisting of progressively more intense alteration about a central upflow of acid water derived from below at least 750 m. The key minerals in the zones were chlorite, montmorillonite, and alunite. A relic alteration pattern of pyrophyllite persisted from an earlier stage of the field's history.

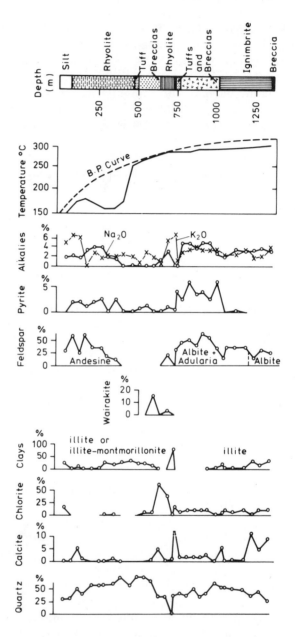

Fig. 3.5 The hydrothermal alteration pattern encountered by well 9, Broadlands. The rock stratigraphy and temperatures are also shown (adapted from Browne and Ellis, 1970).

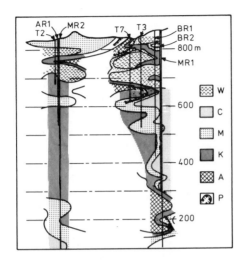

Fig. 3.6 The pattern of hydrothermal alteration of rocks in the Matsukawa geothermal field. W, saponite; C, chlorite; M, montmorillonite; K, kaolin; A, alunite; P, pyrophyllite (from Nakamura *et al.*, 1970).

The chlorite zone included laumontite, hydromica, calcite, quartz, pyrite, anhydrite, and gypsum, while the montmorillonite was accompanied by hydromica, calcite, and anhydrite. Quartz and pyrite were associated with both the kaolin and alunite zones.

The deep central zone water was probably of pH 3-4 and about 250°-280° (Sumi, 1969). Only in the cooler and less acidic outer zones did minerals such as chlorite, laumontite, and calcite occur.

Onikobe, Japan

The hydrothermal alteration of rocks encountered by wells drilled to 700 m in the Katayama area, Onikobe, was described by Seki *et al.* (1969). Within a sequence of andesitic and dacitic lavas and tuffs, the deep and highest-temperature (150°-205°) alteration consisted of quartz, calcite, clay minerals, calcium zeolites, pyrite, and anhydrite. As in other areas, the pyroclastic rocks were most prone to alteration. There was a correlation between the depth of occurrence of various zeolites and the temperature in the wells. The trend was from mordenite, at temperatures up to about 140°; to laumontite, to temperatures of about 170°; to wairakite, at the highest temperatures (about 200°). In addition to the general alteration of host rocks there was a deposition of minerals, including quartz, calcite, anhydrite, and zeolites, in flow fissures.

Salton Sea

The original rocks of the porous aquifer were a series of sands, silts, and clays consisting basically of quartz, calcite, K-feldspar, plagioclase, illite, dolomite, and kaolinite. Mixed-layer montmorillonite-illite formation occurred at about 100° and the conversion to illite was complete at water temperatures of about 210°.

Reaction of dolomite and ankerite to form chlorite, calcite, and carbon dioxide began at temperatures below 180°. At temperatures over about 300°, iron-rich chlorite was an abundant alteration mineral, as was K-feldspar in veins and as a replacement of plagioclase.

The mineral assemblage commonly found at temperatures of about 300° in the sand–clay aquifer at 1200 m was quartz, epidote, chlorite, K-feldspar, albite, K-mica, and pyrite (Helgeson, 1967; Muffler and White, 1969). The deep-water pH was about 4.7 and m_{CO_2} was low (about 0.01).

Muffler and White (1969) considered that the transformation of low-grade hydrated and carbonate sediments to secondary feldspars, epidote, chlorite, carbon dioxide, and water was a metamorphic process. The alteration mineral assemblage, however, is similar to that produced in other geothermal areas by the hydrothermal alteration of glassy acid volcanic rocks.

In both cases, solution–mineral chemical equilibrium can be assumed to have been approached closely from quite different starting materials.

Water Chemistry and Hydrothermal Alteration

The chemical processes of hydrothermal alteration have been qualitatively understood for many years; for example, from studies of cores from Yellowstone Park wells, Fenner (1936) noted that there was a loss of potassium from the waters to form K-feldspar in the rocks. Only in the last decade, however, have sufficient data become available from hydrothermal experiments and from analyses of water compositions from deep levels of hydrothermal areas to enable a quantitative approach to alteration processes. It is now known that high-temperature hydrothermal solutions at deep levels are saturated with silica in equilibrium with quartz (Mahon, 1966; and Fournier and Rowe, 1966). The temperature dependence of the Na/K ratio in solutions in equilibrium with sodium and potassium feldspars has been established from laboratory and field measurements (Orville, 1963; Ellis and Mahon, 1967; Fournier and Truesdell, 1973). (See Chapter 4.)

Helgeson (1971) showed that in general the rate-controlling step in the reaction of aluminosilicates with water was diffusional mass transfer through layers of intermediate reaction products on the reactant mineral surface. Reaction paths, mass transfer, and mineral phase equilibria were predicted from kinetic theory and thermodynamic data. This paper, in conjunction with an earlier paper on mineral–solution equilibria (Helgeson, 1967), has made a considerable contribution to the understanding of the rock alteration reactions which occur in geothermal systems.

Examples of semiquantitative correlations between the chemistry of hot waters and the occurrence of alteration minerals at various levels in hydrothermal systems are now given. These make possible a better understanding of water and steam movement and interactions in the rocks, which is necessary when making such practical engineering decisions as at what level to set the casing of a geothermal well for maximum production. The equilibrium-based concepts which follow can only be used with confidence at temperatures above about 175°, since at lower temperatures the equilibria are not well calibrated by laboratory experiments, nor is attainment of equilibrium certain.

The experimental work of Hemley (1959) in establishing mineral-solution equilibrium conditions in the system $K_2O-Al_2O_3-SiO_2-H_2O$, and of Hemley and Jones (1964) on the equivalent sodium system, are of primary importance in this discussion. Helgeson (1967) gave a detailed analysis of mineral–solution equilibria in the Salton Sea hydrothermal system using Hemley's results in combination with mineralogical and thermodynamic information, and chemical analyses of the waters. He derived a set of phase diagrams expressing mineral stabilities in terms of ion concentrations in solution.

In the presence of quartz, with aluminum assumed to be immobile, the formation of sodium, potassium, calcium, magnesium, etc., aluminosilicates depends on the activity of the respective metal oxides in the rock–water system. At each temperature, the activity of water is assumed to be constant. The activity of Na_2O, CaO, etc., can be expressed as follows:

$$\tfrac{1}{2}Na_2O + H^+ = Na^+ + \tfrac{1}{2}H_2O \tag{3.3}$$

$$K_{Na} = a_{Na^+}/a_{H^+} \cdot a_{Na_2O}^{1/2} \tag{3.4}$$

$$\tfrac{1}{2}\log a_{Na_2O} = \log(a_{Na^+}/a_{H^+}) - \log K_{Na} \tag{3.5}$$

$$CaO + 2H^+ = Ca^{2+} + H_2O \tag{3.6}$$

$$K_{Ca} = a_{Ca^{2+}}/a_{H^+}^2 \cdot a_{CaO} \tag{3.7}$$

$$\log a_{CaO} = \log(a_{Ca^{2+}}/a_{H^+}^2) - \log K_{Ca} \tag{3.8}$$

Variations in Na_2O, CaO, etc., activities can be expressed by variations in the ratios a_{Na^+}/a_{H^+}, $a_{Ca^{2+}}/a_{H^+}^2$, etc. Mineral transformations can therefore be correlated with solution compositions through equations involving ions, as in these examples:

$$3\ KAlSi_3O_8 + 2H^+ = KAl_3Si_3O_{10}(OH)_2 + 6SiO_2 + 2K^+ \qquad (3.9)$$
$$\text{K–feldspar} \qquad\qquad \text{K–mica} \qquad\qquad \text{quartz}$$

$$2\ KAl_3Si_3O_{10}(OH)_2 + 3H_2O + 2H^+ = 3Al_2Si_2O_5(OH)_4 + 2K^+ \qquad (3.10)$$
$$\text{K–mica} \qquad\qquad\qquad\qquad \text{kaolin}$$

Equilibrium constants for these transformations at each temperature and pressure correspond to particular ratios of a_{K^+}/a_{H^+}. Hemley and Jones (1964) gave values for several mineral equilibrium boundaries in these terms over a range of temperatures at 1000 bars pressure. Further results for the K-feldspar–K-mica equilibrium were reported by Usdowski and Barnes (1972). At temperatures under about 300°, the pressure effect on equilibrium boundaries can be calculated from partial molar volumes of the ions and the molar volumes of the minerals involved in the equilibria (Ellis and McFadden, 1972). The illustrations that follow (Figs. 3.7–3.9) are based largely on the cation concentrations in waters tapped by wells in geothermal areas, and they apply to pressures in slight excess of saturated water vapor pressures.

The following reaction involves two different alkali ions:

$$3NaAlSi_3O_8 + K^+ + 2H^+ = KAl_3Si_3O_{10}(OH)_2 + 6SiO_2 + 3Na^+ \qquad (3.11)$$
$$\text{albite} \qquad\qquad\qquad \text{K–mica} \qquad\qquad \text{quartz}$$

The equilibrium constant is $K = a_{Na^+}^3/a_{K^+} \cdot a_{H^+}^2$. On a diagram with $\log(a_{K^+}/a_{H^+})$ as the abscissa and $\log(a_{Na^+}/a_{H^+})$ as the ordinate scale, at constant temperature and pressure, the locus of points conforming with this mineral boundary is a straight line of slope 1/3. Figure 3.7 shows a phase diagram at 260° and saturated water vapor pressures for the system Na_2O–K_2O–Al_2O_3–SiO_2–H_2O. The diagram is derived from experimental phase equilibrium data, a knowledge of phase boundary slopes, and the experimental value for the atomic ratio a_{Na^+}/a_{K^+} in solution at 260° for the coexistence of K-mica, K-feldspar, and albite. The dashed lines show the phase boundaries at 230° for feldspars and K-mica.

Comparable aluminosilicate phase diagrams can be constructed for other combinations of alkali cations, and Helgeson (1967) presented a set of six equilibrium diagrams for 300°, for combinations of sodium, potassium, calcium, and magnesium minerals. Through use of the same principles, mineral phase diagrams for the Broadlands area were developed by Browne and Ellis (1970). A wide range of thermodynamic data on mineral–solution reactions was summarized by Helgeson (1969) and by Kharaka and Barnes (1973).

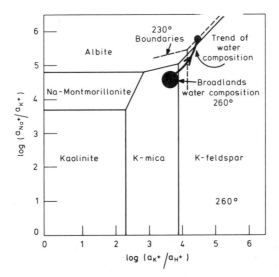

Fig. 3.7 Phase diagram for sodium and potassium in the presence of quartz at 260° in terms of ion activity ratios. Dashed lines show a portion of the diagram for 230° (from Browne and Ellis, 1970).

Phase diagrams for aluminosilicate minerals in the K–Ca and the Ca–Mg systems at 250°–260° are given in Figs. 3.8 and 3.9, drawn to interpret the hydrothermal alteration mineral assemblages for the Wairakei and Broadlands fields. To obtain values of the ratios a_{K^+}/a_{H^+}, $a_{Ca^{2+}}/a_{H^+}^2$, etc., for the deep natural solutions, values of a_{H^+} were calculated from geothermal water and steam analyses (see Chapter 7), and appropriate activity coefficients were applied to convert cation concentrations into activities. Unlike the sodium and potassium mineral phase boundaries, information on the calcium and magnesium silicate boundaries comes from observed mineral assemblages in the field and not from laboratory experimental work.

As an approximation, epidote is treated as zoisite and chlorite as a magnesium mineral (the geothermal variety is iron rich). For simplicity, illite and montmorillonite are treated as separate minerals, although they also may occur as an interlayered mineral. The minerals shown in Figs. 3.8 and 3.9 are assumed to be the only potassium, calcium, and magnesium minerals stable at the temperature (although other minerals—e.g., the calcium zeolites, laumontite, and ptilolite—may become stable at lower temperatures). Each diagram represents a particular element combination plane in a more complex system, and each equilibrium boundary position is associated with particular concentrations of other cations in solution and other cooperative

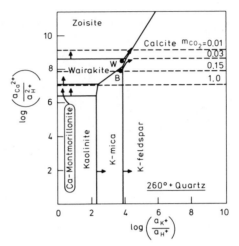

Fig. 3.8 Phase diagram for calcium and potassium minerals at 260°, in terms of ion activity ratios. W and B give the solution ratios for the deep waters in Wairakei and Broadlands fields.

mineral equilibria. Phase boundaries are given as straight lines, although curvature may in fact occur due to intersubstitution of ions in the mineral phases.

The addition of carbon dioxide to a rock–water system results in the

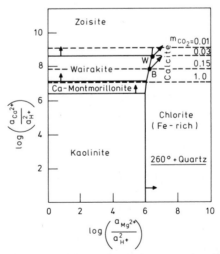

Fig. 3.9 Phase diagram for calcium and magnesium minerals at 260° in terms of ion activity ratios. W and B mark the compositions of the deep Wairakei and Broadlands waters.

formation of bicarbonate and carbonate ions. For a particular value of m_{CO_2} in a hydrothermal solution, a horizontal line can be drawn on a phase diagram with $\log(a_{Ca^{2+}}/a_{H^+}^2)$ as its ordinate, representing the value at which calcite precipitates (see the discussion on calcite solubility in Chapter 4). In the upper part of a diagram like Fig. 3.8, above a dashed line lies values of $a_{Ca^{2+}}/a_{H^+}^2$ which are prohibited at the particular m_{CO_2}. The calcite precipitation line (dashed line, Figs. 3.8 and 3.9) is lowered down the diagram as concentrations of carbon dioxide are increased. On these figures it can be seen that at $m_{CO_2} = 0.01$ (Wairakei) all of the minerals shown have fields of stability. At $m_{CO_2} = 0.15$ (Broadlands), zoisite cannot form at $260°$, while at $m_{CO_2} = 1.0$ the wairakite field would also disappear. The presence or absence of wairakite and epidote in alteration assemblages at these temperatures can therefore be related to carbon dioxide concentrations.

The water compositions at deep levels before any loss of steam and carbon dioxide occurs through boiling are marked as solid points in the diagrams, for the Wairakei (W) and Broadlands (B) areas, and are calculated from well discharge analyses. The Wairakei composition is close to the conditions for equilibrium with wairakite, zoisite, K-feldspar, and chlorite, and the Broadlands water composition for equilibrium with K-feldspar, wairakite, K-mica, calcite, and chlorite.

However, the fact that a geothermal water composition falls in the stability field of a mineral does not require that the mineral be present, nor give any suggestion of the quantity of the mineral. The availability of particular elements from water or rock, the ability of the phase to nucleate, and the rate of crystal growth for particular minerals must also be considered.

In the experimentally based Fig. 3.7 the deep-water compositions for Wairakei and Broadlands fall close to the point of constant a_{Na^+}/a_{H^+}, a_{K^+}/a_{H^+}, and a_{Na^+}/a_{K^+} for coexistence of K-mica, K-feldspar, and albite. This was the situation also at Salton Sea (Helgeson, 1967). The coexistence of these three minerals at deep levels occurs also in many other high-temperature water areas and confirms that this is a common equilibrium point reached in water compositions and in mineralogy when high-temperature waters react with feldspathic quartz rocks. Accepting this as a hypothesis leads to an understanding of several of the differences in water compositions between areas. For example, since the atomic Na/K ratio in geothermal waters is usually eight or more, and since concentrations of other cations are much less than that of sodium (except in concentrated brines), the constant ratio a_{Na^+}/a_{H^+} suggests that there should be an approximate proportionality between the hydrogen ion concentration

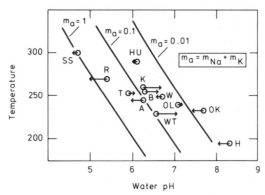

Fig. 3.10   The calculated trends of pH with temperature for waters of three salinities. Open circles show the pH of waters in several geothermal systems and the connected arrowheads indicate the pH calculated for the salinity of the water. SS, Salton Sea; R, Reykjanes; H, Hveragerdi; HU, Hatchobaru; T, El Tatio; OL, Olkaria; A, Ahuachapan; W, Wairakei; K, Kawerau; B, Broadlands; WT, Waiotapu; OK, Orakeikorako.

and the salinity of the deep waters at a given temperature (Ellis, 1969). The relationship can be calculated, and in Fig. 3.10 pH versus temperature trends are given for solutions with $(m_{Na} + m_K)$ equal to 0.01, 0.1, and 1.0 (where calcium and magnesium concentrations are at least a power of ten less than those of sodium and potassium). The solution pHs at deep levels in several geothermal systems are also shown in the diagram and compared with a theoretical point. The general trends of pH values for the waters are as predicted.

Calcite solubility is usually the factor limiting the concentration of calcium in high-temperature waters. Chapter 4 reviews the effects of temperature, pressure, and carbon dioxide concentration on calcite solubility in water and, for the situation where the solution pH is controlled by aluminosilicate mineral equilibria, how calcium and carbonate concentrations are related to the salinity of geothermal waters.

Trends in Hydrothermal Alteration

Mineral phase diagrams (e.g., Figs. 3.7–3.9) can be used to interpret various trends in hydrothermal alteration (Browne and Ellis, 1970).

Boiling and gas evolution occur in high-temperature geothermal areas when the waters rise to a level where the steam and gas pressure becomes equal to the surrounding hydrostatic pressure. With loss of steam, carbon dioxide, hydrogen sulfide, and other gases, the water cools and its pH rises. It is then no longer in equilibrium with the mineral assemblage with which it was in contact. The effects produced in the rocks through which the boiling water rises depend on whether the water rises slowly through homogeneous porosity rock or is channeled. In general, there will be a tendency for the water to impose a new composition on the rocks at levels above the point of first boiling.

The slow upflow of water through homogeneous rock is considered first. There is little quantitative information on how mineral phase boundaries change with temperature, but in Fig. 3.7 the potassium–sodium mineral phase boundaries for 230° are given as a dashed line for comparison with the 260° boundaries. Short arrows qualitatively show the movement of other boundaries with decreasing temperatures.

The Broadlands water composition and mineral assemblages are used as an example (Browne and Ellis, 1970). With increasing steam loss from deep 250°–260° water, and rising pH, the diagrams suggest that the water would tend to impose the following upward sequence of alteration minerals on the rocks. Quartz is present throughout. In Figs. 3.7–3.9, the heavy line with an arrowhead gives the general trend in water composition.

(a)   260° water close to equilibrium with albite, K-mica, K-feldspar, calcite, wairakite, and chlorite.

(b)   A slight steam loss and pH rise moves the water composition point so that K-feldspar, albite, and wairakite, but not K-mica, remain stable (Fig. 3.8), while Fig. 3.9 shows that chlorite remains stable. Due to the slight cooling, the water composition trends into the K-feldspar field of stability (Fig. 3.7). There is a tendency for growth of K-feldspar until the water composition returns to the two-feldspar line. Calcite precipitates as m_{CO_2} decreases, because the dashed horizontal calcite line in Figs. 3.8 and 3.9 rises proportionally to pH change, while the composition point rises proportionally to twice the pH change. Depending on the direction of movement with temperature for the wairakite–K-feldspar and the wairakite–albite phase boundaries, the slight cooling with continuing steam loss would cause growth of one or other of each mineral pair to return the water composition to the equilibrium m_{Ca}/m_K^2 or m_{Ca}/m_{Na}^2 ratios, respectively, for the two-phase boundary conditions.

(c)   With a few percent steam loss, the same trends are followed except that zoisite (epidote) becomes favored rather than wairakite (Fig. 3.8). Calcite and K-feldspar continue to form. The mineral assemblage in contact with the boiling water at this stage trends toward K-feldspar, albite, calcite, epidote, and chlorite.

(d)   After about 3–4% steam loss by boiling, the carbon dioxide concentration in solution falls to about one fifth of the original, or for Broadlands, 0.03 m. The water composition would then approach equilibrium with calcite, K-feldspar, albite, and chlorite.

The comments neglect the buffering action of the rock toward changes. The extent to which the rock mineralogy change moves along the trends suggested is controlled by factors such as the rock/water ratio, the age of the system, and the reactivity of particular minerals.

If the boiling water rises through a simple fissure, the secondary mineralogy in the fissure is dominated by the first minerals formed through the water adjusting its composition toward that for an equilibrium assemblage at a lower temperature and higher pH. Buffering effects from the rock are minimal, and the minerals formed include K-feldspar, calcite, and quartz. Formation of wairakite (or chlorite) in the fissure at the level of first boiling, followed by epidote (or chlorite) at a higher level, would also be consistent with the phase diagrams.

In the case of a water heated by the condensation of steam rich in carbon dioxide, the phase diagrams show that the imposed lower pH conditions cause a trend toward mineral assemblages such as albite, K-mica, wairakite or montmorillonite, K-mica, wairakite.

The phase diagrams also suggest that certain mineral associations are not possible at 260° if chemical equilibrium is attained. For example, Ca-montmorillonite + K-feldspar, epidote + K-mica, and epidote + Ca-montmorillonite should be incompatible mineral pairs at equilibrium.

The differences between the hydrothermal alteration at Wairakei and Broadlands were discussed by Browne and Ellis (1970). As shown by Steiner (1968), the deep alteration mineral assemblage in the western Wairakei field at the highest temperatures is commonly albite, K-feldspar, K-mica, epidote, quartz, and wairakite. The coexistence of K-mica and epidote is contrary to equilibrium requirements in the phase diagram of Fig. 3.8, but there is not a great separation of the water compositions along the feldspar boundary for the stability of each phase. Calcite is a common mineral at Broadlands but much less common at Wairakei, whereas the reverse situation applies with wairakite and epidote. This is as expected from the position of the deep-water compositions with respect to those for wairakite, epidote, and calcite stability (Fig. 3.8). The Wairakei composition point

(0.01 m CO_2) is close to the wairakite, epidote, K-feldspar point, and appreciably below the calcite solubility line. For the first separation of steam, the pH rise brings the water to the epidote–K-feldspar stability line, and an appreciable steam separation is required before the composition point catches up with the calcite line (intersection position not shown in Fig. 3.8).

Broadlands has a deep-water composition point close to equilibrium with K-mica, K-feldspar, and wairakite, as well as with calcite. With steam separation, calcite would precipitate, and although the water composition point would follow the wairakite + K-feldspar boundary, it is unlikely that wairakite would crystallize while calcite was forming.

For a given temperature there is a particular concentration of carbon dioxide in a water above which calcite is likely to precipitate as the calcium mineral phase with first steam separation at depth, and below which the formation or continued stability of a calcium silicate phase is favored. At 260° this carbon dioxide concentration appears to be about the Broadlands value (0.15 m). Calcium silicate phases are unlikely to nucleate on the surface of drill pipes, but calcite can be formed. Reasoning along these lines may make it possible to establish at an early stage of development whether or not a geothermal field is likely to give trouble with calcite precipitation in pipes on the first boiling of water. Calcite deposition occurred in Broadlands wells but it is not important at Wairakei.

The phase diagrams (Figs. 3.7–3.9) indicated that kaolinite formation would be limited to conditions of unusually low water pH.

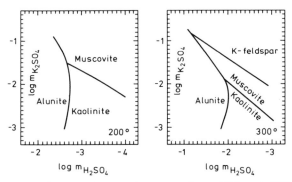

Fig. 3.11 Phase relations of potassium minerals at 200° and 300° in terms of potassium sulfate and sulfuric acid molalities (1000 bars pressure) (from Hemley *et al.* 1969).

In a series of laboratory experiments, Hemley $et\ al.$ (1969) established the stability relationships of kaolinite, alunite, and muscovite
in terms of concentrations of sulfuric acid and potassium sulfate.
Figure 3.11 gives examples of their experimental results for 200° and
300°. It can be seen that with increasing temperature, for a given
potassium concentration, greater concentrations of sulfuric acid are
required for alunite stability, while at 100° it is likely that concentrations of only 10^{-3} or 10^{-4} m are required for dilute geothermal waters.
The acidity required for kaolinite formation in preference to muscovite
also becomes less at low temperatures. These facts are in agreement
with the common formations of kaolinite and alunite as rock alteration
products at the surface where there has been oxidation of sulfides, but
at high temperatures the occurrence is rare and usually in association
with unusually acidic sulfate waters such as those at Matsukawa,
Japan, or Matsao, Taiwan.

Iron Sulfide Phases

The common iron sulfide minerals resulting from the high-temperature hydrothermal alteration of rocks are pyrrhotite and pyrite. The
following reaction enables their relative stabilities to be related to
parameters which can be calculated from analyses, p_{H_2S} and p_{H_2} (the
partial pressures of H_2S and H_2).

$$FeS_2 + H_2 = FeS + H_2S \tag{3.12}$$

Figure 3.12 gives the trend, in values of $\log(p_{H_2S}/p_{H_2})$, with tempera-

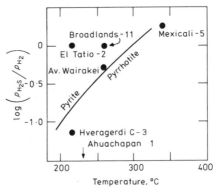

Fig. 3.12   The pyrite–pyrrhotite equilibrium boundary in terms of the P_{H_2S}/P_{H_2} ratio.
The partial pressure ratios for deep waters in several geothermal fields are also shown.

ture for the equilibrium (Browne and Ellis, 1970). Points for the partial pressure ratios for deep waters in several areas (from Table 3.5) are also included. Mexicali well 5 and the Wairakei average points are very close to the equilibrium line, while well 11 Broadlands has a water composition corresponding to pyrite stability, not far from the boundary. Hveragerdi C-3 and Ahuachapan 1 ratios fall in the pyrrhotite field.

Rocks exposed to steam which has boiled from a water body may produce pyrrhotite because of the lower p_{H_2S}/p_{H_2} ratios in the steam (hydrogen is less soluble than hydrogen sulfide). Residual water after steam separation has a high p_{H_2S}/p_{H_2} ratio, favoring pyrite formation (the El Tatio situation). Where organic-rich sediments occur, breakdown of organic material may cause a local low p_{H_2S}/p_{H_2} ratio, and bring the system into the pyrrhotite stability area.

Summary

Many of the mineral–solution equilibria connected with hydrothermal alteration control the concentrations of solutes in deep high-temperature geothermal waters. Silica concentrations correspond to the solubility of quartz. Temperature-dependent aluminosilicate ion exchange equilibria control the ratio of sodium and potassium ions in solution, as well as the ratio of sodium and rubidium ions. As a generalization, the water pH is controlled by the salinity of the waters and aluminosilicate equilibria involving hydrogen and alkali metal ions. The concentration of calcium ions and of bicarbonate ions is related to this pH and the concentration of carbon dioxide present in the waters by solubility product and ionization constant relationships. The concentrations of fluoride ions and of sulfate ions relative to calcium are in turn limited by the solubility of fluorite and of anhydrite, respectively (see Chapter 4).

Magnesium concentrations in solution are controlled at very low levels by silicate equilibria, involving mainly chlorite and montmorillonite minerals. The concentrations are temperature dependent, and as for many other divalent metal cations are approximately proportional to the square of the salinity (see Chapter 4).

The compositions of high-temperature geothermal waters are by no means accidental. The temperature and pressure and the rock compositions and permeability have a major control on the concentrations and ratios of ions. The chemistry of waters sampled at the surface therefore gives important information on the environment of hydrothermal systems at deep levels.

3.6   DEPOSITS FROM WATERS

A brief outline is now given of the deposits which form from natural flows of high-temperature water. An extensive review of this topic was given by Weissberg *et al.* (1976). The deposition of minerals from fluids flowing through geothermal wells is discussed in more detail in Chapter 8.

Siliceous Deposits

Silica is usually deposited at the surface from waters which have flowed through major channels from underground reservoir temperatures in excess of about 180°C. Amorphous silica terraces are commonly formed about boiling hot springs and geysers, and are a good indication of high temperatures underground. The silica may be colored by small concentrations of iron or manganese and may also contain traces of many other elements, including gold and silver (e.g., silica sinter at Whakarewarewa, Rotorua, New Zealand, contains up to 125 ppm silver and 2 ppm gold, and at Champagne Pool, Waiotapu, 8 ppm silver and 5 ppm gold). An interesting sinter containing a high concentration of tungsten occurs at Frying Pan Lake, Waimangu, New Zealand, where slightly acidic (pH 4) sulfate–chloride near-boiling waters deposit a vitreous brown silica sinter containing 3% tungsten, 3% phosphorus, and several percent arsenic.

Where silica is deposited from high-temperature waters within the country rock, quartz is usually formed, but fortunately there are few examples of quartz deposition extending into the interior of drill pipes. Aluminosilicate surfaces are preferred in initiating quartz crystallization. In some geothermal systems quartz formation under natural conditions has been associated with metal sulfide deposition. For example, at Broadlands, mineralized bands of sphalerite, galena, and minor chalcopyrite in association with quartz were encountered over a range of depths and in equilibrium with pyrite and pyrrhotite (Browne, 1971; Weissberg *et al.*, 1976).

At depth in the Salton Sea geothermal system, narrow open veinlets occur containing calcite, pyrite, pyrrhotite, sphalerite, and chalcopyrite, probably deposited from the hot brine (Skinner *et al.*, 1967). A spectacular example of mineral deposition from geothermal waters occurred when several wells in this area produced a siliceous scale containing 10–40% copper, 4–25% iron, 7–23% sulfur, 1–6% silver, 0.5–1% antimony, and 0.1–0.2% arsenic.

Carbonates and Sulfides

Hot spring waters which give rise to extensive travertine (calcium carbonate) deposits usually do not originate at temperatures much in excess of 100°. The occurrence of travertine deposits on the outskirts of boiling hot spring areas may in addition indicate future problems with calcite deposition in well casings during production, as in the zone containing the Kizildere geothermal field of Turkey, where there are extensive travertine deposits (the famous Pamukkale travertine terraces).

Waters from both hot springs and geothermal wells produce a variety of metal- and sulfide-rich precipitates as the waters cool. The precipitates frequently contain high concentrations of arsenic, antimony, and mercury.

Mercury deposition occurs in several geothermal areas of the world (for a review see Weissberg *et al.*, 1976). For example, Sulphur Bank, California, and Ngawha, New Zealand, both are associated with extensive mercury deposits, and in the northern Italian geothermal fields there are many examples of mercury, cinnabar, and stibnite deposition. Associations between extensive mercury deposition and geothermal systems occur mainly in areas of sedimentary rock containing organic-rich shales and sediments. This suggests leaching and mobilizing of mercury from organic-rich sediments (White, 1967; Ellis, 1969). Nevertheless, the mercury concentrations in the coexisting geothermal waters are usually low (about 1 ppb; Weissberg *et al.*, 1976).

At Steamboat Springs, Nevada, both antimony and mercury sulfides are currently depositing from thermal fluids in near-surface rocks as stibnite and cinnabar, and are associated with siliceous antimony-rich precipitates forming as suspensions in hot spring waters. The precipitates contained 400 ppm silver, 10 ppm gold, and 45 ppm mercury (White, 1967). This type of surface deposit is common to several geothermal areas (see Table 3.11), including Broadlands, Rotokaua, and Kawerau in New Zealand. At Broadlands, waters from a major hot spring and from some wells deposited stibnite enriched in thallium and mercury, silver, gold, and arsenic. This was caused by a highly selective adsorption onto the siliceous sulfide precipitate, for the waters contain only very low concentrations of gold, thallium, mercury, and silver (see Table 3.3 and related text).

A comparison of analyses of water from deep wells at Broadlands and local hot spring water showed that only a small proportion of the silver was precipitated as waters traveled to the surface through

TABLE 3.11

Concentration of Metals in Some Precipitates Formed from Geothermal Waters[a]

	Sb	As	Hg	Au	Ag	Tl	Pb	Zn
			Concentration (ppm)					
Ohaki Pool, Broadlands, N.Z.	10%	400	2000	85	500	630	25	70
Well 2, Broadlands, N.Z.	8%	250	200	55	200	1000	50	200
Well 2, Rotokaua, N.Z.	30%	4000	15	70	30	5000	50	100
Champagne Pool, Waiotapu, N.Z.	2%	2%	170	80	175	320	15	50
Spring, El Tatio, Chile	1.5%	12%	50	3	1	10	100	100
Steamboat Springs, Nevada	2000	600	—	60	400	2000	400	200

[a] From the review of Weissberg *et al.* (1976).

natural channels but that most of the thallium precipitated (Weissberg, 1969). The study showed that lead, zinc, and copper precipitates tended to form at depth within the Broadlands system, whereas arsenic, antimony, thallium, mercury, and gold were transported to the surface before precipitation. There is good experimental evidence that strong thio complexes of mercury, gold, antimony, and arsenic are formed at high temperatures (e.g., Dickson, 1966; Seward, 1973) and permit these metals to be transported up through deep zones of changing pH, whereas metals such as lead, zinc, and copper with less stable sulfur complexes are precipitated at depth as sulfides.

Metallic lead is known to deposit from some thermal waters. Lebedev (1972) reported the deposition of metallic lead, sphalerite, and pyrite from highly saline hot waters of the Cheleken Peninsula of the eastern Caspian Sea. Waters from a high-temperature deep geothermal well at Matsao, Taiwan, also produced on outflow pipes a deposite rich in metallic lead and galena.

REFERENCES

Alonso, H. (1966). *Bol. Soc. Geol. Mexico* **29**, 17.
Arnason, B., and Sigurgeirsson, T. (1966). *Isotop. Hydrol. Proc. Symp. Vienna* p. 35.
Arnason, B., and Sigurgeirsson, T. (1968). *Geochim. Cosmochim. Acta* **32**, 807.
Averyev, V. V., Naboko, S. I., and Piip, B. I. (1961). *Dokl. Akad. Nauk SSSR* **137**, 407.
Begemann, F. (1963). *In* "Nuclear Geology on Geothermal Areas" (E. Tongiorgi, ed.), pp. 55–70. Consiglio Nazionale delle Richerche, Lab. di Geol. Nucl., Pisa.
Blattner, P. (1975). *Am. J. Sci.* **275**, 785.
Bodvarsson, G. (1961). *Proc. U.N. Conf. New Sources Energy, Rome* **2**, 82.
Browne, P. R. L. (1971). *Soc. Min. Geol. JPN. (Spec. Issue)* **2**, 64.
Browne, P. R. L., and Ellis, A. J. (1970). *Am. J. Sci.* **269**, 97.
Chelishev, N. F. (1967). *Dokl. Akad. Nauk SSSR* **175**, 205.
Clayton, R. N., and Steiner, A. (1975). *Geochim. Cosmochim. Acta* **39**, 1179.
Clayton, R. N., Muffler, L. P. J., and White, D. E. (1968). *Am. J. Sci.* **266**, 968.
Combs, R. L., and Smith, H. A. (1975). *J. Phys. Chem.* **61**, 441.
Coombs, D. S., Ellis, A. J., Fyfe, W. S., and Taylor, A. M. (1959). *Geochim. Cosmochim. Acta* **17**, 53.
Craig, H., (1963). *In* "Nuclear Geology on Geothermal Areas" (E. Tongiorgi, ed.), pp. 17–53. Consiglio Nazionale delle Richerche, Lab. di Geol. Nucl., Pisa.
Craig, H. (1969). *Am. J. Sci.* **267**, 249.
Craig, H. (1975). *Proc. Int. At. Energy Agency Advisory Group Meeting Appl. Nucl. Tech. Geothermal Stud., Pisa, Italy, Sept.* (in press).
Dickson, F. W. (1966). *Bull. Volcanol.* **29**, 605.
Doe, B. R., Hedge, C. E., and White, D. E. (1966). *Econ. Geol.* **61**, 462.
Elder, J. W. (1965). *In* "Terrestial Heat Flow," pp. 211–239. Am. Geophys. Un., Geophys. Monogr. No. 8.

Ellis, A. J. (1965). *In* "Problemy Geokhimii," pp. 167–179. Inst. Geokhim. Anal. Khim. im V.I. Vernadskogo, Akad. Nauk SSSR, Moscow.

Ellis, A. J. (1966). *Bull. Volcanol.* **29**, 575.

Ellis, A. J. (1967). *In* "Geochemistry of Hydrothermal Ore Deposits" (H. L. Barnes, ed.), pp. 465–514. Holt, New York.

Ellis, A. J. (1968). *Geochim. Cosmochim. Acta* **32**, 1356.

Ellis, A. J. (1969). *Proc. Commonwealth Min. Metall. Congr., 9th, London* **2**, 1.

Ellis, A. J. (1970). *Geothermics (Spec. Issue 2)* **2** *(Pt 1)*, 516.

Ellis, A. J., and McFadden, I. M. (1972). *Geochim. Cosmochim. Acta* **36**, 413.

Ellis, A. J., and Mahon, W. A. J. (1964). *Geochim. Cosmochim. Acta* **28**, 1327.

Ellis, A. J., and Mahon, W. A. J. (1967). *Geochim. Cosmochim. Acta* **31**, 519.

ENEL (1970). Larderello and Monte Amiata: Electric Power by Endogenous Steam. Ente Nazionale per l'energia Elettrica, Rome.

Eslinger, E. V., and Savin, S. M. (1973). *Am. J. Sci.* **273**, 240.

Facca, G., and Tonani, F. (1964). *Bull. Volcanol.* **27**, 1.

Fenner, C. N. (1936). *J. Geol.* **44**, 225.

Fergusson, G. J., and Knox, F. B. (1959). *N.Z. J. Sci.* **2**, 431.

Ferrara, G., Gonfiantini, R., and Pistoia, P. (1963a). *In* "Nuclear Geology on Geothermal Areas" (E. Tongiorgi, ed.), pp. 267–275. Consiglio Nazionale delle Richerche, Lab. di Geol. Nucl., Pisa.

Ferrara, G. C., Ferrara, G., and Gonfiantini, R. (1963b). *In* "Nuclear Geology on Geothermal Areas" (E. Tongiorgi, ed.), pp. 277–284. Consiglio Nazionale delle Richerche, Lab. di Geol. Nucl., Pisa.

Fournier, R. O., and Rowe, J. J. (1966). *Am. J. Sci.* **264**, 685.

Fournier, R. O., and Truesdell, A. H. (1973). *Geochim. Cosmochim. Acta* **37**, 1255.

Giggenbach, W. F. (1974). *N.Z. J. Sci.* **17**, 33.

Giggenbach, W. F. (1977). *In* "Geochemistry 1977" (A. J. Ellis, ed.). New Zealand Dept. Sci. and Ind. Bull. 218, Wellington (in press).

Grindley, G. W. (1965). The Geology, Structure and Exploitation of the Wairakei Geothermal Field, Taupo, New Zealand. Bull. Geol. Surv. New Zealand, No. 75, Wellington.

Gunter, B. D. (1973). *Geochim. Cosmochim. Acta* **37**, 495.

Gutsalo, L. K. (1975). *Proc. U.N. Symp. Develop. Use Geothermal Resources, 2nd, San Francisco, May* **1**, 745.

Helgeson, H. C. (1967). *In* "Researches in Geochemistry" (P. H. Abelson, ed.), Vol. 2, pp. 362–402. Wiley, New York.

Helgeson, H. C. (1968). *Am. J. Sci.* **266**, 129.

Helgeson, H. C. (1969). *Am. J. Sci.* **267**, 729.

Helgeson, H. C. (1970). *Econ. Geol.* **65**, 299.

Helgeson, H. C. (1971). *Geochim. Cosmochim. Acta* **35**, 421.

Helgeson, H. C., Garrels, R. M., and McKenzie, F. T. (1969). *Geochim. Cosmochim. Acta* **33**, 445.

Hemley, J. J. (1959). *Am. J. Sci.* **257**, 241.

Hemley, J. J., and Jones, W. R. (1964). *Econ. Geol.* **59**, 538.

Hemley, J. J., Hostetler, P. B., Gude, A. J., and Mountjoy, W. T. (1969). *Econ. Geol.* **64**, 599.

Hulston, J. R., and McCabe, W. J. (1962a). *Geochim. Cosmochim. Acta* **26**, 383.

Hulston, J. R., and McCabe, W. J. (1962b). *Geochim. Cosmochim. Acta* **26**, 399.

Ivanov, V. V. (1958). *Geochem. Int.* **5**, 600.

Ivanov, V. V., and Nevraev, G. A. (1964). "Klassifikatsiya Podzemnykh Mineral'nykh Vod." Izdatelstvo "Nedra", Moscow.

Kharaka, Y. K., and Barnes, I. (1973). SOLMNEQ: Solution-Mineral Equilibrium Calculations. Nat. Tech. Informat. Serv., Rep. P.B. 215899, Springfield, Virginia.

Kissen, I. G., and Pakhomov, S. I. (1967). *Geochem. Int.* **4**, 295.

Kruger, P., and Otte, C. (1973). "Geothermal Energy," Stanford Univ. Press, Stanford, California.

Lebedev, L. M. (1972). *Geochem. Int.* **9**, 485.

Mahon, W. A. J. (1966). *N.Z. J. Sci.* **9**, 135.

Mahon, W. A. J. (1967). *N.Z. J. Sci.* **10**, 206.

Mahon, W. A. J., and McDowell, G. D. (1977). *In* "Geochemistry 1977" (A. J. Ellis, ed.). New Zealand Dept. Sci. and Ind. Res. Bull. 218, Wellington (in press).

Makarenko, F. A., and Mavritsky, B. F. (1965). *Int. Geol. Rev.* **7**, 1387.

Matsubaya, O., Sakai, H., Kusachi, I., and Satake, H. (1973). *Geochem. J.* **7**, 123.

Mazor, E. (1975). *Proc. U.N. Symp. Develop. Use Geothermal Resources, 2nd, San Francisco, May* **1**, 793.

Mazor, E., and Fournier, R. O. (1973). *Geochim. Cosmochim. Acta* **37**, 515.

Mazor, E., and Wasserburg, G. J. (1965). *Geochim. Cosmochim. Acta* **29**, 443.

Mercado, S. (1966). Aspectos quimicos del approvechamiento de la energia geotermica; Campo geotermico Cerro Prieto, B.C.. Comision Federal de Electricidad, Mexico, D.F.

Mercado, S. (1967). Geoquimica Hidrotermal en Cerro Prieto, B.C., Mexico. Comision Federal de Electricidad, Mexicali.

Meyer, C., and Hemley, J. J. (1967). *In* "Geochemistry of Hydrothermal Ore Deposits" (H. L. Barnes, ed.), pp. 166–235. Holt, New York.

Muffler, L. P. J., and White, D. E. (1969). *Geol. Soc. Am. Bull.* **80**, 152.

Nakamura, H., and Sumi, K. (1967). *Bull. Geol. Surv. Jpn.* **18**, 58.

Nakamura, H., Sumi, K., Katagiri, K., and Iwata, T. (1970). *Geothermics (Spec. Issue 2)* **2** *(Pt 1)*, 221.

Nasini, T. (1930). "I Soffione e i Lagoni della Toscana e la Industria Boracifera." Tipografia Editice Italia, Rome.

Noguchi, T. (1966). *Bull. Volcanol.* **29**, 529.

Orville, P. M. (1963). *Am. J. Sci.* **261**, 201.

Panichi, C. (1975). *Proc. Int. At. Energy Agency Advisory Group Meeting Appl. Nucl. Techn. Geothermal Stud., Pisa, Italy, Sept.* (in press).

Polak, B. G., Kononov, V. I., Tolstikhin, I. N., Mamyrin, B. A. and Khabarin, L. V., (1975). *Proc. Symp. Thermal Chem. Probl. Thermal. Waters, I.U.G.C. 16th General Assembly, Grenoble, Sept.* IAHS Publ. No. 119 p. 15.

Seki, Y., Onuki, H., Okumura, K., and Takashima, I. (1969). *Jpn. J. Geol. Geogr.* **40**, 63.

Seward, T. M. (1973). *Geochim. Cosmochim. Acta* **37**, 379.

Skinner, B. J., White, D. E., Rose, H. J., and Mays, R. E. (1967). *Econ. Geol.* **62**, 316.

Shukolyukov, Y. A., and Tolstikhin, I. N. (1965). *Geochem. Int.* **2**, 617.

Steiner, A. (1968). *Clays Clay Miner.* **16**, 193.

Steiner, A., and Rafter, T. A. (1966). *Econ. Geol.* **61**, 1115.

Sumi, K. (1969). *Proc. Int. Clay Conf., Tokyo* **1**, 501.

Truesdell, A. H. (1975). *Proc. U.N. Symp. Develop. Use Geothermal Resources, 2nd, San Francisco, May* **1**, lii.

Usdowski, H. E., and Barnes, H. L. (1972). *Contrib. Mineral. Petrol.* **36**, 207.

Waring, G. A., Blankenship, R. R., and Bentall, R. (1965). Thermal Springs of the United States and other Countries of the World. U.S. Geol. Surv. Professional Paper 492.

Wasserburg, G. J., and Mazor, E. (1965). *In* "Fluids in Subsurface Environments—a Symposium," pp. 386–398. Memoir Am. Assoc. Petrol. Geologists 4.

Wasserburg, G. J., Mazor, E., and Zartman, R. E. (1963). *In* "Earth Science and Meteoritics" (J. Geiss and E. D. Goldberg, eds.), pp. 219–240. North-Holland Publ., Amsterdam.

Weissberg, B. G. (1969). *Econ. Geol.* **64,** 95.

Weissberg, B. G., Browne, P. R. L., and Seward, T. M. (1976). *In* "Geochemistry of Hydrothermal Ore Deposits" (H. L. Barnes, ed.), 2nd ed. Wiley, New York (in press).

White, D. E. (1957). *Geol. Soc. Am. Bull.* **68,** 1637.

White, D. E. (1965). *In* "Fluids in Subsurface Environments—A Symposium." *Memoir Am. Assoc. Petrol. Geol.* **4,** 342.

White, D. E. (1967). *In* "Geochemistry of Hydrothermal Ore Deposits" (H. L. Barnes, ed.), pp. 575–631. Holt, New York.

White, D. E. (1968). *Econ. Geol.* **63,** 301.

White, D. E. (1974). *Econ. Geol.* **69,** 954.

White, D. E., Barnes, I., and O'Neil, J. R. (1973). *Geol. Soc. Am. Bull.* **84,** 547.

Wilson, S. H. (1963a). *In* Waiotapu Geothermal Field, pp. 87–118. New Zealand Dept. Sci. and Ind. Res. Bull. 155, Wellington.

Wilson, S. H. (1963b). *In* "Nuclear Geology on Geothermal Areas" (E. Tongiorgi, ed.), pp. 173–184. Consiglio Nazionale delle richerche, Lab. di Geol. Nucl., Pisa.

Chapter 4

HYDROTHERMAL SOLUTIONS

A better understanding of the physical and chemical processes occurring in geothermal systems may be obtained through an appreciation of the properties of water and water solutions at elevated temperatures. Only a brief outline of some of the more relevant properties is given, since there are extensive reviews of the topic elsewhere, including those of Franks (1972), Horne (1972), and particularly from the viewpoint of geothermal chemistry, Helgeson (1969) and Helgeson and Kirkham (1974a,b, 1976). The chapter continues with a detailed examination of specific chemical equilibria that are of importance in geothermal chemistry.

4.1 THE NATURE OF WATER

The melting point, boiling point, and critical point temperatures for water are unusually high compared with those for other low molecular weight hydrides. It has an exceptionally high dielectric constant. The high specific heat and latent heat of vaporization make water a highly efficient medium for heat transfer. These and other properties are usually explained as being due to extensive hydrogen bonding between molecules in liquid water. On thermodynamic grounds, Pople (1951) pointed out that it was unlikely that considerable breaking of H—O—H bonds occurred when ice melted.

Reviews of the theories of water structure were given by Frank (1970) and Franks (1972). Most of the discussion in these works is centered on whether the liquid is a mixture of distinct structures of a size considerably greater than a molecular diameter, or a homogeneous single-state medium. There is, however, no real agreement as to the size, geometry, or lifetimes of particular hydrogen-bonded structural entities. Various theories on the structure of nature's most common liquid are summarized.

X-ray diffraction work on water led Bernal and Fowler (1933) to propose that liquid water retained a "broken down" ice lattice. On a statistical average, the network of water molecules took different forms as the temperature was raised. A tridymite-like structure was of most importance below 4°, a quartz-like form from 4° to 200°, and a close-packed arrangement above 200°. This "uniformist" model still influences thinking about water structure, although more recent models consider geometries other than the ice structure, such as exchanging networks of small ring polymers of water molecules (Del Bene and Pople, 1970).

Other theories assume that liquid water is a mixture of two or more distinguishable molecular species. It was suggested by Frank and Evans (1945) that the addition of many simple solutes to water created localized water structure, or "icebergs," about the solutes. Frank and Wen (1957) pointed out that long-lived species were unlikely and that hydrogen bonding was a cooperative phenomenon, the formation of one bond promoting the tendency of each water molecule to hydrogen bond to a neighbor. "Flickering clusters" of icelike structures of varying size form, intermixed with nonhydrogen-bonded molecules. The cluster lifetime is long compared with the molecular vibration period, with the proportions and cluster size varying with temperature and pressure. Nemethy and Scheraga (1962) applied statistical mechanics to calculate the average cluster size at various temperatures.

Several models suggest that the structure of water is an expanded version of the framework of ordinary ice, but that unassociated water molecules occupy some of the interstitial positions in the expanded framework (Samoilov, 1946; Forslind, 1952). Pauling (1959) proposed a more specialized cage structure based on an analogy with the clathrate compounds formed with water by some nonpolar gases such as chlorine, krypton, and xenon.

Marchi and Eyring (1964) regarded water as a combination of tetrahedrally bonded nonrotating molecules in equilibrium with freely rotating molecules. Using appropriate partition functions for these structures, they calculated the thermodynamic properties of water up to about 250° at saturated water vapor pressures (s.w.v.p.).

In balance, at present, the two-species concept, involving a tetrahedrally bonded framework plus interstitial water molecules, appears to be the most satisfactory in explaining the general chemical and thermodynamic properties of water.

At pressures higher than saturated water vapor pressures, modification of the liquid water structure occurs as the density increases. At low temperatures, pressure will cause a collapse of open network structures, first through bending and finally breaking of hydrogen bonds. At high temperatures, pressure may promote hydrogen bonding by providing a favorable molecular density.

The extent to which hydrogen bonding contributes to the structure of water at temperatures of the order of 300°–400° and over and at pressures of several thousand bars is at present open to speculation, but some guidance may be obtained from the electrical conductance of the hydrogen ion in water. At ambient temperatures its conductance is anomalously high because of a chain propagation mechanism made possible by hydrogen bonding. With increasing temperatures at low solution densities, the abnormal conductance rapidly decreases, but if the solution density is maintained near unity, the unusually high hydrogen ion conductance persists to temperatures as high as 800° (Quist and Marshall, 1968).

4.2 PHYSICAL PROPERTIES

Density

Table A.1 in the Appendix gives values for the density of liquid water and of steam at saturated water vapor pressures, and Fig. 4.1

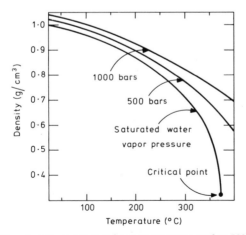

Fig. 4.1 Density of water at saturated vapor pressures and at 500 and 1000 bars.

shows the trend in liquid densities at saturation pressures and at 500 and 1000 bars. At pressures close to saturation values the liquid density drops rapidly at temperatures above about 300° as the critical point is approached, but higher pressures can retain a relatively high density in the fluid at temperatures in excess of the critical point.

Densities of sodium chloride solutions over a wide range of concentrations and temperatures are summarized in Table A.2. In most cases these will be a close approximation to the density of saline geothermal waters of equal ionic strength.

Vapor–Liquid Equilibrium

Table A.1 contains a summary of saturated steam pressures over a range of temperatures of geothermal interest. The presence of salts in solution lowers the vapor pressure of water, and the vapor pressure of sodium chloride solutions may be obtained from the papers of Gardner *et al.* (1963) and Liu and Lindsay (1970, 1972).

The change in boiling point with depth is of particular importance in assessing the characteristics of a geothermal area. In a hot water column at the temperature of boiling throughout its length, the steam pressure at any point is balanced by the hydrostatic head of water. Figure 4.2 shows the boiling point versus depth curve for water and for sodium chloride solutions, calculated from their vapor pressures and densities (Haas, 1971).

The critical point constants for water are $T_c = 374.15°C$, $P_c = 221.29$ bars, and the critical volume $V_c = 3.1$ cm^3/g (Osborne *et al.*, 1937). When considering the mechanisms of geothermal systems it should be noted that when the critical temperature is passed at pressures much higher than the critical pressure there are no sudden changes in physical or chemical properties.

Figure 4.3 shows general pressure–temperature trends in the Wairakei and Broadlands hot water systems and in The Geysers steam-producing field. The vapor pressure curve of water is given, as are the pressure–temperature relationships in water columns existing in geothermal gradients of 30° and 90° per kilometer. As a generalization, the vapor pressure curve is followed up to a limiting temperature, which for a hot water system also relates to the maximum depth at which boiling can occur. For a system with a high carbon dioxide concentration (Broadlands) the vapor pressures are higher than for pure water. Beneath depths at which the limiting temperature (230°–240°) is reached in steam-dominated systems there is little departure from the vapor pressure curve (down to at least 2 km).

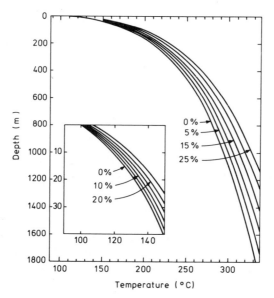

Fig. 4.2 Boiling point-depth curves for water and sodium chloride solutions of various concentrations (from Haas, 1971).

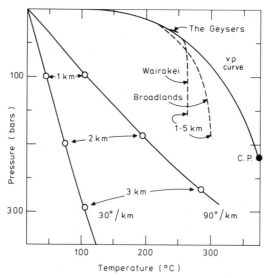

Fig. 4.3 Pressure-temperature trends in the Wairakei, Broadlands, and The Geysers geothermal fields, in comparison with the vapor pressure curve for water and trends in water columns in geothermal gradients of 30° and 90° per kilometer.

In hot water geothermal systems, pressures usually exceed the critical point pressures well before critical point temperatures are reached. The properties of the liquid or liquid-like fluid in the systems should therefore exhibit continuous trends with depth, uninterrupted by critical point phenomena.

Isotope Separation

The isotopes of hydrogen and of oxygen are present in different proportions in water vapor and liquid phases in equilibrium. At ambient temperatures there is an appreciably higher concentration of the heavy isotopes, deuterium, and ^{18}O, in the liquid phase, but the difference between phases becomes less as temperatures rise. Above 224°, deuterium becomes concentrated into the vapor phase.

The equilibrium isotope fractionation factors have been determined by Bottinga and Craig (1968). The results can be expressed as single-stage enrichment factors ϵ^*

$$\epsilon^* = 1000[1 - (r_{vap}/r_{liq})] \tag{4.1}$$

where r_{vap} or r_{liq} is the D/H ratio or $^{18}O/^{16}O$ ratio in the vapor or liquid phases.

Values of ϵ_D^* and ϵ_{18O}^* are given in Fig. 4.4, prepared by Giggenbach (1971) from the data of Bottinga and Craig (1968). Points are added for distribution factors determined from the separate analyses of steam and water samples separated under pressure from geothermal

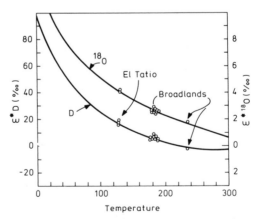

Fig. 4.4 Single-stage enrichment factors for deuterium and ^{18}O (from Giggenbach, 1971).

wells. Even in the very short contact times in the sampling separators, equilibrium isotope distribution was closely approached.

Viscosity

The viscosity of water and of steam at various pressures is summarized in Fig. 4.5, prepared from the data of Dudziak and Franck (1966). The viscosity of water at 250° is only about six times that of steam, yet its density is 40 times that of steam. The formation of steam in a hot water flow therefore lowers mass transfer through a channelway, an effect which becomes more marked at lower temperatures. Since the enthalpy of saturated steam is only two to three times that of water within the temperature range 200°–300°, the extra heat transferred per unit mass of steam is not sufficient to compensate for loss in mass flow, so that heat flow through fluid transfer is also diminished by water boiling in channelways.

Enthalpy

The specific enthalpy of water and of steam at saturation pressures is summarized in the Appendix (Table A.1). More detailed enthalpy information necessary for calculating steam–water mass balances in two-phase flows of geothermal fluids may be obtained from steam

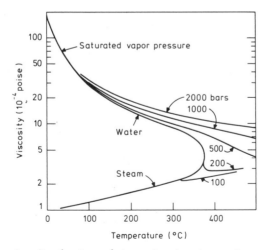

Fig. 4.5 The viscosity of water and steam at various temperatures and pressures.

tables, such as Keenan *et al.* (1969). For sodium chloride solutions, see Silvester and Pitzer (1976).

The thermodynamic properties of water at higher pressures and temperatures are of importance in considering the origins of geothermal fluids. A portion of a Mollier chart relating temperature, pressure, enthalpy, and entropy is given in Fig. 4.6 (Mahon and McDowell, 1977). From this it can be seen that an isoenthalpic expansion of very high-temperature and high-pressure fluid (e.g., 800°, 3000 bars) through porous media intersects the two-phase liquid + steam area of the chart.

The expansion of a magmatic fluid may therefore condense out a small proportion of liquid at a temperature between about 240° and the critical temperature, and this liquid would concentrate within it the salt constituents from the original fluid. There is evidence that concentrated brines exist in some volcanic geothermal areas (e.g., El Tatio; see Chapter 6).

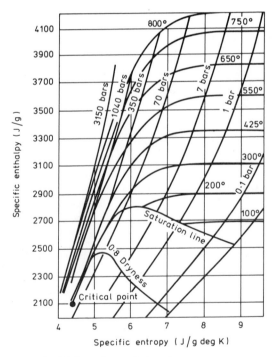

Fig. 4.6 A portion of a Mollier chart relating the temperature, pressure, enthalpy, and entropy of water.

Dielectric Constant

The solubility of salts is the result of a balance between ion–ion and ion–water interactions, and on the simplest electrostatic picture the free energy of ion–ion interactions is inversely proportional to the dielectric constant of the solvent medium. Table A.1 gives values of the dielectric constant of water at saturated water vapor pressures. The rapid decrease in dielectric constant with increasing temperature is reflected in the changing behavior as a solvent. Slightly soluble salts become insoluble, weak acids and bases ionize to a lesser extent, and considerable ion pair complexing occurs in solution.

Increasing pressure raises the dielectric constant, but up to 300° the effect of a 1000-bar pressure increase, for example, is minor in comparison with a temperature increase of a few tens of degrees. Values of the dielectric constant of water over a wide range of conditions were reviewed by Helgeson and Kirkham (1974a).

4.3   IONIC SOLUTIONS

Electrical Conductance

The electrical conductance of rocks containing hot saline fluids is appreciably higher than for rocks that contain cool groundwaters or are impermeable. This fact is widely used in resistivity survey work to outline the area and volume extent of new geothermal areas.

The conductance of many types of salt solutions has been measured at elevated temperatures and pressures, and a review of this work was given by Quist and Marshall (1965). As an approximation, the conductance of sodium chloride solutions can be used as a measure of the conductance of geothermal waters of equal ionic strength, providing that they are of the neutral-pH alkali chloride type. Figure 4.7 gives molar conductance values (λ) for sodium chloride solutions over a wide range of temperatures and molar concentrations (C) (data from Noyes *et al.*, 1907).

The specific conductance of a solution is related to the specific resistivity (R) by the equation $R = 1000/\lambda C$.

For example, at 250° a 0.03 molar NaCl solution has a λ value of about 650 ohm^{-1} cm^{2} mole^{-1}. The value of R is therefore about 0.5 ohm m, while the resistivity of a rock medium of 10% homogeneous porosity containing this solution at 250° would be expected to be

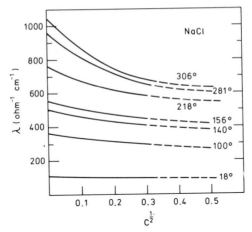

Fig. 4.7 Molar conductance of sodium chloride solutions at various concentrations at high temperatures.

about 5 ohm m. Resistivities about this value were measured in the central part of the Broadlands field where these particular salt concentrations and temperatures exist.

Ionic Equilibria

Equilibrium Constants

In geothermal systems, ionic equilibria are important controls on the chemical nature of hot waters, and on the composition of gases in steam separating from waters. Many solutes in natural waters are weak acid and base systems (e.g., H_2CO_3, H_2S, H_3BO_3, H_4SiO_4, HF, NH_3) which, as temperatures and pressures change, interact with each other, with other solutes in the water, and with minerals and precipitates in contact with waters. Dissolved salts which are extensively ionized at ambient temperatures may be incompletely ionized in low-density high-temperature water, and a wide range of ion complexes may form. A basic requirement in geothermal chemistry is a quantitative knowledge of ionic equilibria in solutions over a wide range of temperatures and pressures, which in turn requires information on the activity coefficients of ions in solutions of various ionic strengths. Because this subject has been extensively reviewed by Helgeson (1969) and Helgeson and Kirkham (1976), only an outline is included here.

For a reaction of the type $MA = M^+ + A^-$, there is an equilibrium constant K.

$$K = a_M a_A / a_{MA} \tag{4.2}$$

$$= m_M m_A \bar{\gamma}_\pm^2 / m_{MA} \bar{\gamma}_{MA} \tag{4.3}$$

where m and a are the molarity and activity, respectively, of an ion or molecule in solution; $\bar{\gamma}_\pm$ is the mean activity coefficient for the ions M^+ and A^- in the solution under consideration and γ_{MA} the activity coefficient for the species MA. With appropriate changes in the equilibrium constant designation, MA may be a molecular solute such as HF, a solid such as $CaCO_3$, a negatively charged ion such as HSO_4^-, or a positively charged ion such as NH_4^+.

$$HF = H^+ + F^- \qquad K_a = a_H \cdot a_F / a_{HF} \tag{4.4}$$

$$HSO_4^- = H^+ + SO_4^{2-} \qquad K_{a_2} = a_H \cdot a_{SO_4} / a_{HSO_4} \tag{4.5}$$

$$NH_4^+ = H^+ + NH_3 \qquad K_a = a_H \cdot a_{NH_3} / a_{NH_4} \tag{4.6}$$

$$CaCO_3 = Ca^{2+} + CO_3^{2-} \qquad K_{CaCO_3} = a_{Ca} \cdot a_{CO_3} \tag{4.7}$$

The reaction of minerals with a solution may be much more complex, but in principle a reaction equation can be written involving each individual mineral and the species with which it reacts in solution. For example, consider the equilibrium between albite $(NaAlSi_3O_8)$ and solution

$$NaAlSi_3O_8 + 4H^+ + 4H_2O = Na^+ + Al^{3+} + H_4SiO_4 \tag{4.8}$$

$$K_{Ab} = a_{Na} a_{Al} a_{H_4SiO_4} / a_H^4 \tag{4.9}$$

The condition of chemical equilibrium in a geothermal solution requires that the concentrations of all solutes adjust so that the equilibrium constants for all interactions between species are satisfied. Computer techniques have been developed to obtain the solution to the very many simultaneous equations which define this unique chemical situation for a given temperature and pressure (Truesdell and Singers, 1971; Kharaka and Barnes, 1973) and these are discussed in Chapter 7.

For dissociation equilibria, such as reactions (4.4) and (4.5), the free energy of dissociation may be separated into an electrostatic contribution arising from the creation or disappearance of ions and the free energy of making or breaking chemical bonds. The electrostatic contribution is responsive to the dielectric constant of water, which at saturated water vapor pressures decreases at an accelerating rate with rising temperature. In many cases, rising temperatures have opposite effects on the electrostatic and chemical energy contributions. A common trend is that at s.w.v.p. simple dissociation constants first increase with temperature to a maximum value and then decrease at an increasing rate as the electrostatic effects become dominant. Examples of this type of behavior are shown in Fig. 4.8, which gives dissociation constants for various weak electrolyte equilibria.

The effect of pressures above s.w.v.p. at a temperature can be estimated from the following equations:

$$d \ln K/dP = -\Delta \bar{V}^\circ/RT \tag{4.10}$$

and

$$d\Delta \bar{V}^\circ/dP = \Delta \bar{K}^\circ \tag{4.11}$$

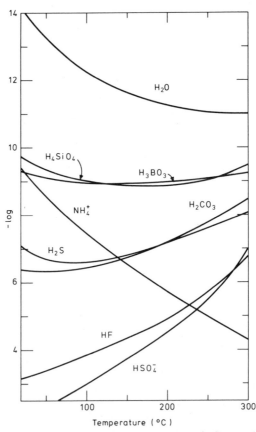

Fig. 4.8 Variations with temperature in various acid dissociation constants at saturated water vapor pressures.

where $\Delta \bar{V}°$ is the change in partial molar volume in the reaction and $\Delta \bar{K}°$ is the change in partial molal compressibility. A sufficient range of partial molal volume and compressibility data for ions is now available (Ellis and McFadden, 1972; Helgeson and Kirkham, 1976) to estimate dissociation constants to pressures of several hundred bars above s.w.v.p.

Laboratory experimental work and theoretical considerations have given values of the equilibrium constants for very many of the ionic reactions of importance in natural hydrothermal systems. Since there are excellent recent compilations of equilibrium constants for temperatures in the geothermal range of interest (0°–350°), no attempt is made in this book to present a full collection of data (Helgeson, 1969, 1976; Kharaka and Barnes, 1973). However, Appendix Table A.3 gives values of equilibrium constants for the ionization of some weak electrolytes.

Note from Fig. 4.8 that the self-ionization constant for water K_w increases markedly from ambient to high temperatures. The neutral pH of pure water is where $a_{H^+} = a_{OH^-}$, that is, $a_{H^+} = (a_{H^+} a_{OH^-})^{1/2}$ or $K_w^{1/2}$. Whereas this is pH 7.00 at 25°, at 300° it is pH 5.5. When discussing whether a geothermal solution is acid or alkaline, the neutral pH for the temperature should be kept in mind.

Activity Coefficients

The mean ionic activity coefficient for a solute forming n cations and m anions is defined as follows:

$$\bar{\gamma}_{\pm} = (\bar{\gamma}_+{}^n \cdot \bar{\gamma}_-{}^m)^{1/(n+m)} \tag{4.12}$$

This coefficient arises largely from electrostatic effects and may be expressed by a form of the Debye–Hückel equation

$$-\log \bar{\gamma}_i = \frac{z_i^2 A I^{1/2}}{1 + \mathring{a} B I^{1/2}} \tag{4.13}$$

where $A = (1.8246 \times 10^6) \rho^{1/2}/(\epsilon T)^{3/2}$ and $B = (50.29 \times 10^8) \rho^{1/2}/(\epsilon T)^{1/2}$; $\bar{\gamma}_i$ is the molar activity coefficient of ion species i, charge z_i; $\mathring{a}$ is the effective ion diameter; ϵ and ρ are, respectively, the temperature- and pressure-dependent dielectric constant and density of water; and I is the true ionic strength of the solution. Values of A and B over a wide temperature and pressure range were given by Helgeson and Kirkham (1974b).

For many univalent ions at temperatures below 300° and at ionic strengths less than about 0.1 molal the Debye–Hückel activity coefficients $\bar{\gamma}_i$ are reasonably satisfactory for use in equilibrium calculations. For divalent ions the coefficients obtained in this way are only a rough approximation.

At high temperatures and higher salt concentrations even the so-called strong electrolytes (e.g., NaCl, KCl, HCl, Na$_2$SO$_4$) are not completely dissociated and the true ionic concentration may be unknown. The stoichiometric mean ion activity coefficient $\gamma_\pm$ conveniently relates the activity of the electrolyte ions to the total dissolved electrolyte concentration. For alkali and hydrogen chloride solutions over about 0.1 m at temperatures above 300°, for example, appreciable concentrations of un-ionized molecules occur and there is an appreciable difference between $\bar{\gamma}_\pm$ and $\gamma_\pm$.

Values of the stoichiometric mean activity coefficients for electrolytes involved in particular mineral and solution equilibria are required to relate thermodynamic dissociation constants to salt concentrations [Eqs. (4.2) and (4.3)]. There is an increasing body of information available on $\gamma_\pm$ values for chloride, sulfate, and carbonate systems from high-temperature electromotive force (emf) measurements, solubility studies on sparingly soluble salts, and vapor pressure measurements. Helgeson (1969) gave a detailed discussion on the problem of activity coefficient estimation and gave values for the

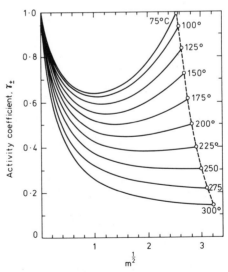

Fig. 4.9 The trend with temperature and concentration in the mean stoichiometric molar activity coefficient of sodium chloride in water (from Liu and Lindsay, 1972).

stoichiometric activity coefficients of many ions of major geological importance in sodium chloride solutions.

Figure 4.9 shows the trends with temperature for the mean stoichiometric molal activity coefficient of sodium chloride in water at s.w.v.p. (Liu and Lindsay, 1972). See also Silvester and Pitzer (1976).

4.4 SOLUBILITIES

The general effectiveness of water as a solvent for many ionic solids is due to both its high dielectric constant and its participation as a reactant through the formation of hydrated and/or hydroxy complex ions. For molecular substances, such as silica or carbon dioxide, it may also function as a reactant (e.g., through the formation of such entities as $Si(OH)_4$ or H_2CO_3). Some of the characteristics of water solutions which are relevant to geochemical work in geothermal areas are now surveyed.

Gas Solubilities

Natural hot waters contain dissolved gases, the most common of which are carbon dioxide, hydrogen sulfide, ammonia, hydrogen, nitrogen, oxygen, and methane. The proportions in which these gases are present reflect the general rock environment, and the concentrations of some gases are connected with the stability and solubility of particular minerals. The concentration of gases in steam derived from hot waters also affects the economics of power production.

The solubility of gases is required for two common types of calculation. From the analysis of well discharges (water and steam) the mass concentration of gas in the deep supply water can be obtained. To correlate the gas concentration and the partial pressure of the gas, knowledge of the Henry's law coefficient K_g is required at the temperature for the particular solution.

$$K_g = f/x \qquad (4.14)$$

where f is the fugacity of the gas (the partial pressure corrected by an activity coefficient) and x is the mole fraction of the gas in solution. At the gas pressures (up to a few bars) in most natural hydrothermal systems, it is sufficient to equate the partial pressure and fugacity of the gases. Himmelblau (1959) gave a convenient summary of the coefficients for the solution of O_2, N_2, H_2, and CH_4 in water. Table 4.1

TABLE 4.1

Values of K_g for H_2S at Various Temperatures[a]

Temp. (°C)	50	100	150	200	250	300	350
K_g (bars)	1000	1580	1860	2000	1900	1620	1000

[a] (Kozintseva, 1965)

gives values of K_g for hydrogen sulfide and Fig. 4.10 has values for carbon dioxide in water and in NaCl solutions (Ellis and Golding, 1963). For an extensive review of the thermodynamics of the CO_2–H_2O system see Malinin (1974).

It is often necessary to estimate the distribution of a gas between water and steam phases at various stages of steam separation from a high-temperature water. For this purpose the mass distribution coefficient A is a useful measure of the gas solubility.

$$A = \frac{n_g^l/n_{H_2O}^l}{n_g^v/n_{H_2O}^v} \tag{4.15}$$

The relationship $K_g = P_w/AZ_w$ connects the two solubility coefficients, where P_w is the water vapor pressure and z_w is the compressibility factor for steam ($z_w = P_wV/RT$). Both P_w and z_w may be obtained from steam tables.

The solubility of a gas expressed as a volumetric coefficient, such as the Henry's law coefficient, passes through a minimum with rising temperature, then increases to infinite solubility at the critical point

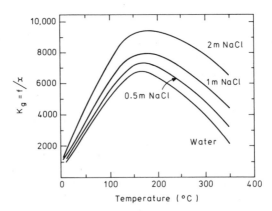

Fig. 4.10 Values of the Henry's law coefficient K_g for carbon dioxide solubility in water and sodium chloride solutions (from Ellis and Golding, 1963).

temperature for the system. Mass distribution coefficients, such as A, have a continuous trend with rising temperature, as shown by Fig. 4.11, which summarizes values of A for most gases of interest to studies of geothermal systems (Ellis, 1967). The values are for solution in water, and ρ^v and ρ^l are the densities of the vapor and the liquid phases, respectively.

It should be noted that for gases such as CO_2, H_2S, and NH_3, only the un-ionized fraction in the liquid phase is used in solubility calculations.

The mass distribution of all the gases mentioned, except ammonia, lies heavily in favor of the vapor phase even at temperatures of 200°–300°. The general order of the solubility of the gases does not change with temperature. The most soluble gas is ammonia, followed by hydrogen sulfide, and then carbon dioxide. Differences between the solubilities of these gases are considerable, especially at lower temperatures, and there is also a considerable difference between carbon dioxide solubility and that of the least soluble group, methane, oxygen, hydrogen, and nitrogen.

The presence of dissolved salts lowers the gas solubilities, and this should be taken into account when hydrothermal solutions have an ionic strength more than a few tenths molal. The "salting out" effect is greatest at higher temperatures, as shown in Fig. 4.10. In this diagram the critical point temperatures are taken from Sourirajan and Kennedy's (1962) work on the $NaCl$-H_2O system.

Fortunately, the salting out effect appears to be of similar magnitude for most simple gases. Table 4.2 gives the relative effects of sodium chloride concentrations on carbon dioxide and hydrogen sulfide solu-

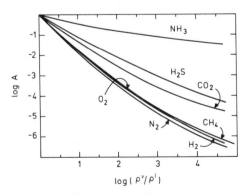

Fig. 4.11 Values of the mass distribution coefficient A for various gases as a function of the relative densities of vapor and liquid phases.

TABLE 4.2

Effect of Sodium Chloride Concentrations on Solubility of Carbon Dioxide and Hydrogen Sulfide[a]

	CO$_2$ solubility				
	100°C	150°C	200°C	250°C	300°C
0.5 m NaCl	0.076	0.070	0.090	0.128	0.172
1.0 m NaCl	0.078	0.076	0.089	0.128	0.176
2.0 m NaCl	0.080	0.073	0.084	0.111	0.151
			H$_2$S solubility		
			202°C	262°C	
0.55 m NaCl			0.10	0.10	
0.12 m NaCl			0.10	0.09	

[a] Expressed as salting out coefficient values $K_x = (1/m) \log(K_g^*/K_g)$, where K_g^* is the gas solubility coefficient in the salt solution.

bilities. Hydrogen sulfide solubility data are from Kozintseva (1965) and those for carbon dioxide from Ellis and Golding (1963).

Solubility of Solids

The solubility of solids in water over a wide temperature and pressure range exhibits a great variety of phenomena. These may include reaction of solid and water, solid–solid reactions, or phase transitions, and the formation of immiscible liquids. Outlines of these types of behavior can be found in Ellis and Fyfe (1957) and Smith (1963). In the temperature and pressure range of interest to geothermal developments, the solubility of solids can be divided into two general classes.

(A)   Solids which have a solubility that increases continuously with temperature at saturated water vapor pressures. At temperatures about the critical point of pure water (374°) the solubility is high, and the critical temperature of the system is raised considerably above that for water.

(B)   Solids which either decrease in solubility with rising temperature at s.w.v.p., or at first increase to a maximum solubility, then decrease. Near the critical temperature region the solubility is usually so low that the critical temperature is raised at most by a few degrees above that for water.

Examples of the first type of behavior are the alkali metal chlorides and bromides; calcium nitrate; potassium nitrate; potassium fluoride; potassium carbonate; and cesium sulfate. Examples of the second type of behavior are lithium and sodium fluorides; lithium, sodium, and potassium sulfates; sodium carbonate; calcium fluoride; calcium carbonate; and silica.

Figure 4.12 shows the solubility of sodium chloride in steam and in water over a wide range of high temperatures and pressures (Sourirajan and Kennedy, 1962). For this A-type system, pressures of the order of 1000 bars have only a minor effect on the solubility at temperatures below about 350°. On the other hand, as shown in Fig. 4.13, the solubility of quartz (from Kennedy, 1950) is markedly dependent on the pressure (and water density). Although at very high pressures the maximum in the solubility curve for this type of system may disappear, it can be seen in Tables 4.4–4.6 that retrograde solubility behavior of some B-type solutes is of major importance at the temperatures and pressures of geothermal systems.

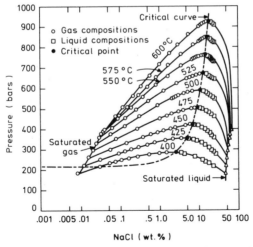

Fig. 4.12 The solubility of sodium chloride in water and steam phases over a wide range of temperatures (from Sourirajan and Kennedy, 1962).

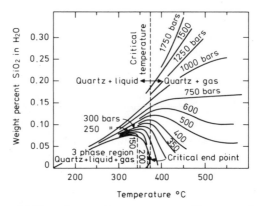

Fig. 4.13 The solubility of quartz in water over a wide range of temperatures and pressures (from Kennedy, 1950).

Halides

The solubilities of simple halide salts such as NaCl, KCl, NaBr, and CaCl$_2$ are considerable, and they increase continuously with temperature. Table 4.3 gives examples of solubilities up to 400° at s.w.v.p. (Seidell and Linke, 1965).

Boric Acid

The solubility of boric acid in water is quite high, increasing from about 5% by weight at 25° to become completely miscible at its melting point in water at 181°. No case is known where the concentra-

TABLE 4.3

Solubility of Halide Salts[a]

	Temperature (°C)							
	25	100	150	200	250	300	350	400
NaCl	26.43	28.2	29.7	31.6	34.1	37.3	41.9	46.4
KCl	26.4	36.0	40.2	44.6	49.0	53.8	58.4	63.4
NaBr	48.61	54.3	56.2	59.3	62.0	66.1	69.8	—
CaCl$_2$	45.3	61.4	67.2	75.7	77.3	—	—	—
MgCl$_2$	35.5	42.3	51.8	56.8	62.7	67.8	—	—

[a] Expressed as grams of salt per 100 g of solution.

tion of boric acid in natural high-temperature waters reaches saturation level.

Sulfates

Most alkali and alkali earth sulfates are solutes which at high temperatures and s.w.v.p. decrease in solubility to low values near the critical temperature of water. The solubilities of Li_2SO_4, Na_2SO_4, K_2SO_4, $MgSO_4$, and $CaSO_3$ in water are given in Table 4.4 (Seidell and Linke, 1965). Calcium sulfate solubilities above $100°$ are from Blount and Dickson (1969) and $MgSO_4$ values above $100°$ are from Marshall and Slusher (1965).

At the sulfate concentrations usually encountered in deep high-temperature geothermal waters, the solubility of lithium, sodium, and potassium sulfates are not exceeded (an exception is the Monte Cimini, Italy, geothermal system; see Chapter 1). Clay mineral equilibria (discussed later) usually maintain magnesium concentrations well below the level for magnesium sulfate saturation, but calcium sulfates (anhydrite or gypsum) are found as secondary minerals in some geothermal areas, and their solubility acts as a limit on calcium and sulfate concentrations in the solutions. The effect of temperature (up to $450°$), pressure (up to 1000 bars), and salinity (up to 6 m NaCl) on anhydrite solubility was given by Blount and Dickson (1969).

Calcite

A detailed review of calcite solubility was given by Holland (1967). The solubility of calcite in water can be expressed as

$$CaCO_3 + CO_2 + H_2O = Ca^{2+} + 2HCO_3^- \qquad (4.16)$$

TABLE 4.4

Solubility of Sulfate Salts in Water[a]

	Temperature (°C)						
	25	100	150	200	250	300	350
Li_2SO_4	25.5	23.5	22.7	22.9	—	—	—
Na_2SO_4	21.9	29.5	29.7	30.6	30.5	19.9	2.3
K_2SO_4	10.8	19.4	23.2	25.6	26.0	25.2	6.2
$MgSO_4$	26.7	33.5	18.5	1.3	0.18	0.06	0.014
$CaSO_4$	0.21	0.073	0.023	0.0074	0.0024	0.0008	—

[a] Expressed as grams of salt per 100 g of solution.

By taking the activity of water as constant, and equating the activity of CO_2 to the molarity m_{CO_2} and its fugacity to the partial pressure p_{CO_2}, the following equation applies.

$$K_{CHC} = a_{Ca^{2+}} a_{HCO_3^-}^2 / m_{CO_2} \tag{4.17}$$

$$= m_{Ca^{2+}} m_{HCO_3^-}^2 \gamma_\pm^3 / B p_{CO_2} \tag{4.18}$$

where B is an inverse Henry's law coefficient. In saline solutions at high temperatures the ionic strength is usually controlled by salts other than bicarbonates, and $\gamma_{\pm Ca(HCO_3)_2}$ is essentially constant for a particular temperature and pressure. As a result there is an approximate proportionality between m_{Ca}^3 and p_{CO_2} ($2m_{Ca} = m_{HCO_3}$), a fact which simplifies interpolation to obtain solubilities of calcite at particular p_{CO_2} values.

Figures 4.14 and 4.15 summarize the solubility of calcite in water and in salt solutions up to temperatures of 300° at s.w.v.p. (from Ellis, 1959, 1963). The solubility decreases rapidly with rising temperature, while carbon dioxide and salt concentrations increase the solubility.

Within the depth range of geothermal wells the usual mechanism of calcite precipitation is through the loss of carbon dioxide from solution as the water rises to a level where the pressure allows boiling to occur [Eq. (4.17)].

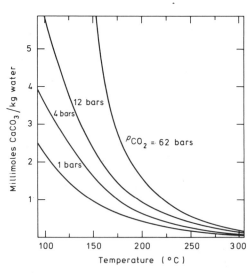

Fig. 4.14 The solubility of calcite in water at different partial pressures of carbon dioxide (from Ellis, 1959).

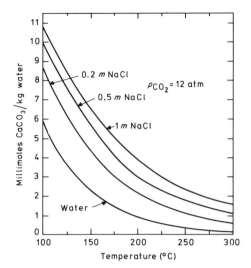

Fig. 4.15 The effect of salt concentrations on the solubility of calcite (from Ellis, 1963).

In contrast, hot waters may dissolve calcium from the rocks after becoming undersaturated with calcite by cooling through conduction or moderate dilution, without loss of carbon dioxide.

In Chapter 3 it was shown that in general the pH of high-temperature geothermal waters is controlled by aluminosilicate mineral equilibria. The consequence of this on calcium and carbonate concentrations in deep waters is now examined, assuming saturation with calcite. From Eq. (4.17), the following applies, where K_{a_1} is the first acid dissociation constant for "carbonic acid."

$$a_{Ca^{2+}} \cdot m_{CO_2} = a_{H^+}^2 \cdot K_{CHC}/K_{a_1}^2 \tag{4.19}$$

For albite, K-mica, K-feldspar equilibrium (Fig. 3.8), at a given temperature,

$$a_{Na^+}/a_{H^+} = K_{fm} \tag{4.20}$$

Therefore,

$$m_{Ca^{2+}} \cdot m_{CO_2} = \frac{K_{CHC}\, m_{Na^+}^2\, \gamma_{Na^+}^2}{K_{a_1}^2 K_{fm}^2\, \gamma_{Ca^{2+}}} \tag{4.21}$$

From a knowledge of the constants and of the ion activity coefficients and by assuming mineral equilibrium, Fig. 4.16 was developed (Ellis, 1970). This gives an approximate relationship between the

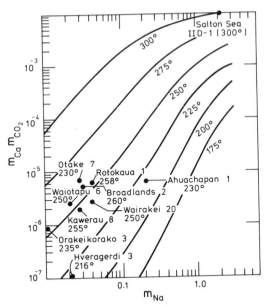

Fig. 4.16 Calculated relationship between the product $m_{Ca} m_{CO_2}$ and m_{Na} in high-temperature geothermal waters, together with points for specific geothermal waters (from Ellis, 1970).

product of calcium and carbon dioxide concentrations, and sodium ion concentrations in high-temperature waters, assuming calcite saturation. There are points in the diagram giving values of $m_{Ca^{2+}} \cdot m_{CO_2}$ for several deep waters, and in general these points fall close to the positions expected from the calculated curves. The calcium concentrations in high-temperature geothermal waters therefore are generally close to the maximum permitted by calcite solubility.

By making the assumption that a geothermal water is saturated with calcite, Fig. 4.16 can be used in an inverse manner to estimate the approximate carbon dioxide (or bicarbonate) concentrations in deep waters from analyses of m_{Na^+} and $m_{Ca^{2+}}$ and a knowledge of the deep-water temperature (also estimated from chemical equilibria; see Sections 4.7 and 4.8).

Pressures in excess of saturated water vapor pressures have an appreciable effect on calcite solubility. The change in K_{CHC} with increased pressure can be calculated from data on the partial molal volume of the solutes involved in the equilibrium, the molar volume of calcite, and estimates of the change in partial molal compressibility for the reaction (Ellis and McFadden, 1972). Table 4.5 gives calculated

TABLE 4.5

Hydrostatic Pressure Increase on Calcite Solubility at Various Temperatures[a]

Temperature	25	50	100	150	200	250	300
Ratio	3.0	2.8	2.7	2.8	3.5	4.6	6.5

[a] Changes calculated for an increase of 500 bars above s.w.v.p. and expressed as ratios of K_{CHC}^P to K_{CHC}° where the latter is the value at s.w.v.p.

changes in K_{CHC} from 25° to 300° for a pressure increase of 500 bars in excess of s.w.v.p. at the temperature.

The positive pressure coefficient shows that calcite would deposit continuously in rock fissures if water saturated with calcite rises rapidly from great depths at constant temperature, salinity, and carbon dioxide concentration but out of equilibrium with aluminosilicate minerals. This could fill fissures and stop the flow of hot water to the surface.

A calculation of the effect of pressures above saturated water vapor pressure on the equilibrium reaction (4.21) shows that by coincidence the effects on $\log K_{CHC}$ and $2 \log K_{a_1}$ are almost equal and approximately cancel out at temperatures up to about 300° (Ellis and Mc-Fadden, 1972). The value of $\log K_{fm}$ increases by about 0.15 unit per 1000 bars, so that the equilibrium concentration of calcium in solution becomes slightly less at high pressures and constant salinity and m_{CO_2}. Geothermal waters rising slowly at constant high temperature from deep, high-pressure conditions, while maintaining equilibrium with K-feldspar, K-mica, and albite, would therefore dissolve a small amount of calcium from the rocks (porous flow). Whether calcite precipitates from hot waters or whether they become undersaturated with calcite on route to the surface clearly is dependent on the rate and type of upflow.

Calcium Fluoride

The solubility of calcium fluoride in water and in salt solutions was reviewed by Holland (1967). It is considerably affected by salt concentrations, as well as by the presence of silica, which may cause reactions of the following type to occur:

$$CaF_2 + H_2O + SiO_2 = 2HF + CaSiO_3 \tag{4.22}$$

In a solution saturated with amorphous silica at s.w.v.p. the fluoride concentrations in solutions in equilibrium with fluorite are as

TABLE 4.6

Approximate Concentrations of Fluoride in a Water Solution in Equilibrium with
CaF$_2$ and Saturated with Amorphous Silica[a]

Temp. (°C)	100	150	200	250	300	350
F (ppm)	14	12	10	9	8	7

[a] At saturated water vapor pressure.

shown in Table 4.6 (Ellis and Mahon, 1964). Fluorite solubility usually determines the maximum concentration of fluoride ions that can exist in natural hydrothermal solutions (Mahon, 1964).

4.5   SULFUR CHEMISTRY

Sulfur chemistry creates some of the more obvious features of geothermal discharges, including colors and smells surrounding natural vents. The oxidation of sulfides creates acidic conditions at the surface in at least parts of most geothermal areas, while the presence of yellow sulfur or purple and black iron sulfides adds tourist appeal to many thermal areas.

In deep high-temperature dilute geothermal waters, sulfate and sulfide are commonly present in comparable concentrations. The equilibrium redox conditions necessary to maintain this ratio can be calculated (Kusakabe, 1974) and they are often more oxidizing than are found by direct measurement from other chemical reactions (see Section 4.6). Some of this anomaly could be explained through sulfate ions being complexed to neutral molecular species in high-temperature water, but sulfate is very unreactive and it may persist metastably for long periods in highly reducing solutions, even at high temperatures. Kusakabe suggested that the sulfate/sulfide ratio in Wairakei waters resulted from a slow and only partial reduction of sulfate derived from deep sediments under the prevailing redox conditions.

Very high sulfate concentrations may be found in unusual situations, for example, in deep acidic waters in volcanic areas or in areas where waters have leached sulfate evaporites.

Where high-temperature water comes in contact with native sulfur the following reaction occurs:

$$4S + 4H_2O = 3H_2O + H_2SO_4 \qquad (4.23)$$

Ellis and Giggenbach (1971) determined the points of equilibrium at saturated water vapor pressures for this reaction experimentally, obtaining typical values of K_S of $10^{-6.4}$ at 300°, $10^{-8.7}$ at 250°, and $10^{-11.6}$ at 200°.

$$K_S = m_{H_2S}^3 m_{H^+} m_{HSO_4^-} \tag{4.24}$$

For the pure water–sulfur system at 250° the solution contains about 0.03 m H_2S and 0.01 m H_2SO_4. The concentrations are in approximate agreement with those found in deep waters in the Matsao, Taiwan, geothermal field, where molten sulfur was encountered at this temperature. Figure 4.17 gives the sulfur–water phase diagram at 250° (from Ellis and Giggenbach, 1971).

Sulfur species other than sulfide and sulfate are unlikely in high-temperature waters. It was shown by Pryor (1960) and by Ellis and Golding (unpublished) that dilute solutions of ions such as $S_2O_3^{2-}$, $S_4O_6^{2-}$, and $S_5O_6^{2-}$ disproportionate rapidly into a mixture of sulfide and sulfate at temperatures of 250°–300°.

At temperatures below 200° and at pH and redox conditions in solution close to the point for equilibrium for sulfur, Giggenbach (1974) demonstrated that thiosulfate and several polysulfide ions can exist, such as the S_4^{2-} ion and its dissociation products $.S_2^-$ or $.S_3^-$. The latter are intense blue radical ions which may account at least in part for the color of the partly oxidized waters in hot springs. For common concentration levels of sulfate and sulfide of about 0.01 m the

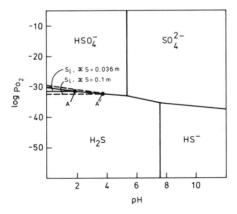

Fig. 4.17 The sulfur–water phase diagram for 250° and total dissolved sulfur concentrations of 0.036 m and 0.1 m. A and A' are the limits of coexistence of liquid sulfur (from Ellis and Giggenbach, 1971).

equilibrium pH for sulfur coexistence is pH 7 at 150°, pH 3.6 at 200°, and pH 0.7 at 250°. For geothermal waters at temperatures over about 200° it would therefore be only in very acid waters, or in waters with very high concentrations of total sulfur, that sulfur solutes other than sulfide and sulfate exist.

4.6   REDOX EQUILIBRIA

The most direct indicator of redox conditions in a deep geothermal water is the partial pressure of hydrogen, which can be estimated from analyses of well discharges. Equilibrium hydrogen and oxygen partial pressures are linked through the gaseous water dissociation reaction $2H_2O = 2H_2 + O_2$, for which the logarithm of the equilibrium constant $\log K_{H_2O}$ has values of -45.9, -43.5, -41.3, and -39.2, at 225°, 250°, 275° and 300°, respectively. The partial pressure of hydrogen in many deep geothermal waters is of the order of 0.1 bar (Chapter 3.2), so that corresponding oxygen partial pressures would be of the order of $10^{-42.5}$, 10^{-40}, $10^{-37.5}$, and 10^{-35} at the four respective temperatures.

Seward (1974) reviewed several ways in which redox conditions can be calculated for deep geothermal waters. For example, independent estimates can be obtained through the use of standard thermodynamic data to calculate the hydrogen pressure to balance the CO_2/CH_4 ratio in the reaction $CH_4 + 2H_2O = CO_2 + 4H_2$ or the N_2/NH_3 ratio in the reaction $2NH_3 = N_2 + 3H_2$. In the Broadlands geothermal field pyrite and pyrrhotite coexist, and a further estimate of redox conditions was obtained through knowledge of the solution composition. Generally good agreement was obtained between the methods for the Broadlands field, the average redox potential of the deep 250° water being about -0.55 V, equivalent to about 0.1 bar partial pressure of hydrogen or $10^{-43.5}$ bar of oxygen at pH 6.2.

For the deep Wairakei system Foster (1959) showed that the redox conditions were close to the point for pyrite, pyrrhotite, and magnetite coexistence at the particular solution pH and sulfide concentration.

4.7   CHEMICAL GEOTHERMOMETERS

Changing temperature or pressure has an effect on the equilibrium concentration of all reactive solutes in geothermal fluids. However, for a particular chemical equilibrium to be of use as a geothermometer to measure temperature conditions at depth it is essential for the reaction rate to be sufficiently slow that further reaction does not occur to alter the relative concentrations of constituents as fluids travel to the surface. The reaction must not be so slow, however, that it fails to respond to the temperatures of the system during its association with a particular depth situation. The elements used as geothermometers must also be present in abundance so that there is no question of the geothermometer failing through a lack of the element in a particular rock type.

Few chemical geothermometers have been calibrated. The silica and the Na/K ratio geothermometers have received most attention and are discussed in detail. Other chemical systems have been used in a less quantitative way and include magnesium, iron, and manganese concentrations, and the Na/Rb ratio. The concentrations of ions such as HCO_3^-, SO_4^{2-}, and F^-, which are limited by the solubility of their calcium compounds, have also been used.

Silica

In hydrothermal areas silica occurs at different depths in various forms: quartz, chalcedony, cristobalite, and amorphous silica (gelatinous silica, sinter, opal). Quartz is the stable form of silica, and it has the lowest solubility. Amorphous silica, cristobalite, and chalcedony should not have an equilibrium solubility, but due to the very slow rate of their conversion to quartz in near-neutral-pH solutions, consistent solubility values are obtained at moderate temperatures. Figure 4.18 presents the solubilities of various forms of silica at s.w.v.p. (Fournier, 1973).

Pressure effects on quartz solubility up to 300° were given by Morey et al. (1962), and over a wider temperature range by Kennedy (1950). Only near the critical temperature of water are the effects of pressure considerable. Quartz solubility is little affected by the presence of dissolved salts, unless they appreciably lower the concentrations of water. However, nearer to the critical temperature of water, high salt concentrations may greatly change the solubilities at s.w.y.p. by increasing the solution density and raising the critical point temperature.

Cristobalite is commonly a first alteration product of volcanic glasses, which may be considered as "contaminated" silica glass, especially in rhyolite volcanic areas. In the experimental reaction of

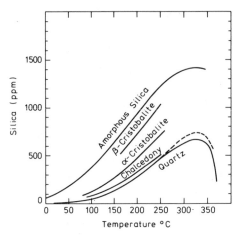

Fig. 4.18   The solubility of various forms of silica in water at saturated water vapor pressures (from Fournier, 1973).

hot waters with volcanic rocks, Ellis and Mahon (1964, 1967) showed that the silica in solution attained a level between amorphous silica and cristobalite solubility, indicating that high-temperature geothermal waters entering fresh volcanic rocks may dissolve high concentrations of silica.

Where measured, high-temperature hydrothermal solutions at deep levels have been found to be saturated with quartz (Mahon, 1966; Fournier and Rowe, 1966). In discharges from high-temperature geothermal wells (180°–260°) they found a close correlation between temperatures estimated from measured silica concentrations (when equilibrium with quartz was assumed) and those measured directly by physical methods. Silica temperatures show a quick response to changing underground temperatures and it is considered that at 250° or above, reequilibrium between quartz and solution takes a few hours or less. For this reason silica temperatures (T_{SiO_2}) much above 200° generally are not estimated from hot spring compositions, due to reequilibration following the cooling of waters by steam loss or by conduction on route to the surface. Furthermore, after appreciable cooling, amorphous silica solubility may be exceeded and opaline silica may be deposited in channelways near the surface (see Chapter 8).

The silica geothermometer is of greatest use in accurate monitoring of the inflow water temperatures of high-temperature wells. Applied to hot spring waters during geochemical survey work, it is also useful in a semiquantitative way to indicate minimum underground water temperatures, but due consideration must be given to the effects of mixing and dilution processes on silica concentrations (see Chapter 6 on mixing models).

The temperature in deep water saturated with quartz can be estimated from the silica concentrations (in parts per million) in the water discharged at the surface, using the following formula, which assumes adiabatic, isoenthalpic cooling (Truesdell, 1975a).

$$t°C = 1533.5/(5.768 - \log SiO_2) - 273.15 \qquad (4.25)$$

Arnorsson (1975) estimated silica temperatures for the waters discharged from geothermal wells in Iceland, covering a wide temperature range (about 50°–300°). He found that at temperatures over 180°, waters were in equilibrium with quartz, and below about 110° with chalcedony. Between these temperatures the silica concentrations did not consistently agree with either quartz or chalcedony solubility (Fig. 4.19). Equilibrium with chalcedony at lower temperatures may be a particular feature of the Icelandic geothermal field.

Two important facts must be considered when applying the silica geothermometer. It is only the un-ionized fraction of total silica in solution that is in equilibrium with quartz (or other silica phase). If the pK_a value for silicic acid minus the pH of the water at its original temperature is less than two, a correction must be made by calculating the concentration of silicate ions and subtracting it from total silica.

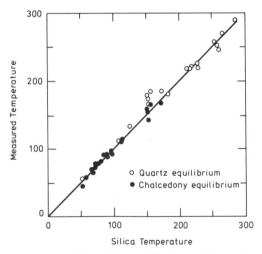

Fig. 4.19 Temperatures calculated for equilibrium of thermal waters with either quartz or chalcedony, compared with measured temperatures (from Arnorsson, 1975).

Fortunately this is a major correction only for the most dilute high-temperature geothermal waters. Means of estimating the pH of deep hot waters are outlined in Chapter 7.

Many wells in hot water geothermal fields do not draw entirely on a liquid phase. Excess steam from boiling and evaporation in the country rock may enter a well and the discharge at the surface then has an enthalpy higher than that of the deep hot water. To estimate downhole temperatures from silica concentrations in waters separated from the discharge at atmospheric pressure, a correction must be made for the evaporation of the water due to steam loss. This evaporation factor varies with the amount of excess steam accompanying the water. Figure 4.20 is a calibration graph for estimating deep-water temperatures from the concentration of silica in waters separated from well discharges at 100°, knowing the enthalpy of the discharge. Accurate results for T_{SiO_2} are obtained with high-temperature wells ($\pm 2°$-$3°$) in both volcanic and nonvolcanic rock areas (see Chapter 7 also).

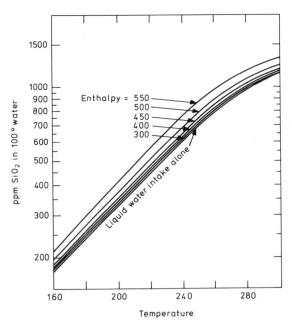

Fig. 4.20 Calibration graph for estimating well water supply temperatures from the quartz geothermometer, incorporating corrections for excess enthalpy discharges. Enthalpy in calories per gram.

Sodium/Potassium Ratio

The partitioning of sodium and potassium between aluminosilicates and solutions is strongly temperature dependent and this fact was utilized from the beginnings of the Wairakei geothermal scheme to measure changing underground water temperatures. There are many discussions of the Na/K ratio geothermometer applied to geothermal solutions (e.g., White, 1965; Ellis and Mahon, 1967; Fournier and Truesdell, 1973) but the general interpretation is that the ratio is related to the exchange reaction

$$K^+ + \text{Na-feldspar} = \text{K-feldspar} + Na^+ \qquad (4.26)$$

Laboratory hydrothermal experiments on rock–water reaction and analysis of waters from high-temperature geothermal wells have been found to give a generally consistent trend in Na/K ratios with temperature. For near-neutral-pH geothermal waters low in calcium ($Ca^{1/2}/Na$) < 1) the simple Na/K ratio gives a good measure of water temperatures in the range of 180°–350° in a wide range of volcanic and sedimentary rock types (Ellis, 1970; Fournier and Truesdell, 1973). A simple formula for calculating temperatures from the Na/K ratio is (Truesdell, 1975a)

$$t°C = [855.6/\log(\text{Na/K}) + 0.8573] - 273.15 \qquad (4.27)$$
$$\text{\small{concentrations in ppm}}$$

Consistent Na/K temperatures were obtained during about 15 years of monitoring at Wairakei (see Chapter 7). The Na/K temperatures ($T_{\text{Na/K}}$) were generally in good agreement with T_{SiO_2} values but sometimes were rather higher. Agreement between geothermometers, or rather higher $T_{\text{Na/K}}$ values (several tens of degrees) was also found with 1000- to 1200-m deep wells at Broadlands. However, later drilling to 2000 m gave measured temperatures more in line with the Na/K values. At Broadlands, cooling by steam loss may have occurred between the two levels, and while there was silica reequilibration, Na/K reequilibration was slower. Ratios of Na to K may in general give an indication of temperatures beneath the level of drilling in a field.

Although the simple Na/K ratio gives good results for many high-temperature geothermal waters, anomalous temperatures have been found for waters high in calcium and of lower temperature. Below about 180°, simple cation exchange between sodium and potassium feldspars may not be the dominant reaction. Taking into account a very wide range of solution compositions for both geothermal and oil well waters, Fournier and Truesdell (1973) proposed an empirical

Na–K–Ca geothermometer which considered the participation of calcium in the aluminosilicate reactions. It is used as follows (Truesdell, 1975a).

$$t°C = \frac{1647}{\log(Na/K) + \beta \log(Ca^{1/2}/Na) + 2.24} - 273.15 \qquad (4.28)$$

where

$$\beta = \tfrac{4}{3} \quad \text{for} \quad Ca^{1/2}/Na > 1 \quad \text{and} \quad t < 100°$$

$$\beta = \tfrac{1}{3} \quad \text{for} \quad Ca^{1/2}/Na < 1 \quad \text{or if} \quad t_{4/3} > 100°$$

Using molal concentrations, $(Ca^{1/2}/Na)$ is calculated. If this is less than 1, a value of $\tfrac{1}{3}$ is used for β and the temperature is estimated. If $Ca^{1/2}/Na$ is greater than 1, a value of $\tfrac{4}{3}$ is used for β. If the temperature estimated in the latter way is greater than 100°, revert to a value of $\tfrac{1}{3}$ for β and use the temperature obtained; otherwise the temperature for $\beta = \tfrac{4}{3}$ is taken as correct.

For dilute high-temperature waters with high carbon dioxide concentrations, where calcium concentrations are very low (a few ppm or less), the term $\beta \log(Ca^{1/2}/Na)$ may become strongly negative and rather variable. In this case the simpler Na/K ratio is more useful.

The mineralogical background of the geothermometers must not be forgotten. It is not correct to apply the Na/K or the Na–K–Ca geothermometers to acid waters, which would not be in equilibrium with feldspars, nor are they applicable to water systems in rocks with unusually high or low concentrations of a particular alkali unless direct correlations with measured temperatures are available.

Magnesium

Low-salinity, high-temperature geothermal waters have very low magnesium concentrations (commonly in the range of 0.001–0.1 ppm), which are controlled by equilibrium reactions with chlorite or montmorillonite (Ellis, 1971). The levels are in marked contrast with higher concentrations in cold or warm waters. Magnesium concentrations in hot spring waters of good flow may be used as a semiquantitative indication whether the waters have a high-temperature origin.

Ellis (1971) showed that the ratio of $\log(a_{Mg}/a_H^2)$ in high-temperature geothermal waters commonly had a value of about 6.2 at temperatures between 200° and 300° and was affected only slightly by partial pressures of carbon dioxide in the range of 1–30 bars. The marked variation with temperature in magnesium concentrations in geother-

mal waters therefore mainly reflects pH trends. At equilibrium for a
fixed salinity there is a marked decrease in water pH with increasing
temperature (see Fig. 3.10). A similar diagram could be drawn relating
magnesium concentrations to salinity and temperature, and examples
of magnesium concentrations that could be expected are shown in
Table 4.7. This type of information can only be used for geothermometry on well discharges while it is certain that the water is still in
equilibrium with the surrounding rocks. The pH of underground
waters may rise when boiling occurs and carbon dioxide is lost, and
the concentrations of magnesium may trend to lower values following
extensive well discharging.

Other Equilibria

Like magnesium, several other cations, such as iron and manganese,
are at low levels in high-temperature water and vary widely in
concentration depending on the water temperature and salinity. The
variation in the equilibrium water pH is again a major control,
through reactions such as $MO + 2H^+ = M^{2+} + H_2O$ or $MS + 2H^+ =
M^{2+} + H_2S$. Concentrations of this group of metals nevertheless
provide a secondary check on water temperatures.

Figure 4.16 shows that the equilibrium product $m_{Ca} \cdot m_{CO_2}$ in underground waters varies widely with temperature for a fixed salinity. For
the initial stages of well discharges in a field this may provide a useful
geothermometer . Alternatively, the product $m_{Ca} \cdot m_{HCO_3}$ estimated for
underground waters can be used in a similar way, again assuming
calcite saturation equilibrium. However, this geothermometer would
not be very reliable for hot springs due to the possible reactions while
waters rise to the surface.

The ratio of bicarbonate to carbon dioxide decreases rapidly with

TABLE 4.7

Approximate Magnesium Concentrations Expected for
Aluminosilicate–Solution Equilibrium with Waters of Three Salinities[a]

	200°C	250°C	300°C
0.01 m	10^{-5}	0.0004	0.01
0.1 m	0.002	0.05	2
1.0 m	0.3	9	300

[a] Concentrations in parts per million; $m = (m_{Na} + m_K)$.

increasing temperature where geothermal waters are in equilibrium with the minerals quartz, feldspar, and mica. This ratio is simpler to determine than the deep-water pH. Table 4.8 gives the m_{HCO_3}/m_{CO_2} ratio for two water salinities, 0.01 and 0.1 m, over a range of temperatures. A high ratio is an indication of low water temperatures, but conclusions can be drawn only where it is certain that there has been no loss of CO_2 from the water.

The solubility of calcium sulfate (Table 4.4) provides a geothermometer which indicates maximum temperatures, since few geothermal waters are saturated with gypsum or anhydrite. High $m_{Ca} \cdot m_{SO_4}$ products in neutral-pH springs of good flow suggest a low-temperature source.

The Na/Rb ratio can be used as an empirical geothermometer parallel to the Na/K ratio, due to the similar chemistry of rubidium and potassium. Good correlation between the two ratios has been obtained with New Zealand waters (R. L. Goguel, personal communication.) but the variation in the Na/Rb ratio with temperature is not as great as for Na/K. The low levels of rubidium in some basic volcanic rocks are also likely to limit its general usefulness.

The reaction $CO_2 + 4H_2 = CH_4 + 2H_2O$ may be used as a geothermometer, since thermodynamic data are readily available for all reactants (Hulston, 1964). The partial pressure of the gases must first be estimated for the depth of equilibrium (i.e., in a deep liquid or steam phase). Because the equilibrium is pressure dependent, interpretation is not simple except for wells in steam-producing fields. For natural steam discharges it may not be known whether the steam arises from the boiling of hot water near the surface or from a large steam reservoir. Since the solubilities of the gases differ, their partial pressures in a separated steam phase are not in the same proportion as in the original water. The slowness of the chemical reaction coupled

TABLE 4.8

Approximate Ratios of m_{HCO_3} to m_{CO_2} in Waters of Two Salinities[a]

Salinity	Ratio				
	200°C	225°C	250°C	275°C	300°C
0.01 m	4.5	1.2	0.3	0.06	0.012
0.1 m	0.65	0.17	0.05	0.010	0.0025

[a] At varying temperatures, for equilibrium with quartz, feldspar, and mica.

with the possibility of separate near-surface origins for at least some of the gases (CH_4, H_2, or CO_2) also complicates the use of the geothermometer.

4.8   ISOTOPE GEOTHERMOMETERS

Isotopes of elements are fractionated in the chemical processes operating in natural rock–water systems. The fractionation is greatest for the lightest elements, and measurements of isotope ratios have been made in geothermal systems mainly for compounds for hydrogen, carbon, oxygen, and sulfur.

Some isotope exchange reactions which achieve equilibrium in the natural system, and for which experimental or theoretical equilibrium constants are available over a range of temperatures, can be used as geothermometers. Alternatively, isotope analyses on solutions and coexisting minerals can comment on the extent of equilibration between phases (see Chapter 3.5).

The isotope exchange reactions of interest may be between gases and a steam phase, a mineral and a gas or solution species, water and a solute, or solute and solute. Although in theory there are many isotope exchange processes which could be used to estimate equilibrium temperatures, a few have been favored because of simplicity of sample collection and preparation, ease of isotopic measurement, a suitable rate of achieving isotopic equilibrium, and knowledge of the equilibrium constants. Isotopic exchange reactions which achieve equilibrium at different rates should be capable of indicating temperatures at various depths within a geothermal system.

Carbon Isotopes

Carbon Dioxide–Methane Exchange

The relative values of the $^{13}C/^{12}C$ ratio in coexisting CO_2 and CH_4 have been used as a geothermometer, assuming that the gases are in isotopic equilibrium through the reaction $CO_2 + 4H_2 = CH_4 + 2H_2O$.

Early work by Hulston and McCabe (1962), Ferrara et al. (1963), and Craig (1963) calculated temperatures of about 250° for Wairakei wells, 215°–315° for Larderello steam, 200° for Yellowstone Park discharges, and 340° for steam at The Geysers. They used the calculated equilibrium fractionation factor of Craig (1953), which was modified by Bottinga (1969). The revised fractionation factor raises the temperatures by 50°–75°.

In most cases the carbon isotopic temperatures shown in Table 4.9 are much higher than those measured in wells in the area; for example, in most Wairakei and Broadlands wells, temperature maxima were approximately 250°–270°, although temperatures slightly over 300° were recorded in one Broadlands well.

The carbon dioxide–methane reaction is very slow to equilibrate, and experimental attempts to achieve carbon isotopic equilibrium between the gases at temperatures of 200°–300° have been unsuccessful, even over periods of months. Better proof is required that isotopic equilibrium is achieved in the natural situations. In most cases the carbon isotopic composition in geothermal methane is similar to that for natural organic carbon, and in combination with average $\delta^{13}C$ values for geothermal carbon dioxide, the temperatures calculated could be fortuitous. As pointed out by Panichi (1975), a wide range of organic and carbonate materials in rocks produce carbon gases at temperatures as low as 70°. If, however, carbon isotope temperatures can be believed, very much higher temperatures exist beneath the depths of drilling in some geothermal areas. Other evidence suggests that this may be true in the Broadlands field (e.g., the sulfur isotope exchange reaction between sulfate and sulfide, discussed later).

TABLE 4.9

Temperatures Calculated from $\Delta^{13}C(CO_2-CH_4)$ Geothermometer [a]

Area	Temperature (°C)
New Zealand	
Wairakei	300–360
Broadlands	380–420
Tikitere	~330
United States	
Yellowstone	240–380
Imperial Valley wells	300–380
Indonesia	
Kamojang well 3	260
Kenya	
Olkaria	420–450
Hannington	300–500

[a] From Craig, 1975; and Hulston, 1975.

Bicarbonate-Carbon Dioxide

The carbon isotope distribution between bicarbonate ions and carbon dioxide has been proposed as a geothermometer by O'Neil *et al.* (1975) and Truesdell (1975b). It was used to estimate temperatures at Steamboat Springs, Nevada, and Yellowstone Park, yielding results in reasonable agreement with other indicators in the temperature range 170°-290°. A potential difficulty with this geothermometer is the tendency for bicarbonate present in deep waters to decompose by reaction with other weak acids as fluids rise to the surface with exsolution of CO_2. The temperature indicated would be for the last stage of bicarbonate reaction.

$$HCO_3^- + H_3BO_3 = H_2BO_3^- + H_2O + CO_2 \qquad (4.29)$$

Oxygen Isotopes

In applying isotope geothermometers involving oxygen or hydrogen exchange reactions with water, the isotopic composition of the water used must be that at the point of equilibrium. An isotopic fractionation occurs when steam separates from hot water (see Section 4.2). For wells tapping a hot water phase, the average hydrogen or oxygen composition of the total discharge must be found by determining the isotopic compositions of both the steam and water phases and by knowing the steam and water proportions. Some authors have used analysis of the water phase alone, which by chance can give almost correct results if steam separation occurs at a temperature of near-zero isotopic fractionation (see Fig. 4.4).

Many reactions involving oxygen isotope exchange occur in hydrothermal systems. However, in some reactions isotopes exchange so rapidly that the reactions are not useful as temperature indicators. For example, the reaction

$$H_2{}^{16}O + C{}^{18}O{}^{16}O = H_2{}^{18}O + C{}^{16}O_2 \qquad (4.30)$$

and probably also the following reaction fall in this category.

$$DHO + H_2S = H_2O + DHS \qquad (4.31)$$

Also, there is probably little point in investigating the equilibrium $HC{}^{16}O_3^- + H_2{}^{18}O = HC{}^{18}O_3^- + H_2{}^{16}O$, since the bicarbonate ion partly decomposes to carbon dioxide through reactions with other weak acids as waters approach the surface.

However, the following exchange reaction has shown great promise as a geothermometer.

$$H_2{}^{16}O + HS^{18}O^{16}O_3{}^- = H_2{}^{18}O + HS^{16}O_4{}^- \tag{4.32}$$

Hulston (1975) reviewed the application of this reaction, noting the pH and temperature dependence of the exchange reaction rate. He estimated for Wairakei water (250°, pH 6.5) that the halftime of the exchange reaction was of the order of 4 months. The reaction therefore has ideal characteristics to show temperatures near the base of wells and to ensure that the temperature record is preserved as waters pass to the surface, even from temperatures as high as 300°. The calibration graphs for the sulfate and bisulfate exchange reactions are given in Fig. 4.21 (Robinson, 1977).

Mitzutani (1972) applied this geothermometer to well discharges in the Otake field, obtaining the results shown in Table 4.10. The results are compared with silica temperatures and Na/K temperatures. For the Wairakei field, deep temperatures of 250°–300° were obtained by the method, rather higher in general than the maximum measured temperatures of 250°–260°. At least in some situations the sulfate oxygen geothermometer appears to give slightly high temperatures.

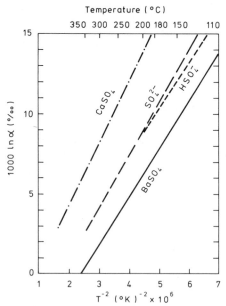

Fig. 4.21 Oxygen isotope distribution between water and various forms of sulfate (from Robinson, 1977).

TABLE 4.10

Underground Water Temperatures in the Otake Field[a]

Well no.	Temperature		
	Isotope	SiO_2	Na/K
7	180	206	195
8	220	220	210
9	280	234	215
10	280	227	200

[a] Calculated from the sulfate–water oxygen isotope geothermometer, from silica concentrations, and from Na/K ratios (from Mitzutani, 1972).

In applying the $\Delta^{18}O(H_2O-HSO_4^-)$ geothermometer to hot spring surveys it should be noted that much of the sulfate may be derived by surface oxidation processes. Truesdell (1975b) used the method for Yellowstone Park springs by plotting $\delta^{18}O$ values for both water and sulfate against chloride concentrations. He used the isotope values corresponding to the highest chloride concentrations to estimate underground temperatures, obtaining values in the range of 260°–314°.

Hydrogen Isotopes

There are many reactions in hydrothermal systems that involve hydrogen exchange. Those occurring in the gas phase are of potential value as geothermometers for steam reservoir systems.

$$HD + H_2O = H_2 + HDO \tag{4.33}$$

This simple exchange reaction has been used as a geothermometer, the calibration being available from the calculations of Bottinga (1969) for the vapor system, in conjunction with the experimental water vapor–liquid separation factor (see Section 4.2).

According to Hulston (1975), the reaction equilibrium is complete in less than a year, and temperatures derived should be those near the base of a well. Laboratory experiments by Arnason (1975) showed that several days or weeks were required for isotopic equilibrium between H_2 and H_2O.

Arnason obtained reasonable agreement between temperatures measured in wells and temperatures from the hydrogen–water exchange reaction, for example, from Hveragerdi, 200°–230°, in comparison with 218° measured. At Reykjanes a temperature of 368° was obtained instead of the measured 290°, but this was considered to be due to late-stage mixing of high-temperature water with cold water. A 260° average temperature was obtained for Wairakei wells by Hulston (1975). However, Craig (1975) obtained temperatures of 140°–220° for Imperial Valley wells in comparison with real temperatures in the vicinity of 300°, suggesting that the equilibrium had readjusted to lower temperatures as waters rose to the surface. Further investigations of the kinetics of this reaction are therefore required.

The following reaction provides a potential geothermometer:

$$CH_3D + H_2 = CH_4 + HD \tag{4.34}$$

This has been used to a limited extent in areas of New Zealand and the United States. The system has been calibrated experimentally (Craig, 1975), and the following equation relates α, the hydrogen isotope fractionation factor, and temperature.

$$10^3 \ln \alpha = -90.888 + 181.264(10^6/T^2) - 8.949(10^6/T^2)^2 \tag{4.35}$$

The results obtained so far have been variable; reasonable temperatures were obtained for the Broadlands field (275°), but rather low temperatures for samples from the Imperial Valley (255°) and Yellowstone Park (70°) (Truesdell, 1975a).

Sulfur Isotopes

The exchange of sulfur isotopes between sulfate and sulfide in geothermal waters has been used as a geothermometer.

$$^{34}SO_4^- + H_2{}^{32}S = {}^{32}SO_4^{2-} + H_2{}^{34}S \tag{4.36}$$

Robinson (1973) determined the isotope exchange equilibrium experimentally up to 320° by sulfur hydrolysis in water. He found that the rates of sulfur exchange with sulfate were very much slower than for oxygen exchange. For example, in the deep Wairakei waters the half-time of the sulfur isotope exchange was considered to be greater than 6 years. Complete equilibrium would therefore take a period of tens of years. Temperatures estimated may therefore be those at levels much deeper than the geothermal wells.

Sulfate may be present as either SO_4^{2-} or HSO_4^- in high-temperature waters, and sulfide may be present as HS^- or H_2S. The equilibrium constant for isotope exchange is slightly different according to

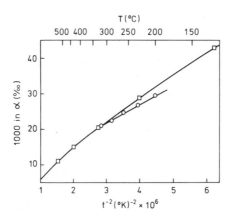

Fig. 4.22 Fractionation factors for the distribution of sulfur isotopes between SO_4^{2-} and H_2S (top line) and between HSO_4^- and H_2S (bottom line) (from Robinson, 1973).

the sulfur species exchanging, and Fig. 4.22 shows the fractionation factor for SO_4^{2-}–H_2S and HSO_4^-–H_2S exchange (from Robinson, 1973).

For Wairakei wells of average discharge, sulfur isotope temperatures of 370°–400° were obtained (Kusakabe, 1974). In interpreting the results, he suggested that the sulfate and sulfide were both derived from basement greywacke rocks, the sulfate being partly reduced to sulfide under the prevailing redox conditions. The results may have represented a kinetic rather than an equilibrium isotope fractionation. Alternatively, if equilibrium was achieved at deep levels, the temperatures at considerable depths beneath the Wairakei field are much higher than at the base of the wells (compare the methane–carbon dioxide carbon isotope temperatures). With further investigation the sulfur isotope exchange reaction may be useful in suggesting very deep temperature conditions in a geothermal field.

Because of oxidation processes at the surface the sulfur isotope exchange reaction is likely to be of limited value applied to hot spring discharges.

REFERENCES

Arnason, B. (1975). *Proc. Int. At. Energy Agency Advisory Group Meeting Appl. Nucl. Tech. Geothermal Stud., Pisa, Italy, Sept.* (in press).
Arnorsson, S. (1975). *Am. J. Sci.* **275**, 763.

Bernal, J. D., and Fowler, R. H. (1933). *J. Chem. Phys.* **1**, 575.

Blount, C. W., and Dickson, F. W. (1969). *Geochim. Cosmochim. Acta* **33**, 227.

Bottinga, Y. (1969). *Geochim. Cosmochim. Acta* **33**, 49.

Bottinga, Y., and Craig, H. (1968). *Am. Geophys. Un. Trans.* **49**, 356.

Craig, H. (1953). *Geochim. Cosmochim. Acta* **3**, 53.

Craig, H. (1963). *In* "Nuclear Geology on Geothermal Areas" (E. Tongiorgi, ed.), pp. 17-23. Consiglio Nazionale delle Richerche Lab. di Geol. Nucl., Pisa.

Craig, H. (1975). *Proc. Int. At. Energy Agency Advisory Group Meeting Appl. Nucl. Tech. Geothermal Stud., Pisa, Italy, Sept.* (in press).

Del Bene, J., and Pople, J. A. (1970). *J. Chem. Phys.* **52**, 4858.

Dudziak, K. H., and Franck, E. V. (1966). *Ber Bunsenges. Phys. Chem.* **70**, 1120.

Ellis, A. J. (1959). *Am. J. Sci.* **257**, 354.

Ellis, A. J. (1963). *Am. J. Sci.* **261**, 259.

Ellis, A. J. (1967). *In* "Geochemistry of Hydrothermal Ore Deposits" (H. L. Barnes, ed.), pp. 465-514. Holt, New York.

Ellis, A. J. (1970). *Geothermics (Spec. Issue 2)* **2**(*Pt 1*), 516.

Ellis, A. J. (1971). *Am. J. Sci.* **271**, 481.

Ellis, A. J., and Fyfe, W. S. (1957). *Rev. Pure Appl. Chem.* **7**, 261.

Ellis, A. J., and Giggenbach, W. F. (1971). *Geochim. Cosmochim. Acta* **35**, 247.

Ellis, A. J., and Golding, R. M. (1963). *Am. J. Sci.* **261**, 47.

Ellis, A. J., and McFadden, I. M. (1972). *Geochim. Cosmochim. Acta* **36**, 413.

Ellis, A. J., and Mahon, W. A. J. (1964). *Geochim. Cosmochim. Acta* **28**, 1323.

Ellis, A. J., and Mahon, W. A. J. (1967). *Geochim. Cosmochim. Acta* **31**, 519.

Ferrara, G. C., Ferrara, G., and Gonfiantini, R. (1963). *In* "Nuclear Geology on Geothermal Areas" (E. Tongiorgi, ed.), pp. 277-284. Consiglio Nazionale delle Richerche Lab. di Geol. Nucl., Pisa.

Forslind, E. (1952). *Acta Polytech. Scand.* **115**, 9.

Foster, P. K. (1959). *N.Z. J. Sci.* **2**, 422.

Fournier, R. O. (1973). *In Proc. Symp. Hydrogeochem. Biogeochem.* **I**, 122-139. Clarke Co., Washington, D.C.

Fournier, R. O., and Rowe, J. J. (1966). *Am. J. Sci.* **264**, 685.

Fournier, R. O., and Truesdell, A. H. (1973). *Geochim. Cosmochim. Acta* **37**, 1255.

Frank, H. S. (1970). *Science* **169**, 635.

Frank, H. S., and Evans, M. W. (1945). *J. Chem. Phys.* **13**, 507.

Frank, H. S., and Wen, W. Y. (1957). *Discuss. Faraday Soc.* **24**, 133.

Franks, F. E. (1972). "Water, a Comprehensive Treatise," Vol. I. Plenum Press, New York.

Gardner, E. R., Jones, P. J., and de Nordwell, H. J. (1963). *Trans. Faraday Soc.* **59**, 1994.

Giggenbach, W. F. (1971). *N.Z. J. Sci.* **14**, 959.

Giggenbach, W. F. (1974). *Inorg. Chem.* **13**, 1724.

Haas, J. L. (1971). *Econ. Geol.* **66**, 940.

Helgeson, H. C. (1969). *Am. J. Sci.* **267**, 729.

Helgeson, H. C., and Kirkham, D. H. (1974a). *Am. J. Sci.* **274**, 1089.

Helgeson, H. C., and Kirkham, D. H. (1974b). *Am. J. Sci.* **274**, 1199.

Helgeson, H. C., and Kirkham, D. H. (1976). *Am. J. Sci.* **276**, 97.

Himmelblau, D. M. (1959). *J. Chem. Phys.* **63**, 1803.

Holland, H. D. (1967). *In* "Geochemistry of Hydrothermal Ore Deposits" (H. L. Barnes, ed.), pp. 382-436. Holt, New York.

Horne, R. A. (1972). "Water and Aqueous Solutions." Wiley, New York.

Hulston, J. R. (1964). *Proc. U.N. Conf. New Sources Energy, Rome* **2**, 259.

Hulston, J. R. (1975). *Proc. Int. At. Energy Agency Advisory Group Meeting Appl. Nucl. Tech. Geothermal Stud., Pisa, Italy, Sept.* (in press).

Hulston, J. R., and McCabe, W. J. (1962). *Geochim. Cosmochim. Acta* **26,** 383.

Keenan, J. H., Keyes, F. G., Hill, P. G., and Moore, J. G. (1969). "Steam Tables." Wiley, New York.

Kennedy, G. C. (1950). *Econ. Geol.* **45,** 629.

Kharaka, Y. K., and Barnes, I. (1973). SOLMNEQ: Solution-Mineral Equilibrium Computations. U.S. Geolog. Surv. Rep. WRD-73-002, Menlo Park, California (NTIS Rep. PB 215 899).

Kozintseva, T. N. (1965). *In* "Geochemical Investigation in the Field of Higher Temperatures and Pressures" (N. I. Khitarov, ed.), pp. 121–134. "Nauka," Moscow.

Kusakabe, M. (1974). *N.Z. J. Sci.* **17,** 183.

Liu, C. T., and Lindsay, W. T. (1970). *J. Phys. Chem.* **74,** 341.

Liu, C. T., and Lindsay, W. T. (1972). *J. Solut. Chem.* **1,** 45.

Mahon, W. A. J. (1964). *N.Z. J. Sci.* **7,** 3.

Mahon, W. A. J. (1966). *N.Z. J. Sci.* **9,** 135.

Mahon, W. A. J., and McDowell, G. D. (1977). *In* "Geochemistry 1977" (A. J. Ellis, ed.). New Zealand Dept. Sci. and Ind. Res. Bull. 218, (In Press). Wellington.

Malinin, S. D. (1974). *Geochem. Int.* **11,** 1060.

Marchi, R. P., and Eyring, H. (1964). *J. Phys. Chem.* **68,** 221.

Marshall, W. L., and Slusher, R. (1965). *J. Chem. Eng. Data* **10,** 353.

Mizutani, Y. (1972). *Geochem. J. (Jpn.)* **6,** 67.

Morey, G. W., Fournier, R. O., and Rowe, J. J. (1962). *Geochim. Cosmochim. Acta* **26,** 1029.

Nemethy, G., and Scheraga, H. A. (1962). *J. Chem. Phys.* **36,** 3382.

Noyes, A. A. (1907). The Electrical Conductivity of Aqueous Solutions. 352 pp. Carnegie Inst. Washington Publ. No. 63, Washington, D.C.

O'Neil, J. R., Truesdell, A. H., and McKenzie, W. F. (1975). *U.N. Symp. Develop. Use Geothermal Resources, San Francisco, May,* Abstract III-71.

Osborne, N. S., Stimson, H. F., and Ginnings, D. C. (1937). *J. Res. Nat. Bur. Std.* **18,** 389.

Panichi, C. (1975). *Proc. Int. At. Energy Agency Advisory Group Meeting Appl. Nucl. Tech. Geothermal Stud., Pisa, Italy, Sept.* (in press).

Pauling, L. (1959). *In* "Hydrogen Bonding" (D. Hadzi and H. W. Thompson, eds.), pp. 1–30. Pergamon, Oxford.

Pople, J. A. (1951). *Proc. Roy. Soc. (London)* **A205,** 163.

Pryor, W. A. (1960). *J. Am. Chem. Soc.* **82,** 4794.

Quist, A. S., and Marshall, W. L. (1965). *J. Phys. Chem.* **69,** 2984.

Quist, A. S., and Marshall, W. L. (1968). *J. Phys. Chem.* **72,** 1545, 2100.

Read, A. J. (1975). *J. Solut. Chem.* **4,** 53.

Robinson, B. W. (1973). *Earth Planet. Sci. Lett.* **18,** 443.

Robinson, B. W. (1977). *In* "Geochemistry 1977" (A. J. Ellis, ed.). N.Z. Dept. Sci. Ind. Res. Bull. 218, Wellington. (in press).

Samoilov, O. Y. (1946). *Zh. Fiz. Khim.* **20,** 12.

Seidell, A., and Linke, W. F. (1965). "Solubilities Inorganic and Metal-organic Compounds," 4th ed., Vol. I and Vol. II. Am. Chem. Soc., Washington, D.C.

Seward, T. M. (1974). *Am. J. Sci.* **274,** 190.

Silvester, L. F., and Pitzer, K. S. (1976). Thermodynamics of Geothermal Brines. I. Thermodynamic Properties of Vapor-saturated NaCl (aq). Solutions from 0–300°C. Nat. Tech. Informat. Serv. Rep. T1D-4500-R64, Springfield, Virginia.

Smith, F. G. (1963). "Physical Geochemistry." Addison-Wesley, Reading, Massachusetts.

Sourirajan, S., and Kennedy, G. C. (1962). *Am. J. Sci.* **260,** 115.

Truesdell, A. H. (1975a). Proc. U.N. Symp. Develop. *Use Geothermal Resources, 2nd, San Francisco, California, May* 1, liii.

Truesdell, A. H. (1975b). *Proc. Int. At. Energy Agency Advisory Group Meeting Appl. Nucl. Tech. Geothermal Stud., Pisa, Italy, Sept.* (in press).

Truesdell, A. H., and Singers, W. (1971). Computer Calculation of Down-hole Temperatures in Geothermal Areas. New Zealand Dept. Sci. and Ind. Res. Rep. C.D. 2136, Lower Hutt, New Zealand.

White, D. E. (1965). *In* "Fluids in Subsurface Environments—a Symposium," *Memoir Am. Assoc. Petrol. Geolog.* **4,** 342.

Chapter 5

SAMPLING AND DATA COLLECTION

5.1 COLLECTION OF SAMPLES OF WATER AND STEAM

Introduction

This chapter discusses field and laboratory methods in use in New Zealand for sampling and analyzing water, steam, and gases from natural thermal activity and from geothermal wells.

The credibility and usefulness of geochemical data depend on the methods used and the care taken in the collection of samples. For this reason it is recommended that geochemists undertake the fieldwork. If the sampling conditions are not well known, the significance of analytical results may not be fully appreciated. A person without chemical knowledge may contaminate samples during collection, or volatile constituents may be lost from samples by faulty handling. It is also unrealistic to expect a geochemist to take a detailed interest in an area with which he has had no practical association.

Field Data

In the field it is essential that complete records be kept of samples collected and features sampled and that this information be carefully transferred to standard sampling forms on arrival at base. There is constant difficulty over the identification of springs and fumaroles sampled, particularly when a change of personnel occurs. The solution is careful mapping and numbering of features. When chemical sampling precedes mapping, local spring names, descriptions, and sketch maps are used. A convenient field record card for springs is reproduced here and a similar card is used for fumaroles. For wells, the number, wellhead pressure, sampling pressure, and separating pressure are recorded as applicable, for both water and steam samples.

Chemical Analysis

                                                                Lab. No.:

Location in regard to other springs:

Map No.:   Grid Ref.:   Area:
Collected by:
Size l, b, d, (m) & Sketch:

  Appearance: Spring Name:
                                                   Spring No.:

  Deposits: Date:

  Spouting and Periodicity: Temp.:

  Description of spring: Weather:
                                                   Outflow:

  Surroundings: Inflow:

  Gas:

  Remarks:

Natural Activity

Hot Springs

In the preliminary investigations of a hot water geothermal area the most useful chemical data for interpreting the underground conditions are obtained from boiling springs with flows over 0.5 l/sec. These are less likely to have suffered dilution, surface evaporation, and extensive interaction with surface rocks. In the absence of boiling springs, springs of lower temperatures, with small flows, or even stagnant springs, may give some useful information.

In areas with many boiling or close to boiling springs it is unnecessary to sample more than a representative number. In sampling other springs, consider the distribution of springs throughout the area, the size of springs, springs in valleys or on hillsides, geological associations such as faults and rock types, proximity to steam discharges, and distance from cold surface water such as streams, rivers, and lakes. Field analysis of constituents in the waters, such as chloride, may delineate particular zones of interest and aid in planning the sampling program. A portable conductivity meter may also be of use but the results it yields are less specific and should be treated with caution.

Measurements of temperature and flow, and of the shape and size of the feature, are made when the samples are collected for analysis. A grid reference is calculated and a number allocated to the feature. A wooden or plastic stake (or other acceptable form of marking) with the number inscribed on it should be set into the surrounding sinter or rock, for permanent recognition of the sampled feature. Details of the extent, thickness, and characteristics of sinter and deposits surrounding the spring are recorded.

Water is sampled at the hottest part of the spring, usually directly above the inlet. In a large spring the temperature variation across the surface area may be appreciable, perhaps 10° or more, due to evaporative heat loss, conductive heat loss (minor), and possibly dilution. No simple correction factor can be applied to the results for samples collected below boiling point to obtain concentrations at boiling point. Direct comparison of solute concentrations in spring waters is thus often difficult.

Water sample volumes should be minimal, since there are frequently transport difficulties in isolated thermal areas. Water samples of 500 ml are a good compromise between size and convenience in analysis, although at times smaller samples are used.

Sampling techniques for boiling and warm springs vary according to the constituents that are being investigated. Samples for unreactive constituents (alkalis, halogens, boron, arsenic) are collected in a stainless steel or polyethylene beaker attached, if necessary, to an extendable handle. The sample is transferred as quickly as possible to bottles made of unfilled polyethylene which have been rinsed at least twice with spring water. The bottles are capped as air free as possible. Glass containers are less preferable, since minor reaction between the glass and water occurs and precipitates tend to form more readily in glass. With continued use, glass containers become frosted from silica precipitation and some of the minor, but important, constituents in the waters, such as magnesium, may be increased or decreased in concentration to a major extent by glass–water reaction.

Certain constituents, such as silica, magnesium, calcium, iron, and manganese, may precipitate as the waters cool. Acidification to a pH of less than 3 with a measured amount of pure nitric acid immediately after collection prevents precipitation for lengthy periods.

Water samples for constituents affected by exposure to air (pH, SO_4, H_2S, NH_3, HCO_3) are collected with the simple apparatus shown in Fig. 5.1. The aim is to fill a sample container with water that has not been exposed to the atmosphere (as would happen if a bottle were placed directly within the hot spring). Two glass, or polyethylene, sample bottles with 15-cm lengths of wide-bore butyl rubber tubing over their necks and with a screw clip placed halfway up the tube are used. Rubber stoppers fitted with stainless steel tubing are placed in the ends of the wide-bore tubing. The assembly is submerged so that the shorter tube is under the surface and the long tube is exposed. Both bottles fill, with one bottle being purged with an extra 500 ml of water. The screw clamp on the tube is closed, allowing the water in the purged bottle to cool in the absence of air. Immediately after cooling, the sample is analyzed for pH, carbonates, and hydrogen sulfide, and a precise chloride value for the spring water can be obtained, since the water is little affected by evaporation. Sulfate is determined after acidification of the sample and removal of hydrogen sulfide by blowing nitrogen through the water.

Samples are collected by this method for the determination of hydrogen and oxygen isotopes (D/H, $^{18}O/^{16}O$). Where this is impossible, collection by dipper is made, with rapid transfer to a glass bottle and closure by clipping a butyl rubber tube. To preserve the isotope water samples during transportation and storage (often for long

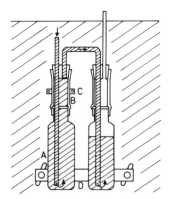

Fig. 5.1 Apparatus for collecting air-free water samples from hot springs. A, sample bottle; B, butyl rubber hose; C, clip; D, clamp.

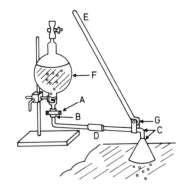

Fig. 5.2 Apparatus for collecting gas samples from springs. A, clip; B, butyl rubber seal connector; C, stainless steel tubing and funnel; D, butyl rubber flexible connector; E, handle; F, double-ended Pyrex flask; G, swivel.

periods), aliquots are transferred when cold to glass ampoules which are drawn off in a hot flame (butane burner).

Samples for trace metal analysis should be collected in all-polyethylene or teflon equipment which has been thoroughly acid washed and rinsed before use. Bottles made of polyethylene containing filler compounds, usually recognizable by the opaqueness of the bottles, should not be used. If the hot water shows any trace of suspension, or particles, it is passed through a medium filter paper held in a polyethylene or Teflon filter funnel. Extreme care regarding cleanliness is taken with this sample to ensure that the water does not contact hands that have touched metal equipment.

Discharges of gases commonly occur in hot springs and gas analysis can produce useful information. This type of sample is appropriate when natural steam discharges are absent in the thermal area. Gas is collected by the displacement of spring water from a double-ended thin-walled glass flask (Fig. 5.2). A polyethylene funnel fitted into a butyl rubber hose attached to the flask is placed below water level above a stream of gas bubbles. Carbon dioxide and hydrogen sulfide are the predominant gases in hot springs and may be absorbed into solution by the addition of a caustic soda solution to the flask. This allows the collection of larger volumes of nonacidic gases. The nonacidic gases are analyzed by gas chromatography, while carbon dioxide and hydrogen sulfide are determined by acid and by iodine titration, respectively.

All collection bottles must be clearly marked for identification. Stick-on labels or various ink markings, which are easily removed or erased during transportation and storage, are inadequate. Markings should be permanently etched or cut into the glass or polyethylene. At the time of collection, a record of the sample number, bottle number, and reasons for collection is made.

Facilities for carrying out simple chemical analyses should be made available in the field. Unstable constituents, such as hydrogen sulfide, carbon dioxide, and bicarbonate, and the water pH are best determined on the day of collection. A portable solid-state electronic pH meter, powered with mercury cell batteries, is useful for field determinations of pH and redox potential (eH), and for titrations for carbonates. Some chloride analyses should be made in the field, since the results aid in programming subsequent sampling. Hydrogen sulfide can be determined by a simple iodine titration, and the Mohr method for determining chloride is simple and convenient (after sulfide is removed from the water by nitrogen purging).

It is convenient to have a vehicle equipped as a field laboratory available for the sampling program. A long-wheelbase Landrover, fitted with a water tank for cooling water (see p. 179) and other field equipment, has been used in the New Zealand geothermal areas.

Fumaroles

Natural steam discharges occur in many forms: as gentle discharges over a large area of hot ground (steaming ground), as major discharges from large orifices, or commonly from hydrothermal explosion craters. Superheating of the vapor is common, and the outlet temperatures of large fumaroles can reach 160°C. (High-temperature active volcanic centers are not considered here.)

In New Zealand the most useful information has been obtained from steam discharged from the large fumaroles. This frequently originates from deep levels, and the large mass flow and velocity prevents excessive condensation of steam, or reaction between the gases and the confining rocks. These fumaroles are also the easiest to sample without contamination.

Figure 5.3 shows an apparatus for sampling a large fumarole with a concentrated discharge. A stainless steel tube with a gravity flap valve on the top and a side outlet for collecting samples is inserted into the fumarole outlet. The flap valve controls the flow of vapor through the sampling outlet and can be weight adjusted according to the velocity of the discharge. For sampling smaller steam discharges, such as those at the bottom of explosion craters or ill-defined outlet points in an area of steaming ground, a stainless steel or aluminum cone attached to the stainless steel tube is firmly placed over the discharge and the surrounds are packed with earth and sacking to reduce air contamination and heat losses to a minimum. The vapor discharged through the

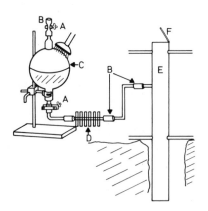

Fig. 5.3 Apparatus for sampling large fumaroles. A, clamps; B, butyl rubber seal/ connector; C, double-ended Pyrex flask; D, air-cooled condenser; E, stainless steel probe; F, flap valve.

sampling tube is passed into an evacuated thin-walled Pyrex flask containing 100–200 cm^3 of a concentrated solution of potassium or sodium hydroxide (50–70 vol %). Minor air contamination is purged from the apparatus by steam flow before the flask is opened and collection commenced. The flask is positioned so that the entering steam and gas bubble through the alkaline solution. The flask is cooled by water or by air when water is not available. Carbon dioxide and hydrogen sulfide frequently contribute 60–99% of the total gases present and are absorbed into the alkali solution, while the residual gases (N_2, CH_4, H_2, etc.) remain undissolved. The pressure of the residual gases is initially low and a pressure gradient is maintained between the steam inlet and the flask. Collection ceases when the residual gas pressure equals the steam inlet pressure. Samples for ammonia, fluoride, boron, and silica determinations are collected in evacuated flasks containing no alkali solution.

Variations of this technique are used. For example, an air condenser consisting of a long narrow tube of stainless steel fitted with large-area cooling vanes can be inserted between the outlet pipe and the flask. This increases the efficiency of air cooling and condensation and increases the rate of collection.

When the concentration of the nonacidic gases is high, or when air contamination occurs (a more frequent occurrence), pressure in the flask builds up rapidly, and collection ceases. To obtain sufficient quantities of carbon dioxide, hydrogen sulfide, and condensate for analysis, a vacuum pump is attached to the outlet of the flask and the residual gases or air removed to allow collection to continue.

The size of sampling flask used for each collection depends on the quantity of sample required and the proportion of gas in the steam. The flask capacities are 0.5–10 l. Every flask is numbered and its volume etched on the glass. If water is used for cooling, a simple recirculation system to conserve water can be set up using a hand pump.

The alkaline condensates are analyzed as soon as possible after collection, if necessary in the field. Alkaline sodium sulfide oxidizes rapidly in the presence of small amounts of air. Aliquots of the sample are withdrawn from the flask by means of a probe similar to that shown in Fig. 5.4. The probe is evacuated by a hand vacuum pump or filled with distilled water before use. Sodium carbonate solutions are stable, and may be retained for long periods before analysis.

The partial pressure of residual gases can be measured in a simple field laboratory and the gases transferred to small glass ampoules for storage, using the apparatus shown in Fig. 5.5. All connections are

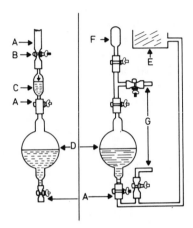

Figs. 5.4 and 5.5 Apparatus for transferring steam condensate and gas samples in the field. A, butyl rubber seal/connectors; B, clips; C, double-ended Pyrex ampoule; D, double-ended Pyrex flask; E, reservoir of distilled or thermal water as transfer fluid; F, Pyrex ampoule; G, connection to vacuum pump.

made of thick-walled (o.d. 22 mm; i.d. 7 mm) butyl rubber, which is impervious to gases. The pressure of residual gas in the storage ampoule is kept as near atmospheric pressure as possible. The residual gases are analyzed in the laboratory using standard PVT techniques or gas chromatography. The composition of the steam discharged from a fumarole is assessed from the analyses of carbon dioxide and hydrogen sulfide, the partial pressure and analysis of the residual gases, the temperature of the flask during residual gas pressure measurement, and the known volumes of the flask, condensate, and sodium hydroxide.

During a survey the grid reference of each fumarole is noted, the feature plotted on a map, and clear records kept of physical characteristics, such as outlet temperatures and approximate flows, since the nature of fumarolic areas can change dramatically over short periods of time. A maximum thermometer, or thermocouple, and a simple Pitot tube should be available.

Geothermal Wells

The collection of representative samples from a geothermal well discharging steam and water is a complex procedure. High-temperature water (150°–350°) from deep levels partially flashes into steam as pressures decrease toward the surface and the steam–water mixture discharges into the atmosphere at velocities approaching the speed of sound. The volume ratio of steam to water in the mixture at the atmospheric discharge point is about 500 to 1000, with the water phase traveling as a thin layer on the inside pipe surface, and steam plus water droplets in the pipe center. One-stage sampling of this rapidly moving mixture to obtain the correct steam/water proportions in a sample is almost impossible. It is necessary to sample each phase separately, at a known pressure, and subsequently combine the analyses of the separate phases. Knowledge of the discharge enthalpy, or the steam-to-water weight ratio at the sampling pressure, is necessary.

The method used for sampling a geothermal well depends on the surface pipework. During the early development work in New Zealand it was not feasible to fit standard surface equipment to the exploration wells, and several different techniques were used for sampling. However, standard pipework systems are now used on both exploration and production wells (Smith, 1958). An exploration well is fitted with a 10- to 20-m horizontal length of 20-cm diameter mild steel discharge pipe, which enters a concrete, steel, or wooden vertical twin-tower silencer (Dench, 1961). The wellhead of a production well is fitted with a Webre cyclone separator, which connects to a steam

Fig. 5.6 Surface equipment on a typical production well at Wairakei.

Fig. 5.7 Surface equipment on a typical exploration well at Broadlands.

reticulation line and to a horizontal water discharge pipe which enters a twin-tower silencer. Figures 5.6 and 5.7 show these features on wells at Wairakei and Broadlands, New Zealand.

Representative sampling of steam, gas, and water from wells is made using one or more of the following techniques.

(a) A sampling device is lowered down the well and samples are collected from various depths. Samples collected are cooled within the sampling device and are directly representative of the reservoir fluid at the point of sampling.

(b) Samples of steam, water, and gas are collected under pressure from sampling points on the surface discharge pipes.

(c) Steam, water, and gas samples are collected from the vertical twin-tower silencer at atmospheric pressure and boiling point.

Before satisfactory pressure sampling was achieved, many sampling positions on the discharge pipes and many types of sampling probes and points were tested. Various types of miniature steam–water separators, for processing samples discharged from the sampling points, were designed and tested. Downhole sampling devices capable of withstanding the temperatures, pressures, and chemical conditions encountered in hydrothermal systems were also experimented with over many years before a satisfactory instrument was found.

Atmospheric Pressure Sampling

The method employed for sampling a well at atmospheric pressure is determined by the equipment or silencer attached to the end of the horizontal discharge pipe. A vertical twin-tower silencer (cf. Figs. 5.6 and 5.7; Dench, 1961) is frequently used in New Zealand for reducing discharge noise to a tolerable level. The silencer is fitted with a titanium-tipped steel splitter plate to spin the discharge on entry and it thereby acts as an atmospheric pressure separator. Water leaves the unit at its base and passes through a weir box to a drainage channel. Steam is emitted at low velocity into the atmosphere from the top of the two towers. Where the end of the horizontal discharge pipe fits into the pipe inlet of the silencer, the annular space is closed with a wooden or steel frame to prevent debris (or people) from being drawn into the silencer. A partial vacuum is created here by the rapid expansion of the discharge and air is drawn into the discharge at the silencer inlet.

Boiling water discharged from the outlet weir of the silencer is sampled with a stainless steel or polyethylene beaker, or alternatively by the dual bottle arrangement described previously. For routine sampling of wells at Wairakei, where 60 or 70 are processed in a day, the sampler shown in Fig. 5.8 is used. It consists of a bottle holder, into which a different bottle is fitted for each sampling, attached to a long handle. Connected to the handle is a spring-operated rubber block which covers the top of the bottle during its immersion and withdrawal from the weir box. This prevents evaporation of the sample.

Steam and gas samples can be collected from the silencer by the equipment shown in Fig. 5.9. A stainless steel or aluminum cone is suspended over the top of the silencer, or alternatively, a stainless steel probe is inserted through the silencer wall near the top of the structure. Steam and gas entering the cone (or probe) pass to a water condenser by stainless steel tube or butyl rubber hose. An evacuated double-ended thin-walled glass flask partially filled with a solution of sodium hydroxide is attached to the condenser.

Air entering the steam at the silencer inlet is collected in the flask with the geothermal gases. Air rapidly fills the flask, and must be

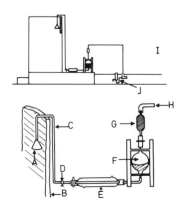

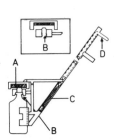

Fig. 5.8 Sampler for collecting water samples from the weir box of a geothermal well. A, neoprene seals; B, screw thread loosely matched to bottle; C, valve return spring; D, valve lift push rod. Inset shows seal in open position.

Fig. 5.9 Apparatus for collecting steam and gas samples from the silencer of a geothermal well. A, funnel for collecting steam and gas; B, silencer wall; C, stainless steel tubing; D, valve for regulating flow; E, water-cooled condenser; F, double-ended Pyrex flask; G, capsule filled with sand soaked in lead acetate; H, connection to steam pump; I, apparatus installed on typical well; J, steam pump.

removed to allow carbon dioxide and hydrogen sulfide to be collected in sufficient quantities for analysis. Removal is by a hand vacuum pump, or by a simple pump operated by steam or water discharged from a convenient tapping on the horizontal discharge pipe. A capsule filled with sand saturated with lead acetate is fitted between the flask and pump to detect hydrogen sulfide escaping through the alkaline solution. The method is not applicable for collecting residual gas samples.

Twin-tower silencers are expensive and although portable units have been used at Wairakei, they may not be generally available. Exploratory wells, or wells with poor production, are frequently fitted with less expensive horizontal-type silencers consisting of a series of concrete cylinders of increasing diameter through which the discharge expands. In some cases no silencing equipment is fitted.

Water sampling from a horizontal silencer is difficult, and the results are in many cases unreliable. A double-walled stainless steel bucket is used, so as to minimize heat losses to the atmosphere. At the silencer outlet, water is distributed mainly around the sides and bottom of the concrete cylinder and is collected from this point. Water temperature is measured to allow analyzed concentrations to be corrected for evaporation, to give results at boiling point temperature. Although concentrations of constituents present in the water are not particularly reliable, ratios between ions (e.g., Na/K, Cl/F) are accurately obtained. This method of sampling should not be used when other techniques are possible.

Water sampling from the end of an unsilenced discharge pipe is dangerous, but with care can be successful. Necessary safety devices are earmuffs or earplugs and heavy-duty insulating gloves. Water can be collected in two ways. A stainless steel bucket (a polyethylene bucket has insufficient strength) attached to a long steel handle is held at the bottom of the discharge. The bucket is lifted into the discharge very slowly, with the lip tipped slightly toward the discharge. Failure to do this results in the sampler being pulled into the discharge and then violently out of the hand. When the sampler is incorrectly positioned, the velocity of the water creates a vortex which swirls the water out of the bucket again.

An alternative sampler is shown in Fig. 5.10. This is welded to the discharge pipe or held in the hand if the discharge is small. If held in the hand, the sampler should be inserted very slowly into the discharge, preferably toward the bottom of the pipe. Evaporative heat losses are generally small and samples at close to boiling temperature are obtained.

It is very difficult to collect representative steam and gas samples from the end of a discharge pipe; hence, alternative sampling positions should be used.

Pressure Sampling

Outline of Requirements Sampling prospecting wells at pressures above atmospheric pressure requires (a) a sampling point or probe on

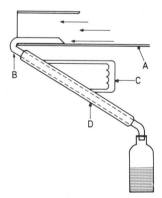

Fig. 5.10 Apparatus for collecting water samples from the end of the horizontal discharge pipe of a geothermal well. A, horizontal discharge pipe; B, stainless steel probe (1- to 1.5-cm stainless steel tube); C, handle; D, lagging to prevent heat loss.

the pipe from which steam and water samples can be taken; (b) sampling positions on the piping which provide representative samples; (c) a small separator unit for separating the steam and water mixture efficiently. Mahon (1961) discussed each of these requirements in detail and a summary is now given. Sampling of production wells is simpler and is described subsequently.

The standard type of sampling probe (A.S.T.M., 1966) used for steam–water mixtures, which extends across the diameter of a discharge pipe with the sampling ports facing upstream, is unsatisfactory for routine sampling of geothermal discharges. The probe is susceptible to breakage from vibrations set up by the impact of the two-phase mixture on it and from small pieces of rock brought up by the discharge. Satisfactory robust sampling points, which combine longevity with adequate sampling of the discharge, are 2.5-cm sockets welded onto the discharge pipe, with a 1.25- to 1.9-cm hole drilled through the pipe. A gate valve is attached to the socket. This type of sampling point is readily serviced or replaced with the well closed or discharging.

Figures 5.11 and 5.12 show a separating unit designed in New Zealand that has been used successfully for sampling wells in New Zealand and several other countries for some years. The stainless steel unit consists of two Webre separators (Pollak and Work, 1942) operating in series. Flow through the separator is controlled by three valves, located on the two separated steam lines and on the water outlet of the second separating unit.

When the equipment in Fig. 5.11 is used, the discharge from the sampling point enters the first unit at 1 and valves 2 and 3 are adjusted to obtain dry steam from outlet 4. The second separator is adjusted with the valves at 3 and 9 to operate slightly flooded, a condition which is recognized by the presence of water in the steam discharged from the steam outlet 5. This ensures that water entering the cooling

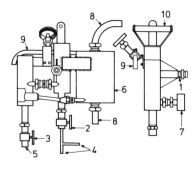

Fig. 5.11 Webre cyclone separator for collecting steam and water samples under pressure from a discharging geothermal well. 1, steam–water inlet; 2, valve for controlling steam discharge on first separator; 3, valve for controlling steam discharge on second separator; 4, dry steam outlet on first separator; 5, dry steam outlet on second separator; 6, separated water cooler; 7, pressure gage fitting; 8, cooling water inlet and outlet; 9, outlet for cooled separated water; 10, handle.

Fig. 5.12 Photograph of Webre cyclone sampling separator used in New Zealand.

coils in cylinder 6 contains no condensed steam. The unit must be operated with a minimum pressure drop (0.2 bar) through the separators, the gage at 7 ensuring that this condition can be controlled through the sampling period. If this condition is not met, the steam sample is diluted by extraneous steam formed from water boiling as it enters lower-pressure zones in the separators. The unit is heavily lagged to prevent heat losses and to reduce the possibility of water samples being diluted with condensed steam.

The unit operates effectively over a sampling pressure range of 1.5–30 bars and a discharge enthalpy range of 190–475 cal/g. At lower or higher enthalpies, there is difficulty in obtaining both dry steam and undiluted water samples at the same valve settings. In these cases the valves are adjusted to collect either steam or water. The presence of condensed steam in a water sample is recognized from gas bubbles in the water emerging from the water cooler. The occurrence of boiling within the separators is more difficult to recognize. Some indication may be obtained if large fluctuations in pressure (±0.3 bar) are recorded on the pressure gage.

Mahon (1961) outlined routine procedures for sampling geothermal wells and detailed the positions of sampling points on the surface discharge pipes from which representative steam and water samples are obtained. The procedures are satisfactory for wells with discharge enthalpies ranging from 190 to 475 cal/g, mass outputs of 20,000–500,000 kg/h, and discharge pipe diameters of 7.5–25 cm. The standard discharge pipe arrangement used on New Zealand geothermal wells is shown in Fig. 5.13. Sampling points are located at three positions: on the vertical wellhead pipe 1–1.5 m upstream of the wellhead-T (position A); on the horizontal discharge pipe 1.5 m downstream of the wellhead-T and at least 1.5 m upstream of a constriction (position B); and 1.5 m from the end of the pipe and at least 7.5–9 m downstream of a constriction (position C).

Water samples are obtained reliably from positions A and B at the high-pressure end of the discharge pipe, while steam and gas samples are obtained from position C at the low-pressure end. If the enthalpy of a well discharge is unknown, it is unwise to collect steam and gas samples at high pressure, since the percentage of steam can be very small and collection of representative samples is difficult.

In a new geothermal area, these sampling techniques are tested on the first available well to ensure that they are satisfactory. Chloride concentrations in water samples collected under pressure are corrected to atmospheric pressure boiling point conditions and compared with results obtained from samples collected at atmospheric pressure. Similarly, gas concentrations in pressure samples are compared with

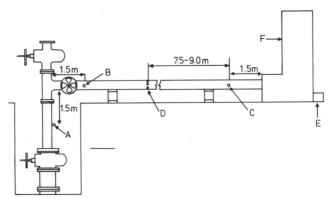

Fig. 5.13 The positions of sampling points on the surface piping of a geothermal well from which representative steam and water samples are collected. A, B, and C, sample points; D, constriction such as back pressure plate or valve; F, silencer; E, weir box of silencer.

each other by correcting results to a common basis (e.g., concentration in total discharge or for atmospheric pressure separation conditions).

Sampling Procedure. The sampling equipment is shown in Fig. 5.14. A Webre separator is attached to the sampling point by a quick release union, with the three control valves in a closed position. The gate valve on the sampling point is fully opened, and the sampling point pressure registers on the gage attached to the separator unit. The three valves on the separator are opened and adjusted to give dry steam from the first separating unit and gas-free water from the water cooler. Dry steam can be recognized by a blue tinge taken on by the discharge, or tested for by running the hand quickly through the fluid. Valve adjustment is made so that the pressure drop through the separator is less than 0.2 bar.

An evacuated glass flask containing a measured volume of sodium hydroxide solution is fitted to the dry steam outlet by a stainless steel T-piece and heavy-wall butyl rubber tubing. The air in the butyl tube is removed by steam purging or with a hand vacuum pump. Flasks are enclosed in wooden crates covered with fine mesh netting, in case of breakage during sampling. With excess steam passing from the T-piece, the clip closing the butyl tube is slowly opened and steam allowed to enter the flask. The flask is then positioned so that the steam and gas entering bubble through the sodium hydroxide. To aid cooling and condensation, cold water is played onto the exterior of the flask.

After adequate gas and condensate are collected, or when the flask goes cold (a condition reached when internal gas pressure equals the pressure of steam from the T-piece), the flask is closed by closing the clip on the connecting hose.

Fig. 5.14 Apparatus for collecting steam and water samples under pressure from a geothermal well.

Whenever possible, water sampling is carried out simultaneously with steam and gas sampling. Alternatively, the separator is attached to a point where water samples are obtainable. After separation adjustment, the flow of water from the cooler is controlled so that the outlet temperature is below 25°C. A stainless steel or plastic tube attached to the water outlet is placed at the bottom of the sample bottle. A water volume approximately twice that of the container is allowed to flow through it during collection.

During gas and water sampling, a continuous check is kept on the separating pressure, the dryness of the separated steam, and the presence of gas in the separated water. At the completion of sampling, the control valves on the separator are closed and the pipeline pressure checked to see whether it has changed. A record is made of well number, bypass pressure, separating pressure, wellhead pressure, flask numbers, and bottle numbers.

Production wells at Wairakei are fitted with standard wellhead equipment (Smith, 1958) in which steam and water are separated by a large Webre centrifugal separator or by a U-bend separator and a Webre separator connected in series. Other hot water fields have similar arrangements. Steam and gas samples are collected directly from the separated steam line, without additional separation, by the method described earlier. Water samples are collected from a high-pressure water-cooling unit attached to a sampling point on the bypass pipe. The stainless steel cooling unit is fitted with needle valves which allow controlled collection of water at approximately 25°C.

It is sometimes impossible during sampling to assess whether samples are being contaminated. This is particularly true when the well discharge enthalpy is very high or very low. The reliability of the sampling should be checked periodically in the field by carrying out simple analyses, such as chloride determinations, and correcting concentrations to a standard collection condition basis for comparison.

Physical Method. The gas concentration in a steam or steam–water discharge from a well can be measured directly without the collection and analyses of samples. McDowell (1973) described a simple means of measuring and monitoring gas concentrations. The method utilizes the partial pressure of gases present in the discharge in relation to the vapor pressure of water at the sampling pressure.

The apparatus is shown in Fig. 5.15. A mixture of steam, water, and gas, discharged from a standard sampling point on the horizontal or vertical discharge pipe of a well, is passed into the heavily lagged,

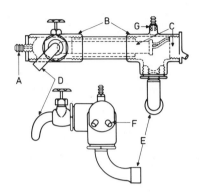

Fig. 5.15 Apparatus for measuring the gas content of the discharge from a geothermal well. A, pressure gage fitting; B, heat-insulated body of instrument; C, stainless steel capsule; D, outlet port; E, inlet port for well fluids; F, differential pressure tap; G, socket for differential tap.

heat-insulated vessel B. A small stainless steel capsule C, which is partially filled with degassed distilled water during measurements, is sealed into the center of B. The interiors of B and C are connected to a standard mercury manometer or differential pressure gage capable of accurately recording pressure differences down to 0.006 bar. The temperature of the water in C rapidly attains the temperature of the discharge, and a pressure equivalent to the vapor pressure of water at the temperature is exerted on one leg of the manometer or gage. The pressure in B, which is equal to the vapor pressure of water plus the sum of the partial pressures of the gases in the discharge, is exerted on the second leg of the gage, and a reading of the total gas pressure is obtained.

The percentage by volume of gas in the discharge is readily obtained by assuming ideal gas behavior. To obtain the percentage by weight of gas in steam, the gas composition is required, but since in New Zealand and many other areas carbon dioxide frequently makes up between 90% and 98% by volume of the gases present, only a small error is produced by assuming it is the only gas present.

Convenient formulas for converting the measured partial pressure of gas into pounds of gas per pound of steam or kilograms of gas per kilogram of steam are

$$W(\text{lb}) = \frac{M144P_g V_s}{1545T} \quad \text{and} \quad W(\text{kg}) = \frac{MP_g V_s}{83.13T}$$

where W is pounds of gas per pound of steam or kilograms of gas per kilogram of steam; P_g partial pressure of gas in pounds per square inch or bars; V_s specific volume of steam at the vapor pressure of the water discharged in cubic feet per pound or cubic centimeters per gram (vapor pressure = total pressure − partial pressure of gas); T absolute temperature in degrees Fahrenheit or degrees Kelvin; and M the molecular weight of the gas.

The method depends on an accurate measurement of partial pressure. Excellent results have been obtained from wells which have 1% by weight or more of gas in the discharge. The method is capable of continuously monitoring the gas concentration in a given discharge and gives rapid results.

Calculation of Results to a Standard Basis

The concentrations of constituents in samples collected from steam and water discharges by the methods described depend on the pressure of sampling, the enthalpy of the discharge, the concentrations in the deep-water supply, and the initial temperature of the water supplying the well. To compare the results obtained from individual wells, results are calculated to a standard basis, such as concentrations in the total discharge, concentrations for atmospheric pressure separation, or concentrations in the deep intake fluid (see also Chapter 7).

To illustrate the calculations, a well with the following characteristics is considered.

(a) Temperature of deep supply water to the well is 250°C.

(b) Chloride concentration in supply water, 1000 ppm.

(c) Carbon dioxide concentration in supply water, 250 mmoles per 100 moles of water.

(d) Enthalpy of discharge, 259.3 cal/g (Cases 1 and 4). Enthalpy of discharge, 344.4 cal/g (Cases 2 and 3).

Separated samples are taken at 6.88 bars, 3.44 bars, and 1.013 bars. The following information is also required in the calculations

Enthalpy of water at 250°	259.3 cal/g
Latent heat of evaporation at 250°	409.5 cal/g
Water temperature at 6.88 bars	164.3°
Enthalpy of water at 164.3°	165.8 cal/g
Latent heat of evaporation at 164.3°	493.8 cal/g
Water temperature at 3.44 bars	138.3°
Enthalpy of water at 138.3°	138.9 cal/g
Latent heat of evaporation at 138.3°	513.3 cal/g
Enthalpy of water at 100°	100 cal/g
Latent heat of evaporation at 100°	539.2 cal/g

The enthalpy of the discharge is given by the relationship $E = H + qL$ where E is enthalpy (in calories per gram), H the enthalpy of water, L the latent heat of evaporation, and q the dryness fraction.

(a) Water Results

Case 1   No production separation of steam and water at wellhead, and sampling pressures are 6.88, 3.44, and 1.013 bars (atmospheric pressure); discharge enthalpy, 259.3 cal/g. The chloride concentrations in water samples at these pressures are calculated as follows.

(a)   At 1.013 bars sampling pressure

$$E = H_{100°} + qL_{(100°)} \quad \text{or} \quad 259.3 = 100 + 539.2q_{100°}$$
$$q_{100°} = 0.2954$$

The chloride concentration in water at 100° and atmospheric pressure $= 1000/(1 - q_{100°}) = 1000/0.7046 = 1419$ ppm.

(b)   At 6.88 bars sampling pressure

$$E = H_{164.3°} + qL_{(164.3°)} \quad \text{or} \quad 259.3 = 165.8 + 493.8q_{164.3°}$$
$$q_{164.3°} = 0.1894$$

The chloride concentration in water at 6.88 bars $= 1000/0.8106 = 1234$ ppm.

(c)   Similarly, at 3.44 bars pressure $q_{138.3°} = 0.2346$, and the chloride concentration is 1306 ppm.

Case 2   No production separator at wellhead, and sampling pressures are 6.88 and 1.013 bars; enthalpy of discharge, 344.4 cal/g. The chloride concentration in water sampled can be calculated as follows.

(a)   At 1.013 bars pressure (100°), excess steam in well supply,

$$E = H_{250°} + qL_{(250°)} \quad \text{or} \quad 344.4 = 259.3 + 409.5q_{250°}$$
$$q_{250°} = 0.2078$$

The steam fraction at 100° is then obtained.

$$E = H_{100°} + qL_{(100°)} \quad \text{or} \quad 344.4 = 100 + 539.2q_{100°}$$
$$q_{100°} = 0.4532$$

Chloride concentration at $100° = 1000 \,(1-0.2078)/(1-0.4532) = 1449$ ppm.

(b)   Similarly, at 6.88 bars sample separation the steam fraction is 0.3616 and the chloride concentration in water is 1241 ppm.

Case 3   Production separator at wellhead operating at 6.88 bars;

enthalpy of discharge is 344.4 cal/g. The chloride concentration at 6.88 bars is 1241 ppm, as calculated in Case 2(b). At 6.88 bars the enthalpy of water from the production separator is 165.8 cal/g.

$$E = H_{100°} + qL_{(100)} \qquad \text{or} \qquad 165.8 = 100 + 539.2q_{100°}$$
$$q_{100°} = 0.1220$$

Chloride in water at 1.013 bars = 1241/(1-0.1220) = 1413 ppm.

(b) Gas Results

Data on the solubility of gases in high-temperature water and saline solutions (carbon dioxide, Ellis, 1959; Ellis and Golding, 1963; hydrogen sulfide, Selleck *et al.* 1952; Kozintseva, 1964) indicate that after several percent of steam have separated from hot water, the gases should be almost entirely in the steam phase. At steam–water equilibrium at 200°C, in a well having the characteristics described earlier, approximately 0.27% of the CO_2 and 0.9% of the H_2S present in the discharge is in the water phase. For equilibrium at 100°, 0.5% of the CO_2 and 0.1% of the H_2S is present in the water phase. In the pressure range over which most steam and gas samples are collected from wells (1–4 bars) carbon dioxide and hydrogen sulfide are considered to occur entirely in the steam phase for the purposes of calculating steam and gas results. At higher sampling pressures, corrections for gas retention, particularly of hydrogen sulfide, in the water phase must be considered. Most other gases found in natural hot waters, such as nitrogen, hydrogen, methane, and inert gases, are less soluble than CO_2.

Ammonia is more soluble than either carbon dioxide or hydrogen sulfide (Clifford and Hunter, 1933) and considerable quantities remain in the water after steam formation. In a steam–water mixture at 200°C at equilibrium, about 30% of the total ammonia is present in the water, and at 100°, 8.6%.

When the carbon dioxide or hydrogen sulfide concentration in the steam fraction of a discharge is known for one pressure, it can be calculated for other pressures when the enthalpy of the discharge is known. When the enthalpy of the discharge is equal to the enthalpy of the supply water (taking into account uncertainties in measurements), the gas concentration in the deep supply water can also be calculated (see Chapter 7). If the discharge enthalpy is higher than that of water at the supply temperature, the gas concentration in the supply water is difficult to estimate. In this case assumptions have to be made about the gas concentration in the excess steam entering the well.

A method for determining the enthalpy of a well from the gas concentrations in the discharge is described in Chapter 6.

Case 4 The carbon dioxide concentration in the supply water is 0.25 mole per 100 moles of water. Steam and gas samples are collected at 6.88, 3.44, and 1.013 bars; enthalpy of discharge is 259.3 cal/g. Carbon dioxide concentrations in steam samples collected at the different pressures are calculated as follows:

(a) At 6.88 bars, from Case 1(b) the steam fraction at 6.88 bars is 0.1894 and the CO_2 concentration in the steam is 0.25/0.1894 = 1.320 moles per 100 moles of water.

(b) At 3.44 bars, from Case 1(c) the steam fraction at 3.44 bars is 0.2346 and the CO_2 concentration is 1.065 moles per 100 moles of water.

(c) At 1.013 bars pressure, from Case 1(a) the steam fraction at 1.013 bars is 0.2954 and the CO_2 concentration is 0.846 moles per 100 moles of water.

Similarly, the carbon dioxide or hydrogen sulfide concentrations in the total discharge can be calculated.

For ammonia, water and steam samples collected at the same pressure are analyzed and the ammonia concentration in the total discharge is obtained by addition: concentration of ammonia in water × water fraction + concentration of ammonia in steam × steam fraction.

Downhole Sampling

Many geothermal systems, such as Wairakei, consist of large reservoirs of hot water of relatively constant chemical composition. Beneath near-surface dilution zones chemical gradients are slight and supply waters to wells are practically homogeneous. Other geothermal systems consist of a series of hot water flows, each of slightly different chemical characteristics.

The relationships of rock compositions and secondary hydrothermal alteration mineralogy to water chemistry can only be understood when they can be clearly related to specific depths in wells. A well which penetrates several inflow zones may be supplied mainly by one and little information on the chemistry of waters at other levels is obtained from the well discharges. In these cases downhole sampling can determine the chemistry of the minor flows.

The downhole sampling bottle now used in New Zealand (Klyen,

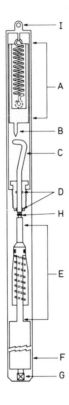

Fig. 5.16 Downhole sampling bottle.
A, inertia mechanism; B, striker; C,
break-off tube; D, seal gland for break-off
tube; E, nonreturn valve (valve stem is of
triangular cross section, allowing transfer
of sample fluids); F, sample vessel; G,
sample release valve; H, filters; I, wire
suspension.

1973) is shown diagrammatically in Fig. 5.16. The stainless steel bottle
of 600-ml capacity is lowered down a well with a winch, the unit
passing through a wellhead recovery tube before entering or leaving
the well.

A glass break-off tube is inserted into the top of the valve assembly
and sealed into the unit by O rings (D). The tube is located immediately below the striker arm (B) fitted to the inertia mechanism (A). The
valve (E) is pressed into the valve seat by a spring with a pressure of
approximately 1.2 bars. A needle valve (G) closes the lower outlet of
the device.

On reaching the sampling depth, the suspension wire is jerked two
or three times, the movement activating the inertia mechanism and
allowing the striker arm to break the glass tube. The fluid pressure in
the well depresses the return spring and water flows through a series
of filters into the sample chamber. When the pressure in the bottle
plus the spring pressure equals the external fluid pressure, the valve
closes. A cone valve seat and an O-ring seal on the valve head ensures

effective sealing. After recovery the bottle is cooled and the sample removed by loosening the top valve and opening the needle valve.

The sampling bottle can hold a high vacuum, preventing the sample's becoming contaminated with air. This is most important when collecting samples for gas analysis. A fitting attached to the needle valve allows a sample to be removed from the sample chamber under vacuum without contact with air.

A large pressure difference always exists between the exterior and interior of the bottle during sampling and this causes hot water entering the bottle to boil. The velocity of fluid entry ensures that the entire flashing steam–water mixture goes into the bottle. When the external fluid pressure is greater than the vapor pressure of the water, the steam within the bottle is subsequently compressed and condensed.

The worst sampling condition occurs when the external fluid pressure is comparable to the vapor pressure of the water. Steam boiled from the water during entry may not condense and remains as vapor after the valve has closed. Condensation of the steam during ascent decreases the internal pressure and reduces the sealing efficiency of the valve. The return pressure exerted by the retaining spring is appreciable, and exhaustive tests have not revealed leakage under these conditions, even when ascent is slow or when the bottle passes through cold water. However, the volume of the bottle occupied by uncondensed steam can be as high as 90% and as little as 50 g of liquid water collected.

The bottle has many advantages. The valve is simple and its sealing efficiency is not affected by the presence of small suspended particles in the well fluids. It can be operated by untrained personnel and loading of the glass tubes is readily carried out in the field. The O-ring seals, which are either neoprene or Viton, are stable at temperatures as high as 300°C and pressures of at least 200 bars. In a series of tests carried out in the El Tatio geothermal area in Chile, the bottle operated with 97% success.

Sample Storage

The majority of samples are analyzed within a month and little deterioration occurs over this period. Samples collected for future reference, stored in stoppered or capped glass bottles, tend to undergo concentration by evaporation and "breathing" over a period of years. Correction for evaporation is made by redetermining chloride on the sample at the time of analysis. Loss of constituents from solution by precipitation or by algal growth is overcome to some extent by dispersing the solid matter before taking an aliquot, and then treating it to get the solids back into solution. Precipitates and algal growth, which are promoted by very fine sediment, can be reduced by filtration of the cold sample before storage. A sample can also deteriorate by reaction with the walls of the storage vessel; for example, over long periods magnesium dissolves from glass surfaces. Losses of solutes to old types of polyethylene containers were greater than to glass in some circumstances (Milkay, 1954), but this is not the case with modern polyethylene containers. No soluble ions are picked up from unfilled polyethylene.

5.2   ANALYTICAL TECHNIQUES FOR GEOTHERMAL FLUIDS

The major constituents in geothermal waters are frequently sodium and potassium chlorides, boron, calcium, silica, carbonates, and sulfate. Important minor constituents include lithium, cesium, rubidium, magnesium, ammonia, fluoride, bromide, iodide, arsenic, antimony, and hydrogen sulfide. The major gases in steam from fumaroles and wells are usually carbon dioxide and hydrogen sulfide, with hydrogen, nitrogen, ammonia, methane, other hydrocarbons, and inert gases making up the residual gases.

In the exploration and initial production stage of a geothermal system, the majority of the foregoing constituents are determined. Analyses for trace metals, including copper, lead, tungsten, zinc, thallium, gold, silver, molybdenum, and mercury are made on a representative number of samples. Tritium, deuterium, and ^{18}O are determined on a selection of both hot and cold waters and of steam condensates from wells and fumaroles.

During chemical monitoring of the well discharges in the production phase, constituents frequently determined on a routine basis are sodium, potassium, lithium, cesium, calcium, boron, silica, chloride, sulfate, carbon dioxide, bicarbonate, and hydrogen sulfide.

Although some of the analytical methods quoted in this section are original, many are well-established procedures. The methods are not described in detail. Only a general outline of the procedures is given, together with likely interferences from other constituents. References are noted for each method. Three particularly useful collections of methods which have been used are "Methods for Collection and

Analysis of Water Samples" Water Supply Paper 1454 (U.S. Geological Survey, 1960); "Standard Methods for the Examination of Water and Waste Water," 13th ed., American Public Health Association: American Water Works Association: Water Pollution Control Federation, New York, (A.P.H.A., 1971); and "Manual on Industrial Water and Industrial Waste Water," 2nd ed., American Society for Testing and Materials, Philadelphia, Pa. (A.S.T.M., 1966).

Alkali Metals: Lithium, Sodium, Potassium, Rubidium, and Cesium

(a) Atomic Absorption Spectrophotometry

Standard procedures are available for the determination of the alkali metals in acid and near-neutral-pH thermal waters by atomic absorption spectrophotometry (Goguel, 1973). They are mainly applicable to Perkin-Elmer 303 or 403 atomic absorption spectrophotometers or similar instruments, but many can be applied with less sophisticated instruments. Suggested flame types, wavelengths of measurement, and other details taken from Goguel (1973) are shown in Table 5.1.

(b) Emission Flame Photometry

A Beckman D.U. spectrophotometer with photomultiplier, flame attachment, and recorder, or an instrument with an equivalent quality monochromater, can accurately determine sodium, potassium, lithium, rubidium, and cesium. Use of a simpler instrument with interference filters to isolate the various emission lines limits accurate determinations to sodium, potassium, and possibly lithium.

The light emissions of the alkalis are enhanced to a greater or lesser extent by the presence of other alkalis, and depressed by high concentrations of silica. Standard solutions are made up which contain concentrations of alkalis and other major constituents similar to those in the hot waters of an area. Sodium emission is measured at 5893 Å, potassium at 7665 Å, lithium at 6708 Å, rubidium at 7800 Å, and cesium at 8521 Å, or with the appropriate filters.

Convenient concentrations to work with are sodium up to 2000 ppm, potassium up to 250 ppm, lithium up to 25 ppm, and rubidium and cesium up to 5 ppm. Higher sodium concentrations tend to block the burner and cause fluctuations in readings.

The concentrations of rubidium and cesium in natural hot waters are often low, and considerable care is required in determinations. The emission of both elements is enhanced by the presence of sodium

TABLE 5.1 Analysis of Near-Neutral pH Thermal Waters with Perkin-Elmer Atomic Absorption Spectrophotometer[a]

	Conc. Range (ppm)	Dilution[h]	Wavelength (Å)	Monochromator[b]				Lamp		Burner		Flame[c]	Sampling height (mm)
				I	II	III	IV	Type[d]	Current (mA)	Slot length (mm)	Alignment[e]		
Li	0.5–30	(O)	6708	335	vis	in	3	HCL	10	70–100	⊥	C₂H₂–air (fuel lean)	8
	1–150	(D)	6708	335	vis	in	3	HCL	10	70–100	∥	C₂H₂–air (fuel lean)	8
Na	100–1000	(D)	5890 5896	295	vis	in	4	ODL	900	70–100	⊥	C₂H₂–air (fuel lean)	8
K	4–400	(D)	7699	385	vis	in	4	ODL	400	70–100	∥	H₂–air (fuel rich)	8
Rb	0.1–10	(B)	7800	390	vis	in	4[f]	ODL	250	70–100	∥	H₂–air (fuel rich)	8
Cs	0.2–20	(B)	8521	426	vis	in	5[f]	ODL	400	70–100	∥	H₂–air (fuel rich)	8
Mg	0.001–1	(B)	2852	285	uv	out	4	HCL	6	70–100	∥	C₂H₂–air (fuel lean)	8
	0.002–2	(O)	2852	285	uv	out	3	HCL	8	70–100	∥	C₂H₂–N₂O (RZ 3)	8
Ca	2–200	(D)	4227	211	vis	out	3	HCL	15	50–70	∥	C₂H₂–N₂O (RZ 5)	13
Mn	0.005–5	(B)	2795	279[g]	uv	out	3	HCL	20	70–100	∥	C₂H₂–air (fuel lean)	8
Al	0.04–50	(B)	3092.7 3092.8	309	uv	out	3	HCL	25	50–70	∥	C₂H₂–N₂O (RZ 12)	8
Fe	0.01–0.1	(B)	2483	248	uv	out	3	HCL	30	70–100	∥	C₂H₂–air (fuel lean)	8
	0.1–20	(O)	2483	248	uv	out	3	HCL	30	50–70	∥	C₂H₂–N₂O (RZ 3)	8
SiO₂	50–1000	(B)	2507	251	uv	out	3	HCL	40	50–70	∥	C₂H₂–N₂O (RZ 25)	7
B	20–1000	(B)	2497	250	uv	out	4	HCL	30	50–70	∥	C₂H₂–N₂O (RZ 25)	7

[a] Operating conditions are for models 303, 305, and 403; data are from Goguel, (1973).

[b] I, wavelength counter setting; II, grating; III, filter which excludes wavelengths shorter than 550 nm (5500 Å); IV, slit setting: (numbers correspond to spectral bandwidths) for the uv grating, $3 = 2.4$ Å, $4 = 7$ Å; for the visual grating, $3 = 5$ Å, $4 = 14$ Å, and $5 = 50$ Å.

[c] RZ denotes height of the red zone in millimeters.

[d] HCL, hollow cathode lamp; ODL, Osram discharge lamp.

[e] Of the flame in the light path of the spectrophotometer: perpendicular alignment produces 0.1 or less of the sensitivity of the parallel alignment. An extra filter must be inserted into the light beam near the window of the spectrophotometer to the left of the flame in order to exclude modulated scattered Na light which reaches the photomultiplier via the reference beam. Filter holder 040-0251 (model 403) or 303-0828 (models 303–305) and a red filter Schott RG 5 or Corning 2-58 (Perkin-Elmer No. 290-1518) are required.

[f] An extra filter must be inserted into the light path of the spectrophotometer: perpendicular alignment produces 0.1 or less of the sensitivity of the parallel alignment.

[g] Shortest wavelength of the triplet.

[h] Dilutions: (B) 1.1 fold. Add 5 ml of a buffer solution (2.5% recrystallized Sr (NO₃)₂, 4% KCl, 1 N HNO₃) to 50 ml of each sample of standard.

and potassium, the enhancement being nonlinear but gradually leveling off with increasing sodium and potassium concentrations. Standards are made up containing fixed amounts of sodium (2500 ppm) and potassium (250 ppm) and the samples adjusted to these values by the addition of *pure* sodium and potassium chlorides. A new or recently adjusted burner should be kept aside for the determination of these elements, since the response is poor with an old or badly adjusted burner.

A wavelength scanning device associated with the spectrophotometer is of great advantage. The flame emission of a sample is scanned from 5893 to 8521 Å, recording the concentrations of the five alkalis.
References: Goguel (1973), Berl (1956).

Ammonia

The following methods are applicable to both water and steam condensate samples.

(a) Ammonia is distilled from an alkaline buffered solution, and an aliquot of the distillate is examined by colorimetry using the Nessler reaction. Because sulfide interferes, it is precipitated in the distillation flask by the addition of lead carbonate. High calcium concentrations, over 500 ppm, can interfere by reacting with the buffer unless excess buffer is added. In some natural hot waters the ammonia is retained strongly in solution and care must be taken to ensure that all the ammonia is distilled off.

(b) Ammonia reacts with a solution of pyridine–pyrazalone at a pH of 3.7, in the presence of chloramine-T, to form a violet-colored complex. This is extracted by carbon tetrachloride and its absorption measured photometrically at 4500 Å. The color is stable in carbon tetrachloride.

Sodium concentrations of 1000 ppm lower the color intensity by about 2%, and iron and aluminum in acid waters form precipitates and have to be separated. Solutions containing more than 4 ppm ammonia should be diluted to bring them into the 0.5–4 ppm range.

(c) Ammonium ion selective electrode. The Orion ammonia electrode is a gas-detecting electrode sensing the level of dissolved ammonia in aqueous solutions. Ammonia concentration can be read directly on a specific ion meter using calibration curves, or can be determined by the known addition method (Orion, 1974).

Antimony

Antimony is concentrated by extraction with dithiocarbamate and determined photometrically with rhodamine B for concentrations less than 100 ppm, or as iodoantimonite for higher concentrations. Acid waters require pretreatment to remove iron. Copper and arsenic are coextracted with dithiocarbamate but do not interfere in the determination.

Reference: Wyatt (1955).

Arsenic

Arsenic is liberated from solution as arsine and absorbed in sodium hypobromite solution. The oxidized arsenic is reacted with ammonium molybdate in acid solution, in the presence of hydrazine sulfate as a reducing agent, to produce a blue color (8400 Å) which is measured for standards and samples. The rate of color development is very dependent on temperature and should always take place at constant temperature.

An alternative method is to absorb the generated arsine in a solution of silver diethyldithiocarbonate in pyridine. A soluble red complex is formed suitable for photometric measurement at 5350 Å.

Sulfide interferes, and must be removed before the arsine is absorbed into the sodium hypobromite or silver salt. A high proportion is initially removed from the test solution by acidification and bubbling with nitrogen. The residual sulfide and any sulfur dioxide present are removed by passing the generated gases through an absorption column filled with lead acetate.

Reference: A.P.H.A. (1971).

Boric Acid

(a)   For concentrations greater than approximately 5 ppm a modified pH titration is used. The addition of a polyhydroxy organic compound such as mannitol to boric acid solutions forms an acid complex which can be titrated with sodium hydroxide.

Carbonates cause buffering in the titration and are removed as carbon dioxide beforehand. The method cannot be applied directly to acid waters containing iron or aluminum, or to waters with any constituent which causes a buffering action in the pH range 4.5–7.3. Preliminary treatment by precipitation or by ion exchange is necessary in these cases to remove interfering ions, or by isolating the boron in the sample by distillation as trimethyl borate.

(b)   Waters containing over 10 ppm boron can be analyzed by atomic absorption spectrophotometry.

References: Goguel (1973), Foote (1932), Muto (1957).

Calcium and Magnesium

(a)   Calcium and magnesium are best determined by atomic absorption spectrophotometry. The procedures are shown in Table 5.1.

(b)   *EDTA titration.*   Standard disodium ethylenediaminetetraacetate is used to titrate calcium at pH 13, with cal-red indicator, and calcium plus magnesium is titrated at pH 10, with solochrome black as indicator. Magnesium is obtained by difference.

Because iron and aluminum interfere in the titration, acid thermal waters containing these elements are pretreated before titration. For calcium in acid waters add 5 ml of 30% v/v triethanolamine to 50 ml of the sample, followed by NaOH to neutralize the solution if it is still acid. The standard titration is then carried out. For the calcium plus magnesium titration, iron and aluminum are separated by double precipitation with ammonia in the presence of ammonium chloride. The combined filtrates are then titrated in the normal manner.

Most high-temperature natural waters contain appreciable quantities of silica, which precipitates when the waters are made alkaline (pH 8–12) and tends to mask the titration end point. Addition of silica to a standard solution of calcium plus magnesium allows the end point in the samples to be seen more clearly.

Low concentrations of magnesium (<0.5 ppm) in the presence of calcium cannot be determined accurately by this method.

Reference: Goguel (1973), Patton and Reader (1956).

Carbon Dioxide, Bicarbonate, Carbonate, and Effective Bicarbonate

When the pH of a dilute water sample is adjusted at ambient temperature to 8.25, the total carbonate in the water is essentially in the form of bicarbonate. This pH for maximum bicarbonate concentration varies from 8.1 to 8.3, according to the ionic strength and temperature. By adding a known amount of acid (0.02 m HCl, titer A), the pH is adjusted to pH 3.8, at which all the carbonate is essentially carbon dioxide. The solution is further acidified (pH 3) and bubbled with nitrogen to remove carbon dioxide. The pH of the solution is readjusted back to pH 8.25 with sodium hydroxide (0.02 m) followed by a further acid titration from 8.25 to 3.8 (titer B). The total carbonate in the sample is given by the difference between titer A and titer B. The titer B gives a measure of interfering acid–base equilibria which operate as a buffer in the pH range of the titration.

The concentrations of bicarbonate, carbonate, and free CO_2 in the samples are estimated from the total carbonate by graphs relating the proportions of these ions to pH and the ionic strength of the sample.

Low-pH waters containing high concentrations of iron and aluminum give precipitates on pH adjustment. For these waters a determination of carbonate usually has little significance and the analysis may be omitted. Low carbonate concentrations in waters containing high boron concentrations cannot be determined by this method. Other methods must be employed, such as the liberation of CO_2 followed by absorption into barium hydroxide solution.

Reference: A.S.T.M. (1966)

Effective Bicarbonate

The bicarbonate and carbonate concentrations in hot waters at the surface are not the same as those in the deeper hot water (see Chapter 7). Surface waters which have cooled by loss of steam, and possibly by dilution, have lost CO_2 and H_2S to the atmosphere. The deep waters ($200°-300°$) contain in solution high concentrations of free carbon dioxide and hydrogen sulfide. Borate and silicate ions, which are present in waters of above-neutral pH at the surface, were formed by their acids' reacting with bicarbonate ions to give carbon dioxide while steam was flashed from the high-temperature water. Titration of a surface water sample from its original pH to pH 3.8 gives a measure of the "effective bicarbonate," which includes silicate, borate, and other constituents that have a buffer capacity over this pH range. This approximates the bicarbonate concentration titer for the deep water (after allowing for concentration by steam loss).

Chloride

(a)   *Mohr Method*

An aliquot of sample is titrated against standard silver nitrate solution using potassium chromate as indicator. The solution must be near neutral pH, acid samples being neutralized with calcium carbonate before titration. Addition of calcium carbonate to all samples has the advantage of giving a white background, which allows the end point to be seen more clearly.

Bromide and iodide are also included in this titration but their concentrations are relatively low and can be subtracted if true chloride concentrations are required. Sulfide is titrated and must be removed from the samples by acidification with nitric acid and bubbling with nitrogen. For chloride concentrations less than 10 ppm, solutions are preconcentrated.

Reference: A.S.T.M. (1966)

(b) Potentiometric Method

The chloride in a supporting sulfuric acid electrolyte is titrated with standard silver nitrate, the end point being detected by a silver electrode and a reference mercurous sulfate electrode in association with a standard expanding-scale pH meter. The titration can also be made using potassium nitrate as a supporting electrolyte, and a silver–silver chloride electrode. Bromide, iodide, and sulfide also react.

The method is used for chloride concentrations as low as 0.5–1 ppm. For lower concentrations, waters are concentrated before analysis.

Reference: Kolthoff and Kuroda (1951).

Fluoride

(a) Direct Titration

This method is possible with waters containing 0–10 ppm of fluoride and low concentrations of sulfate. Highly mineralized waters with more than 50 ppm of sulfate, and waters with iron and aluminum, which form a purple complex with the titration indicator, must be distilled to isolate the fluoride.

Fluoride in solution at pH 3.5 is titrated with thorium nitrate in the presence of the indicator chromazurol-S. The equivalence point is determined by matching the color of the sample being titrated to a blank containing no fluoride.

Sulfate interferes with the titration and also chloride to a lesser extent. Between 0.5 and 10 ppm of fluoride may be determined, within ±10%, by direct titration if chloride does not exceed 1000 ppm and sulfate 30 ppm. Corrections for the effects of these ions can be determined and applied.

(b)   Distillation

An aliquot of sample is steam distilled in a Claisen flask in the presence of powdered quartz, and silver perchlorate is added to precipitate chloride.

The distillate is kept slightly alkaline by the addition of sodium hydroxide. An aliquot, or the total distillate, is titrated against standard thorium nitrate with chromazurol-S indicator.

(c)   Ion Selective Electrode

A fluoride-specific ion electrode gives a direct reading of fluoride contained in samples buffered with a citrate–citric acid buffer.

Direct millivolt readings on an expanded scale pH meter are compared against those for standard fluoride concentrations. Alternatively, samples can be titrated with standard lanthanum nitrate, and the equivalence point obtained from the inflection point in a graph of milliliters of titrant against millivolts.

References: Milton *et al.* (1947), Milton R. F. (1949), A.P.H.A. (1971).

Iodide and Bromide

Iodide in a buffered sample is oxidized to iodate with bromine water. Excess bromide is removed and, by addition of potassium iodide, iodine is liberated for titration with sodium thiosulfate. Iodide and bromide in a second aliquot of the sample are oxidized to iodate and bromate with hypochlorite, the excess hypochlorite then being decomposed. Iodine liberated from both the iodate and bromate on the addition of potassium iodide is titrated with sodium thiosulfate. Bromide is obtained from the difference between the two titrations.

Iron, manganese, and organic materials interfere with the method, and in their presence aliquots of sample are shaken with anhydrous calcium oxide and filtered, the first 25 ml of filtrate being discarded.

Reference: A.S.T.M. (1966).

Iron and Aluminum

The concentrations of iron and aluminum are generally low (<1 ppm) in near-neutral or slightly alkaline pH waters, but can be high in acid waters. They are analyzed by atomic absorption spectrophotometry using the nitrous oxide–acetylene flame. The iron line at 3720 Å and the aluminum line at 3961 Å are utilized.

In the absence of an atomic absorption spectrophotometer, silica is removed from the test solutions and iron and aluminum are precipitated with ammonia. The precipitate is filtered off, ignited, and weighed. Iron is determined photometrically with o-phenanthroline and aluminum by difference.
References: Goguel (1973), Koga, (1967).

pH

The temperature of the pH buffer standards and the samples should be known within $\pm 2°C$ when they are measured with a good quality pH meter. For reliable results the pH of a sample should be taken as soon as it is cooled after collection and before it is unduly exposed to the atmosphere. The pH of water which is only slightly buffered and not in equilibrium with the atmosphere is subject to interference by carbon dioxide from the air.

Silica

(a) Photometric

Silica in samples is initially converted into the ionic form by digestion with alkali, then reacted with an acid solution (pH 1.2–1.5) of ammonium molybdate in the presence of oxalic acid, which stabilizes the yellow silicomolybdic acid that is formed. The light absorbance of the sample is measured at 4100 Å and compared against a standard calibration curve. Concentrations are adjusted to lie between 50 and 200 ppm, where there is a linear relationship between absorbance and concentration.

Color or turbidity interfere, and if sulfide or sulfur dioxide are present, the silicomolybdic acid turns green.

For low concentrations of silica, the silicomolybdic acid is reduced to molybdenum blue with aminonaphthol sulfuric acid solution, and the absorbance is read at 8400 Å.
Reference: A.P.H.A. (1971).

(b) Atomic Absorption

Silica is determined directly, using an acetylene–nitrous oxide flame, a high-temperature burner, and a silica hollow-cathode lamp. Absorbance measurements are made at 2516 Å. This is a delicate method, and optimum conditions of instrument operation and re-

sponse are required for consistent and accurate results. The best results are obtained by using a slightly enriched acetylene flame and standards which approximate the total composition of the samples under test.
Reference: Goguel (1973).

Sulfate

Sulfate in water is reacted with barium chromate to form insoluble barium sulfate and to liberate chromate ions, which are measured photometrically at 3850 Å. The method is applied to samples containing 10–120 ppm sulfate; waters higher in sulfate are diluted.

Sulfide and bicarbonate interfere and must be removed from the waters. Other ions affect the absorption to some extent, and in proportion to the total amounts present. For low sulfate (<25 ppm) in a highly mineralized water, the error could reach ±40%, but since high sulfate frequently accompanies high mineralization, the error will seldom exceed ±10%. The error is minimal if standards approximating the composition of the water are used.
Reference: Iwasaki *et al.* (1958)

Sulfide

In acid solution, hydrogen sulfide is quantitatively oxidized to sulfur by iodine. The quantities of hydrogen sulfide present in surface hot waters are generally low, and large aliquots are frequently necessary for analyses. Excess iodine is back titrated with sodium thiosulfate, using sodium starch glycollate to indicate the equivalence point.

Other reducing sulfur compounds such as sulfite and thiosulfate also react quantitatively with iodine. In the collection of high-temperature fumarole steam where SO_2 and H_2S are present, a complex mixture of sulfur compounds is formed in the condensate. For these, a detailed separation and analysis scheme is necessary (Giggenbach, 1975). Heavy metals may react with iodine, or metals in low valency states may become oxidized. Interference by oxidizing agents other than air is unlikely in natural hot waters.
Reference: Giggenbach (1975).

Trace Metals

(a) A rapid assessment of the trace metals present in waters can be made by spectrographic analysis of the residue obtained by evaporation. Because the least contamination of the sample may constitute gross interference with such a sensitive technique, scrupulous cleanliness is required. The possible loss of constituents by absorption on the sample bottles must always be borne in mind.

A volume of sample is evaporated in a platinum basin to give at least 50 mg of residue as sulfates. Enough sulfuric acid is added to constitute a small excess over that required to combine with the cations. After evaporation to dryness, the residue is placed in a cool muffle furnace and heated to 350°. Spectrographic analysis is made of the residue. Easily volatile constituents (e.g., As, Sb, Hg) are lost, wholly or in part, during ignition.

(b) Successive chloroform extractions with mixed solutions of diethyl dithiocarbamate, oxine, and dithizone—at pH 3, 5, 7, and 9—give a catchall concentrate that is later analyzed spectrographically. Many heavy metals and rare earth metals are included in the extractions.

The method is applicable to all clear waters, the sample size being adjusted according to the level of metal concentration sought, a 1-l sample sufficing for levels of 0.5–1 ppb and higher. Suspended matter causes emulsification during the extractions, while large amounts of iron and aluminum in acid waters require many repeated extractions and may swamp trace metal extraction. Some heavy metals can coprecipitate with the silica and must be tested for if silica precipitates form in the sample.

(c) Copper, lead, zinc, silver, tungsten, nickel, and cadmium are concentrated by one of two techniques: (1) absorption onto Chelex-100 cation exchange resin and elution with 2 N nitric acid; (2) extraction with a benzene solution of mixed diethyl and pyrrolidine dithiocarbamic acids.

The concentrates obtained by both methods are evaporated to dryness, and later dissolved for analysis by atomic absorption spectrophotometry.

References: Pohl (1953), Tennant (1967), Ritchie (1973).

Total Dissolved Solids

A measured volume of water is evaporated and the dry residue is weighed to indicate the total amount of mineral matter in waters. Reweighing after ignition gives the ash content. The determination is empirical, since dissolved matter may either volatilize (e.g., bicarbonate) or remain with the residue.

Reference: A.S.T.M. (1966).

Gas Analysis Methods

Samples of steam condensate and gas from wells and fumaroles are frequently collected in flasks to which sodium hydroxide is added before collection. The alkaline condensate contains the carbon dioxide and hydrogen sulfide from the steam and these are estimated by the methods that follow. If a nonalkaline collection is made, the carbon dioxide and hydrogen sulfide are brought into solution in the laboratory by the addition of sodium hydroxide, as described earlier. When the nonacidic or residual gases are to be analyzed, they are subsequently transferred from the collection flask to small glass ampoules.

(a) Hydrogen Sulfide

An aliquot of alkaline condensate is added to a beaker containing a known volume of 0.01 N iodine and sufficient sulfuric acid to ensure that the solution remains acid. Addition of the sample is made slowly with constant stirring to minimize the effect of the heat of neutralization of the alkali, and to prevent a local buildup of alkalinity. Failure to do this will result in inconsistent results. The excess iodine is titrated with standard sodium thiosulfate, using sodium starch glycollate as indicator.

The analysis is carried out immediately the alkaline condensate is available, due to its rapid oxidation on exposure to air. During repeat determinations, the condensate is kept as air-free as possible.

(b) Carbon Dioxide

Silver nitrate is added to an aliquot of the alkaline condensate to precipitate sulfide. The solution is then adjusted to pH 8.25 with 2 m HCl to convert all the carbonate to bicarbonate, and is titrated with standard acid (0.1 m HCl) to a pH of 3.8, where all the bicarbonate is converted to CO_2. Addition of acid near the end point is made slowly and accompanied with vigorous stirring to ensure the removal of carbon dioxide.

Providing that sulfide is precipitated, the method is relatively free of interference except in samples from high-temperature fumaroles when SO_2 and H_2S are present. Special methods must be applied for these cases (Giggenbach, 1975).

Reference: Tinsley *et al.* (1951).

(c) Ammonia

A separate nonalkaline condensate sample is usually collected for the specific purpose of estimating ammonia by the same methods used for ammonia in waters.

(d) Residual Nonacidic Gases

The residual nonacidic gases present in geothermal steam are commonly H_2, N_2, CH_4, and NH_3, with minor ethane and air (usually from contamination of the sample). In high-temperature fumaroles or volcanic emissions, other sulfur gases, hydrochloric acid, and hydrofluoric acid may also be present. The collection and analysis of these gases are not considered here and reference should be made to Naughton *et al.* (1963), Sigvaldasson and Elisson (1968), Finlayson (1970), and Giggenbach (1975).

The residual gases can be analyzed by standard PVT gas analysis equipment, such as a Bone and Newitt gas analyzer, or more simply, a portable Orsat gas analyzer.

Reference: Nash (1950).

Alternatively, an aliquot of the residual gases from an ampoule is analyzed in a gas chromatograph. A molecular sieve 5A column is employed for separation of the gases, and thermal conductivity and flame ionization detectors determine the amounts of the separated components.

For calibration purposes, aliquots of pure gases must be run through the gas chromatograph to calibrate the detectors. A plot of peak area versus moles of gas at standard temperature and pressure is constructed for each gas. At least three aliquots containing similar quantities of the residual gases should be run for each gas. The size of the gas aliquots used will be dictated by the sensitivity of the gas chromatographic system.

The accuracy of the analysis system is determined by analyzing known mixtures of the gases prepared by means of a gas burette or purchased from commercial sources. Air provides a standard for determining the accuracy with respect to O_2 and N_2.

Any acidic gases remaining in the residual gas samples lead to a gradual deterioration in the separating efficiency of the molecular sieve column. This can be prevented by trapping out acidic gases in a liquid air bath before injecting the sample into the gas chromatograph, or by means of a precolumn of concentrated sodium hydroxide solution absorbed on Chromosorb W.

References: Jeffery and Kipping (1964), Klein (1966) Hodges and Matson (1965), Deitz (1967).

Isotope Analysis

The specialized techniques required for the collection and analysis of samples for stable isotope determinations can be obtained through the review by Gonfiantini (1975).

Analytical Results

Analytical results from water samples are expressed as parts per million (ppm) or equivalents per million (epm), or more correctly as milligrams per liter and milliequivalents per liter, since most analyses are done on a volumetric basis. Ideally the constituents should be reported as the species present in the fluid sampled. In the majority of cases it is most convenient to report constituents as simple atoms or ions, total carbon dioxide as one particular carbonate species (bicarbonate), and silica as SiO_2.

Results on spring waters can be referred simply to the water. With well discharges, where two phases are sampled separately (except during downhole sampling), results are referred, according to the purpose of the analysis, to the respective phase, separated at a particular (stated) pressure (often wellhead pressure or atmospheric pressure) or by computation to the total discharge. For comparison between wells, gas contents are most usefully expressed on a complete discharge basis, as is at least one constituent, usually chloride, in the water phase.

Where a complete analysis of a water sample is made, a check of the accuracy of the results is obtained by comparing the sum of the anion equivalents with that of the cations. The concentration ratios of pairs of many constituents are used in interpreting water and steam compositions. Molecular or atomic ratios are preferred to the weight ratio, since they reflect the proportions in which the constituents react with rock minerals.

REFERENCES

A.P.H.A. (1971). "Standard Methods for the Examination of Water and Waste Water," 13th ed. Amer. Public Health Assoc., Am. Water Works Assoc., Water Pollut. Contr. Federation, New York.

A.S.T.M., (1966). "Manual on Industrial Water and Industrial Waste Water." 2nd ed. Am. Soc. for Testing Mater., Philadelphia, Pennsylvania.

Berl, W. G. (1956). "Physical Methods in Chemical Analysis," Vol. 3, pp. 220–276.

Clifford, I. L., and Hunter, E. (1933). *J. Phys. Chem.* **37**, 101.

Deitz, W. A. (1967). *J. Gas Chromatogr.* **5**, 68.

Dench, N. D. (1961). *Proc. U.N. Conf. New Sources Energy, Rome* **3**, 134.

Ellis, A. J. (1959). *Am. J. Sci.* **257**, 217.

Ellis, A. J., and Golding, R. M. (1963). *Am. J. Sci.* **261**, 47.

Finlayson, J. B. (1970). *Geothermics (Spec. Issue 2)* **2** (Pt 2), 1344.

Foote, F. J. (1932). *Ind. Eng. Chem. (Anal. Ed.)* **4**, 39.

Giggenbach, W. F. (1975). *Bull. Volcanol.* **39** (No. 1), 132.

Goguel, R. L. (1973). Rep. C.D. 2151, Chem. Div. D.S.I.R., N.Z.

Gonfiantini, R. (ed.) (1975). *Proc. Int. At. Energy Agency Advisory Group Meeting Appl. Nucl. Tech. Geothermal Stud., Pisa, Italy, Sept.* (in press).

Hodges, C. T., and Matson, R. F. (1965). *Anal. Chem.* **37**, 1065.

Iwasaki, J., Hagino, K., and Utsumi, S. (1958). *J. Chem. Soc. Jpn.* **79**, 38.

Jeffery, P. G., and Kipping, P. J. (1964). "Gas Analysis by Gas Chromatography." Pergamon, Oxford.

Koga, A. (1967). *N.Z. J. Sci.* **10**, 979.

Kolthoff, I. M., and Kuroda, P. K. (1951). *Anal. Chem.* **23**, 1304.

Kozintseva, V. L. (1964). *Geokhimiya* **8**, 758.

Klein, A. (1966). *In* "Poropak Columns," Varian-Aerograph Tech. Bull. 128, Walnut Creek, California.

Klyen, L. E. (1973). *Geothermics* **2**, 57.

Mahon, W. A. J. (1961). *Proc. U.N. Conf. New Sources Energy, Rome* **2**, 269.

McDowell, G. (1974). *Geothermics* **3**, 100.

Milkey, R. G. (1954). *Anal. Chem.* **26**, 1800.

Milton, R. F. (1949). *Analyst* **74**, 54.

Milton, R. F., Liddell, H. F., and Chivers, J. E. (1947). *Analyst* **72**, 43.

Muto, S. (1957). *Bull. Chem. Soc. Jpn.* **30**, 881.

Nash, L. K. (1950). *Anal. Chem.* **22**, 108.

Naughton, J. J., Heald, E. F., and Barnes, I. J. (1963). *J. Geophys. Res.* **68**, 539.

Orion (1974). Orion Research Inc., Cambridge, Massachusetts. Ammonia electrode model 95-10.

Patton, J., and Reeder, W. (1956). *Anal. Chem.* **28**, 1026.

Pohl, F. A. (1953). *Z. Anal. Chem.* **139**, 241.

Pollak, A., and Work, L. F. (1942). *Mech. Eng.* **64**, 27.

Ritchie, J. A. (1973). Rep. C.D. 2164, Chem. Div., D.S.I.R., N.Z.

Selleck, F. T., Carmichael, L. T., and Sage (1952). *Ind. Eng. Chem.* **44**, 2219.

Sigvaldasson, G. E., and Elisson, G. (1968). *Geochim. Cosmochim. Acta* **32**, 797.

Smith, J. H. (1958). *N.Z. Eng.* **13**, 354.

Tennant, W. C. (1957). *Appl. Spectrosc.* **21**, 282.

Tinsley, J., Taylor, T. G., and Moore, J. H. (1951). *Analyst* **76**, 300.

U.S. Geological Survey (1960). "Methods for Collection and Analysis of Water Samples. Water Supply Paper 1454, U.S. Geolog. Surv., Washington, D.C.

Wyatt, P. F. (1955). *Analyst* **80**, 368.

Chapter 6

PREDEVELOPMENT INVESTIGATIONS

6.1 OBSERVABLE NATURAL FEATURES

Introduction

The extent and shape of a geothermal area can sometimes be outlined by visual surface or aerial surveys, or infrared and black and white photography, but more elaborate survey work is required for a quantitative assessment. Some of the characteristic features of geothermal areas are summarized in this section.

Geothermal areas commonly have well-defined springs and fumaroles, but hot water seepages, areas of steaming ground, mud pools, and zones of high gas emissions can form a major part of the total natural heat flow. The dominant type of discharge varies considerably from area to area. Where the topography is uneven, hot springs are usually found at lower levels close to the cold water table, whereas fumaroles and steaming ground occur at higher altitudes. Surface activity is not a direct guide to deep conditions and a steam-heated system may have as many surface springs as a hot chloride water system. Geysers are common in areas with high water temperatures and where the subsurface rock permeabilities are high.

Large exposures of hydrothermally altered rock, frequently multicolored by iron oxides and sulfides, inevitably occur in association with steaming or hot ground. Significant sinter and salt deposits may be found in hot spring areas, but because sinters are very stable to weathering, they may remain as permanent features even when spring activity disappears. In areas where snow occurs, such as at Yellowstone Park or in the high Andes in Chile, melted snow can give a line of demarcation between hot and cold ground. Trees, shrubs, and grasses will not survive above certain ground temperatures (40°–60°) and many hot zones are outlined by dead vegetation. For example, the Karapiti fumarole area at Wairakei is shown in Fig. 6.1. Characteristic types of flora found in areas of elevated ground temperatures can define the limits of hot and warm ground. Varieties of algae grow prolifically in hot seepages and in rivers and streams receiving hot waters.

Major discontinuities in subsurface rocks may sometimes be recognized from the surface pattern of water, steam, or gas emissions. Hot springs or fumaroles aligned or forming semicircular patterns on the surface may outline faults or old caldera structures. Outflows may also occur along the surface exposures of contacts between dissimilar rock formations, possibly marking the perimeter of a geothermal system.

The natural heat escaping from a hot spring area is an approximate guide to the minimum energy available for development, and a rough

Fig. 6.1 The Karapiti fumarole area, Wairakei, New Zealand, showing hot areas with dead vegetation.

assessment of this can be made by measuring the temperature or chloride increment in local waterways receiving the hot water drainage from springs (Ellis and Wilson, 1955).

Initial Assessment of a Geothermal Area

In the preliminary investigation of a geothermal area, simultaneous geological, geophysical, and geochemical surveys are advisable. Results from each discipline are continually correlated to produce guidelines for the investigations and to update the model of the system.

Geology

Information may be available from published aerial photographs or geological and topographical maps of the region. The first investigations are directed toward obtaining a broad picture of both local and regional geology, examining the general structure and fault patterns, and determining their association with the natural thermal activity. The local stratigraphy is examined for the possible occurrence of permeable aquifer beds or cap rocks, and porosities and permeabilities are noted. From the topography of the surrounding country, the general movement of cold groundwater in formations of the area, and evidence of permeability from faults and particular rock types, a local hydrology model is developed. Possible recharge zones are indicated.

Surface exposures of hydrothermally altered rock give a measure of the size and intensity of the system but a decision must be made as to whether the alteration is caused by current activity or is a relic from the past.

Geophysics

Initial observations in an area include the extent and nature of the activity, surface temperatures, and measurement of shallow temperature gradients. Rough estimates of hot water and steam discharges and of the area of steaming and hot ground lead to a crude estimate of the natural heat flow. A certain amount of meteorological information, such as precipitation and evaporation rates and maximum and minimum air temperatures, is necessary for accurate assessments.

The geophysical prospecting techniques most likely to be successful in the area are assessed. Hot water systems with deep saline solutions are most commonly surveyed by electrical resistivity techniques, but other methods commonly applied are aerial infrared, electromagnetic, seismic, and gravity techniques. Steam-heated systems may best be investigated by measurement of temperature gradients in shallow (100-m) slim wells, together with electrical resistivity methods.

A high proportion of geophysical prospecting techniques require a grid system of measurements. A prerequisite for geophysical studies is therefore an accurate survey of the area and establishment of benchmarks. Once reference points are established and related to the national grid, a local grid is constructed for the area, and a common base map agreed to for all the studies in the area.

Geochemistry

Some geochemical information is frequently available at the outset of a survey but its quality and significance must be evaluated. The natural surroundings can give comment on the chemistry of the geothermal system. Deposits and sinters around the hot springs may indicate the nature and extent of mineralization of the waters. Terraces of amorphous silica are commonly found about neutral-pH chloride water springs originating from deep high-temperature waters. Antimony and arsenic sulfides are frequently found in sinters formed from these waters, and are recognized by their color. Calcareous deposits are often found around springs where the deep-water temperatures are low (less than 150°), or where hot waters are considerably diluted with surface cold waters. Waters rich in sodium chloride may deposit salt about their drainage channels when evaporation rates are high.

Acid waters high in sulfate may be associated with sulfur in or about the springs, and are usually turbid. The rocks surrounding the springs are highly altered to kaolin-type minerals and calcium sulfate deposits may occur. Acid springs of this type are often associated with fumaroles in the elevated parts of an area, and there may be vigorous ebullition of gases through the waters.

Sulfur is commonly found in small quantities around fumaroles, as are alums such as ammonium potassium alum. Boron compounds are also found about fumaroles, particularly those with steam originating from a high-temperature steam reservoir. A mercaptan smell, characteristic of carbon–sulfur gases, is frequently present in areas where steam passes through zones of decomposing vegetation.

Steam arising from active volcanism is readily distinguishable from steam boiling from underground hot water systems. Volcanic steam usually contains sulfur dioxide, hydrochloric acid, and hydrofluoric acid, which are uncommon compounds in steam formed from underground hot water. Sulfur dioxide is detectable by its acrid smell and its irritation to the throat, while gaseous hydrochloric acid and hydrofluoric acid rapidly produce skin and eye irritation. The majority of high-temperature volcanic steam vents are surrounded by appreciable deposits of sulfur.

Areas may contain up to several hundred hot springs or fumaroles. The first reconnaissance survey is used to plan the sampling of springs and fumaroles. Areas are initially separated into high-temperature (over 80°) and low-temperature (under 80°) discharge zones, and a decision is made on the number of samples required from each zone to get representative sampling, taking into account the areal distribution of springs, spring alignments, and the association with geological features. An investigation of the compositions and distribution of cold surface waters in the area aids later interpretations.

6.2 ROLE OF SCIENCES IN PRELIMINARY INVESTIGATIONS

The initial reconnaissance is frequently sufficient to give a preliminary assessment of the area's potential for development. More detailed confirmatory information is then necessary from scientific surveys.

Since the bulk of the cost of geothermal exploration occurs in drilling investigation wells, preliminary scientific surveys should be aimed at siting wells to obtain the maximum amount of information about the system with the minimum drilling costs. The investigation drilling in many geothermal areas has been to depths of only 200–700 m, often producing inconclusive data. It is considered that the drilling rigs used in exploration should be capable of reaching depths of at least 1200 m. The high cost of drilling to this depth, however, requires very careful selection of sites.

The initial wells are aimed at determining the areal and vertical extent of the system, the geological structure, and aquifers present and their permeability, major upflow zones, and the temperatures, pressures, and compositions of the deep fluids. The wells also give information on the mineral deposition potential of the deep waters and the chemical gradients within the system. This information is used for estimating the potential of the area for exploitation.

This section briefly outlines the prospecting techniques employed in determining the sites of exploratory wells. All the techniques are not necessarily employed in investigating every area.

Geology

Careful mapping of the surface activity is completed (hot springs, fumaroles, and areas of hot ground), and detailed fieldwork carried out on local and regional geology with the aid of aerial photographs. These help to define areas of hydrothermal alteration, surface activity, lava flows, eruption centers, rock formation contacts, and surface traces of faults. If it has not already been done, survey work should be carried out at the same time as the geological mapping, since accurate base maps are required. Where there is no geological record, a mapping program may require one or two years.

The natural thermal activity is examined for preferential associations with particular geographic or structural zones, or with particular rock types (Healy, 1970). From the general characteristics of the area it is tentatively classified into one of the system types described in Chapter 2.

Measurements of rock porosities and permeabilities (joints, fissures) and the degree of alteration and mineralization give an improved assessment of possible aquifer and confining formations. The recognition of volcanic centers and the distribution of various types of volcanic rocks can help to identify zones where subsurface permeability may be high.

The directions of cold and hot water movement through the system are assessed from local hydraulic gradients. Pressure measurements made in shallow wells (30–50 m) are sometimes useful. Cold water movements can suggest whether the surface thermal activity is likely to be displaced laterally from the center of hot water upflow at deep levels.

Different catchment areas are defined by surface water analyses, and estimates made of rainfall, evaporation, transpiration, and surface flows. The estimate of subsurface flows then gives an assessment of the recharge capacity of the system.

Maps of surface and subsurface geology are produced from the investigations and suggestions made on the nature and locality of the heat source for the system and on productive aquifers and permeable flow zones.

Geophysics

The geophysical exploration methods used are determined by the type of geothermal system. In New Zealand where hot chloride water systems predominate, heat flow measurement and electrical resistivity surveys have proved to be the most useful techniques available (Dawson and Dickinson, 1970; Risk *et al.*, 1970). In steam systems, heat flow measurements and a combination of gravity, magnetic, and seismic surveys have been used to outline the systems. In recent investigations of steam systems in the United States and Indonesia, the electrical resistivity measurements have proved useful in outlining hot subsurface fluids (Zohdy *et al.* 1973; Hochstein, 1975).

Heat Flow Measurements

The total heat and mass flows of steam and water from a geothermal system give a rough minimum measure of its utilization potential and of the enthalpy of the deep fluid. Heat escapes from an area through conduction and convection processes in the top rock formations, by evaporation, and by direct discharge of steam and water. A breakdown of the natural heat loss at Wairakei (from Dawson and Dickinson, 1970) is shown in Table 6.1.

Heat loss by conduction is estimated by measuring the temperature gradient in areas of warm ground. Temperatures at 15 cm and 1 m are measured and the gradient calculated from the equation $Q = K \, \Delta T$, where K is the thermal conductivity (calories per centimeter per second) and ΔT is the temperature gradient (degrees per centimeter). The value of K is estimated from measurement of daily or annual temperature variations at different depths (Ingersoll *et al.* 1955).

Heat loss from steaming ground is measured using a calorimeter

TABLE 6.1

Total Natural Heat Loss from the Wairakei Geothermal Area [a]

Discharge type	Heat loss by	Percentage of total heat loss
Warm ground	Conduction	3
Steaming ground	Convection, evaporation	36
Fumaroles	Evaporation, direct discharge	10
Geysers	Direct discharge	0
Hot pools	Evaporation	32
Hot springs	Direct discharge	16
Seepages	Direct discharge	3

[a] From Dawson and Dickenson (1970).

(Benseman, 1959a,b) or a venturi meter (Thompson *et al.*, 1961). Measurement of steam flows from fumaroles is made with a traversing Pitot tube (Thompson *et al.*, 1961). Black and white photography as well as reflection infrared photography is also used for qualitative assessment of heat flows (Dawson and Dickinson 1970; Dickinson, 1975). Heat flow measurements using individual snowfalls as calorimeters were described by White (1970).

Heat loss due to evaporation is calculated from the temperature of the water surface and the ambient air temperature, using semiempirical equations presented in the International Critical Tables. These were validated over the temperature range 20°–98° for calm water surfaces, and for low wind speeds, by Banwell *et al.* (1957) and Dawson (1964). Pool areas are measured or taken from aerial photographs. The heat loss due to ebullition from vigorously boiling pools is significant and Dawson (1964) estimated that the additional heat loss is approximately proportional to the ebullition height. Similar heat losses occur from geysers.

Heat outputs of hot springs and geysers are estimated from temperatures and flow measurements using standard hydrological methods employing flumes or V-notch weirs in association with flowmeters or Pitot tubes. Frequently, hot water seepage occurs into streams or rivers and measurements of mass flows and water temperatures upstream and downstream of the seepage entry points give the heat output.

Natural heat flows from various geothermal areas in New Zealand are listed in Table 6.2. The productive capacity of a field may, however, greatly exceed the natural heat output; for example, the production from wells at Wairakei is several times the natural heat

TABLE 6.2

Natural Heat Flow from Some Geothermal Areas in New Zealand[a]

Area	Date of measurement	Natural heat flow (kcal/sec)
Wairakei	1951–1952	103,000
Rotokaua	1951–1952	50,000
Waiotapu	1957	134,000
Orakeikorako	1957–1964	82,000
Kawerau	1962	18,000

[a] Values taken from Dawson and Dickinson (1970).

flow of 103,000 kcal/sec. Heat flow measurements and estimates of geothermal gradients have been carried out in many geothermal areas of the world. A selection of the results from various countries is available in the Proceedings of the U.N. Symposia on the Development and Use of Geothermal Resources at Pisa (1970) and San Francisco (1975).

Electrical Resistivity Techniques

Electrical prospecting techniques have been used mainly for hot water systems and are based on the high conductivity of hot sodium chloride solutions. Recently, similar methods have been used to explore steam-heated systems, using the conductivity of the sodium sulfate, sodium bicarbonate, or sulfuric acid solutions which may be present (Zohdy *et al.*, 1973; Hochstein, 1975). The techniques can outline the areal and vertical extent of hot water in subsurface rocks and delineate certain structural features, such as faulting, within an area. In the past it has not been possible to obtain deep rock permeabilities from resistivity measurements. However, recent advances in the technique enable this to be done qualitatively (Risk, 1975).

Meidav (1970) discussed the factors determining the electrical conductivity of fluid-saturated rocks. The resistivity is dependent on the concentrations of conducting ions in solution, concentrations of conducting ions in the rocks, porosity of the rocks, and the temperature. Resistivity is approximately inversely proportional to salt concentration, and Hochstein (1971) showed that at constant salinity and temperature, a change in the porosity of rocks from 10% to 35% caused a twelvefold change in resistivity. Rock conductance is usually small compared to that of the solution, although certain rocks such as clays, have a significant conductivity. To interpret resistivity soundings, the salinity and temperature of the deep waters and the porosity of the aquifer rocks are required. The first two are provided by the geochemist and the third by the geologist. Shallow slim wells approximately 100 m deep drilled at points around the area can supply information on these parameters and can be used for thermal gradient measurements.

For resistivity surveys to depths of approximately 30 m, the electromagnetic method, employing the horizontal loop technique, is both quick and inexpensive (Lumb and MacDonald, 1970). It can replace near-surface temperature measurements for locating hot waters close to the surface. Cold groundwater movement or impermeable surface layers can mask the presence of shallow hot water to temperature surveys, but are recognized by resistivity measurement. Details of the theory and interpretation of electromagnetic induction methods were given by Wait (1962) and Keller and Frichknecht (1966). The application of the method to geothermal surveys was discussed by Lumb and MacDonald (1970).

A number of direct current resistivity techniques are available for investigating hot water systems down to depths of at least 3 km. These can be roughly classified under five headings, each technique having a number of variations: (a) Wenner; (b) Schlumberger; (c) polar dipole; (d) equatorial dipole; (e) random dipole.

The first two techniques use a linear spread of source and detection electrodes, whereas the dipole techniques employ a nonlinear arrangement of electrodes. The depth of current penetration in the linear electrode arrangement is dependent, among a number of parameters, on the electrode separation. In a homogeneous body of rock, of constant porosity and electrical resistance, the depth of current penetration is approximately equal to the distance of electrode separation.

To obtain deep resistivity soundings, the linear electrode arrangement of the Wenner and Schlumberger methods is frequently impracticable. It is often difficult to lay long lengths of electrical cable across country, and extraneous electrical currents may occur, particularly in populated areas (wire fences, overhead power lines). In a dipole survey, current passes through the ground from a source dipole and the electrical field strength produced is measured at various receivers in the survey area. The dipole methods offer the advantages that the logistic requirements are almost independent of the probing depth, and that lateral resistivity discontinuities can be more accurately located.

The theory and application of the methods are given in standard texts of electrical prospecting techniques, such as Keller and Frichknecht (1966), and their applications to geothermal investigations in many countries were described by Bodvarsson (1951), Mazzoni and Breusse (1954), Studt (1958), Alfano (1960), Battini and Menut (1961), Banwell and MacDonald (1965), Hayakawa *et al.* (1967), Risk *et al.* (1970), Meidav (1970), and McEuen (1970). The Proceedings of the 1975 U.N. Symposium contains many more references to electrical prospecting techniques.

The results for each probing depth (in New Zealand, electrode spacings of 180, 550, and 1100 m have often been used) can be superimposed on a map to give a three-dimensional model, or plotted separately to show the boundary of the hot water at each depth. Some geothermal areas are found to have almost vertical resistivity boundaries, values rising from between 5 and 20 ohm m to over 50 ohm m in short lateral distances. Thus discontinuities such as faults which contain hot water flows are recognized in dipole surveys as low resistivity zones.

A resistivity contour map of the Broadlands geothermal field in New Zealand, taken from Risk *et al.* (1970), is shown in Fig. 6.2 (the probing depth was 250 m). The small separation between the 20- and 50-ohm m contours is illustrated. The abrupt boundary between high and low resistivities may represent a pressure boundary.

Schlumberger resistivity techniques have been used with considerable success in some of the steam-heated systems in Indonesia (Hochstein, 1975). Sulfate and free hydrogen ions formed from hydrogen sulfide oxidation or sulfur hydrolysis have significant conductance, as do sodium and bicarbonate ions present in pore waters of subsurface rocks.

Seismic, Gravity, and Magnetic Studies

These methods are used to obtain information on the structure of the subsurface rocks in relation to several basic concepts of geothermal areas (Hochstein and Hunt, 1970).

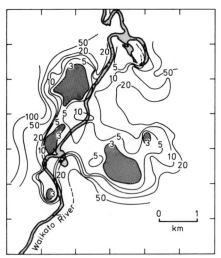

Fig. 6.2 Apparent resistivity contours, in ohm meters, over the Broadlands geothermal area, New Zealand, from Wenner survey; $a = 180$ m ($p = 250$ m) (from Risk, MacDonald, and Dawson, 1970).

(a) Geothermal systems in both sedimentary and volcanic rocks are located above fissure zones or major faults, both presumably related to basement tectonics.

(b) Secondary fissures and minor faults at shallow depths are often tapped for exploitation.

(c) Geothermal systems in a volcanic environment are associated with nonmagnetic rocks.

Seismic, gravity, and magnetic surveys can in principle detect basement and aquifer structures. Studies in geothermal areas of Italy were described by Cassinis (1961), Vecchia (1961), Battini and Menut (1961), and Mouton (1969); in geothermal areas of Japan by Hayakawa and Mori (1962), Noguchi (1966), Hayakawa *et al.* (1967), Hayakawa (1970), and Baba (1975); in New Zealand areas by Beck and Robertson (1955), Modriniak and Studt (1959), Studt (1959), and Hochstein and Hunt (1970); and in the United States by Isherwood (1975).

Hochstein and Hunt (1970) suggested that the success of structural investigations was limited. Although a generalized picture of the basement structure was obtained from gravity and seismic surveys, fissures and faults at shallow depths could not be detected. Magnetic survey anomalies could be explained by the presence of nonmagnetic rocks associated with geothermal fluids, but similar anomalies often occurred nearby. Nevertheless, the techniques have had some success and should be attempted where more direct methods, such as electrical resistivity soundings cannot be applied. Thompson (1975) successfully outlined geothermal areas in New Zealand by magnetic surveys.

Microgravity Surveys

Although not directly a prospecting technique, microgravity measurements may be made with advantage during the exploration stage. The results are used for monitoring the net mass loss from a geothermal area after exploitation commences, and Hunt (1970) described its application to the Wairakei area, where measurements of gravity at particular benchmarks are made at intervals of time. Gravity differences after corrections for known changes in elevation reflect the net mass of fluid lost from the aquifer, and may also outline the recharge areas.

Other Methods

The following four methods are less common than those already discussed but have been applied in some geothermal areas.

The magnetotelluric method (Keller, 1970; Strangeway, 1970) makes use of the natural fluctuations of magnetic micropulses in the ionosphere at subaudio frequencies, and thunderstorms at audio frequencies. Sufficient ground penetration is available in the lower part of the available frequency range to enable electromagnetic soundings of the crust and upper mantle, while the audio frequencies are used for shallow penetration. Keller (1970) applied the method in New Zealand in investigating possible active magma chambers of the Taupo Volcanic Zone. Recent advances in the technique were described by Whiteford (1975), Wilt and Combs (1975), Hoover and Long (1975), and Hermance *et al.* (1975).

The distribution of microearthquakes in a geothermal area can be used to determine movement and activity on major faults. The method was successfully applied in one of the Kenyan geothermal areas by Hamilton *et al.* (1972) and more recently in other geothermal areas by Combs and Rotstein (1975) and by Maasha (1975).

Geothermal ground noise investigations have been made in New Zealand by Whiteford (1970, 1975), in Mexico by Banwell and Gomez (1970), and in the United States by Iyer and Hitchcock (1975). A conspicuous noise pattern was found to occur near low resistivity zones in the geothermal areas.

Black and white photography and aerial infrared photography are becoming important in locating geothermal areas and for monitoring changing boundaries during exploitation (Hochstein and Dickinson, 1970; Palmason *et al.*, 1970; and several papers at the 1975 U.N. Symposium: Dickinson, 1975; Hodder, 1975; Marsh *et al.*, 1975; Yuhara *et al.*, 1975). In black and white photography vertical photographs are usually taken at heights of 600–1200 m. The infrared method depends on imagery of emissive radiation in the 4.5- to 5.5-m band, the minimum heat flow detectable ranging from 150 to 350 cal $cm^{-2} sec^{-1}$.

Geochemistry

In the initial exploration stages, geochemical investigations of hot springs, fumaroles, steaming ground, and cold surface waters can supply information on:

(1) the range in compositions and homogeneity of the hot fluids;
(2) subsurface fluid temperatures and pressures;
(3) the type of system present—steam-heated or hot chloride water;
(4) the subsurface rocks associated with the hot fluids;

>(5)   the origin of the hot fluids, the direction of flow through the area, and the turnover times of water in the system;
> (6)   the mineral deposition potential of the fluid;
> (7)   the natural heat flow;
> (8)   the zones of high upflow permeability;
> (9)   fluids constituents which could have economic value;
> (10) the feasibility of reinjecting fluid back into the system to eliminate local thermal and chemical pollution problems.

Interpretative methods are outlined briefly in this chapter. Details of the methods used in the New Zealand geothermal areas are available from Ellis (1970), Mahon (1970, 1975), Glover (1970), and Finlayson (1970); in the United States from White (1970), Fournier and Truesdell (1970, 1975), Bowman *et al.* (1975), Mazor (1975), McKenzie and Truesdell (1975), McKenzie *et al.* (1975), O'Neil *et al.* (1975), Truesdell *et al.* (1975); in Iceland from Arnorsson (1970), and Arnorsson *et al.* (1975); in Italy from Tonani (1970), Baldi *et al.* (1975), Fancelli *et al.* (1975), and Panichi and Tongiorgi (1975); in Turkey from Dominco and Samilgil (1970); in Japan from Koga (1971), Koga and Noda (1975), and Fujii and Akeno (1970); in Mexico from Mercado (1970); in India from Gupta *et al.* (1975); in the Soviet Union from Gutsalo (1975); in Indonesia from Kartokusumo *et al.* (1975); and in El Salvador from Sigvaldasson and Cuellar (1970).

The majority of known geothermal areas have near-boiling springs discharging neutral-pH or slightly alkaline sodium chloride waters, but the extent of mineralization varies widely between areas. In some areas the chloride concentrations in near-neutral-pH boiling springs are rather low (<100 ppm); for example, about 40 ppm at Beowawe Geysers, Nevada (Nolan and Anderson, 1934), and Carboli, Italy (Cataldi *et al.* 1969). These were nevertheless still considerably higher than those found in local surface waters.

Steam rising into surface waters from steam-heated reservoir systems may contain high concentrations of ammonia, which neutralizes the acidity caused by the oxidation of sulfide gases. At Larderello, there were boiling springs of small output, low in chloride but high in boron and ammonia. At Kamojang, Indonesia, where a steam-heated reservoir exists to depths of 600–700 m, there were neutral-pH springs low in chloride ($\approx$5 ppm) but with high concentrations of sulfate and bicarbonate. Acids in the waters were neutralized by rock–water reactions.

Since water compositions vary widely, the type of system present may not always be clear. Discharges from shallow wells (about 100 m) may assist (e.g., from wells drilled to measure the local geothermal gradient).

Chloride is possibly the most important chemical tracer. Its role in assessing whether a system is water or steam based was discussed by White (1970). However, exploration of Indonesian geothermal systems showed that the absence of chloride in hot surface waters does not necessarily indicate a steam reservoir beneath. Also, alteration of surface rocks can give rise to moderate concentrations of chloride in spring waters.

The deep hot water in the center of an area is often of near-constant chloride concentration. For example, at Wairakei the concentration is about 1500 ppm over a wide area of the deep reservoir, while at El Tatio, Chile, the concentration is nearly constant at 6200 ppm. In many areas chloride concentrations in boiling hot springs of high flow have been found to be similar to those in the deep water, after allowance for concentration by boiling on route to the surface.

A map of isochloride contours in a geothermal area may show the position of the deep-water source, the extent of dilution of hot waters rising to the surface, and possibly the boundary of the hot water area. Lower chloride concentrations near the edge of a field often follow the higher resistivity contours. Figure 6.3 shows isochloride lines drawn across the Waiotapu area, New Zealand (Lloyd, 1959).

Maximum chloride concentrations indicate permeable zones from deep levels to the surface, and siting of the first exploratory wells near

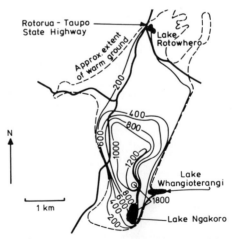

Fig. 6.3 Isochloride lines drawn from hot spring analysis at Waiotapu, New Zealand (from Lloyd, 1959).

the least diluted springs generally produces good results. However, it should be noted that major spring activity can frequently occur on the perimeter of a system.

In geothermal areas with activity occurring at higher altitudes, the nature of the discharges may suggest the system to be a steam-heated reservoir type. However, a survey of chloride concentrations of streams and rivers draining the areas should be made, particularly at lower altitudes, to detect any leakage of chloride from a deep hot water aquifer. Such surveys in Indonesian geothermal areas enabled the depths of steam-heated reservoirs and of hot water in the systems to be assessed.

Other soluble constituents such as boron, cesium, arsenic, and bromide may be used in association with chloride to determine whether the system contains one or several hot water aquifers. The use of atomic Cl/B, Cl/Cs, Cl/As, and Cl/Br ratios overcomes dilution and evaporation effects.

Chemistry gives information about the nature of the rocks contacting the deep hot waters. High boron, ammonia, iodide, bicarbonate, and carbon dioxide concentrations and low Cl/B ratios often indicate the presence of sedimentary rocks. Appreciable concentrations of lithium, rubidium, and cesium suggest the reaction of water with rocks such as rhyolites, ignimbrites, and pumice breccias, whereas many andesites and basalts liberate into solution appreciable concentrations of boron and fluoride, respectively.

The concentrations of silica, magnesium, fluoride, sulfate, and calcium in waters, the Na/K and Na/Rb ratios, and several light element isotope exchange equilibria are controlled by temperature, and enable deep-water temperatures to be assessed from spring water analysis (see Sections 4.7 and 4.8). Of the isotopic reactions, H_2O–HSO_4^- oxygen exchange is probably the most reliable.

On route to the surface, hot water often undergoes dilution and reaction with rock. In estimating deep temperatures, near-boiling-point springs of large flow and with near-neutral-pH water should be chosen. The estimates of underground temperatures are generally minimum values.

Chloride concentrations in boiling springs can be used to estimate minimum underground water temperatures (Mahon, 1970). The highest chloride value in flowing springs of boiling temperature is assumed to be similar to that in the deep water (after allowing for concentration as the water rises to the surface), while boiling springs with lower chloride concentrations (but lacking steam-heating characteristics) are assumed to be diluted by low chloride surface waters.

The amount of dilution in going from the highest to lowest chloride concentrations allows a minimum deep temperature to be calculated.

Fournier and Truesdell (1973) and Truesdell and Fournier (1975) developed improved methods of calculating deep temperatures and hot water fractions for mixed and diluted springs that issue at boiling temperatures. They used the equations

$$X = \frac{Cl^m H_{ws}^n (H_s^m - H_{sil}^m) + Cl^n H_{ws}^m (H_{sil}^m - H_w^c) - Cl^c H_{ws}^n H_{ws}^m}{Cl^n H_{ws}^m (H_s^n - H_w^c) - Cl^c H_{ws}^n H_{ws}^m}$$

$$H^h = \left(\frac{H_{sil}^m - H_w^c}{X}\right) + H_w^c$$

where Cl is the chloride content; H_w, H_s, and H_{ws} are specific enthalpies at the surface temperature of water, steam, and evaporation, respectively; H_{sil} is the enthalpy of liquid water at the temperature indicated by the silica geothermometer, and superscripts m, n, and c denote mixed, nonmixed, and cold springs, respectively. H^h and X are the specific enthalpy and mass fraction of the hot component of the mixed water. The boiling spring mixing model assumes conservation of chloride and enthalpy and reequilibration with quartz after mixing. Similar deductions can be made graphically, as shown in Section 6.4.

Knowing the approximate chloride concentration and temperature of the deep water, we can estimate the minimum natural heat flow and energy potential of the area. The springs probably discharge into a stream or river, allowing the total amount of chloride discharged from the area to be determined accurately. The minimum natural heat output is then calculated, based on the ratio of heat to chloride in the deep water (Ellis and Wilson, 1955). The method ignores conductive heat loss and steam flows to the surface unaccompanied by water.

From estimates of the subsurface water temperature together with concentrations of calcium and sodium, or calcium and bicarbonate, in spring waters, approximate carbon dioxide concentrations in the deep water can be calculated (Ellis, 1970). Areas with high carbon dioxide concentrations are most likely to be troubled with calcite deposition in drill pipes and fissures.

A survey of the hydrogen and oxygen isotope ratios in spring waters and in local surface waters will probably indicate the recharge area for the system (see Chapter 3). A knowledge of the recharge area rainfall enables the long-term output capacity of the system to be estimated. Determinations of the tritium concentrations in the least diluted spring, in steam condensates from active fumaroles, and in local meteoric waters allow the residence time or flow time of the hot waters to be assessed.

The analysis of gases accompanying steam flows enables underground temperatures to be estimated and separation and flow patterns of steam established. The chemical and isotopic exchange reactions involving carbon dioxide, methane, hydrogen, and water provide several cross-checking geothermometers (see Sections 4.7 and 4.8). A comparison of the ratios between different gases (e.g., CO_2/H_2S, CO_2/NH_3, CO_2/H_2) from a representative selection of fumaroles may indicate the direction of steam or hot water flow in the system. The assessment is based on the different solubilities of the gases in hot water, their distribution between a liquid and a vapor phase, and their reaction with rocks (see Chapter 3). It is only with major fumaroles, discharging large volumes of steam at temperatures of at least 100°, that quantitative significance can be attached to the results.

The use of volatile constituents (such as mercury, boron, and ammonia), trace metals (such as arsenic, antimony, and uranium), and noble gases in geochemical prospecting has received considerable attention in the last few years, but the methods are still under development (see, e.g., Koga and Noda, 1975; Mazor, 1975; Celati *et al.*, 1975; Gutsalo, 1975).

At the completion of the geochemical survey, maps are prepared with contour lines for constant chloride, Na/K, Na/Ca, etc. A typical set of data from the El Tatio geothermal area is shown later in this chapter.

It is important that all chemical data be utilized to make predictions (i.e., the results from one particular type of measurement should not be overemphasized).

Estimating the Potential of an Area

The data accumulated from geological, geophysical, and geochemical work are used to assess the utilization potential of an area, often on the basis of electricity generation alone. For example, Wairakei was conservatively estimated to have a 5000-MW yr potential, or 250 MW for 20 yr. However, a full assessment should include both electrical generation potential and possible utilization of low-grade heat for community heating, desalination, drying, refrigeration, and timber processing.

Estimating the recoverable heat in a system requires knowledge of the temperature of the deep fluid, the size of the system, the porosity of the rocks, and the specific heats of the fluid and rock at the temperature. Approximate information of these parameters is available from surface surveys.

Banwell (1963) gave data on the energy content of rock, saturated water, and saturated steam for temperatures above 50°. As an example (Turner, 1969), an ·estimate is given of the electricity-generating potential of a geothermal system 14 km^3 in area, with deep water at 225° and an average rock porosity of 16.5%. At a temperature of 225°, saturated water and reservoir rock, have heat contents of 5000 and 3000 MW yr per cubic kilometer, respectively. For a reservoir of 14 km^3 the total thermal energy is 46,000 MW yr, or assuming 10% efficiency in electricity generation, 4600 MW yr of electrical energy. Other factors affecting the suitability of an area for development must also be considered. Low permeability may inhibit fluid withdrawal to an extent where development is impracticable. Recharge to an area may be minor and this could eventually limit production. Mineral deposition could also be troublesome and result in expensive cleaning operations. Distribution costs and local energy requirements are other important considerations. The minimum recoverable energy flow from the system is represented by the natural heat discharge, assuming that this represents an equilibrium state between discharge and recharge.

Areas with temperatures less than 200°, or with low rock permeability, are at present regarded as marginal for electricity production, but as pointed out in Chapter 1, recent advances in geothermal technology, particularly binary fluid cycle systems, may change this concept.

Siting Exploratory Wells

The expense of drilling is only warranted if surface surveys suggest favorable conditions for development. Siting the first well in an area is a challenging task, and there will be conflicting opinions. For example, stratigraphic problems important to the geologist may influence his choice of site, but be less important to others. The first well should supply as much information as possible about the deep system, but it should also be capable of discharging. Three wells is usually the minimum to ensure an acceptable model. Frequently it is argued that if the first well produces successfully, the next wells should be located within the same geological feature or zone. This may prove the production from one portion of an area but it is unfortunate if the possibilities of immediate production cloud the issue of a proper assessment of the whole system. In a number of systems exploratory and production wells were drilled in the same general locality and there is still little known about other parts of the areas. However, in some cases where power is urgently required only a small part of the area may be assessed immediately and production wells drilled to produce a small quantity of electricity (20–30 MW). This may also lend confidence in investing in further developments in the area.

Geological evidence will indicate a well site where high permeability or porosity at depth is indicated. In some New Zealand areas success was also obtained by drilling to intersect the down-thrown side of a fault zone at a depth where safe exploitation was possible.

In hot water systems, the geophysicist selects a site in a zone with the lowest resistivity values at all probing depths, and where major discontinuities or fault zones are predicted. In many (but not all) areas the low resistivity zones include the surface hot spring areas. In steam-heated reservoir systems, resistivity surveys may be of less value, and sites are selected in zones of high heat escape and temperature gradient. Gravity and magnetic anomalies may also assist.

Well site selection from geochemical data depends on the type of surface activity. In hot water systems, sites are selected close to high-temperature springs, where temperature and dilution indicators suggest the most direct hot water supply from high temperatures at depth.

Chemical information from fumaroles may lead to well sites away from hot spring activity if high subsurface temperatures are indicated from chemical and isotope equilibria. Assessment is made of the distance that steam has traveled from its underground source by using ratios of gas constituents (CO_2/H_2S, CO_2/H_2, CO_2/CH_4, CO_2/NH_3).

6.3 PROSPECTING WELLS

Physical Characteristics of Discharges

The methods of testing the physical characteristics of wells and their discharges vary from country to country. Some common methods are outlined.

When drilling is completed, slotted liner is set into the uncased portion of the well and the drilling mud is removed with water. Zones of high rock permeability occurring within the slotted casing level (recognized during drilling by loss of mud circulation) are investigated by cold water injection and temperature measurement. These tests, commonly referred to as completion tests (Wainwright, 1970), also give information on the overall rock permeability by measuring the increase in well pressure at three water-injection rates (Wainwright, 1970).

The pressure and temperature gradients in a well are measured under static conditions immediately after the water-injection runs and then at intervals for about one month. This is generally long enough for near attainment of temperature and pressure equilibrium between the fluids in the well and the surrounding rocks. Final pressure and temperature equilibrium is approximated by extrapolation of incremental increases with time. Wells filled with vapor or with a vapor-water mixture are readily distinguished from a well filled with water by downhole pressure and temperature measurements.

The pressures measured in wells may not be representative of the pressures in the rocks at particular depths before drilling, since individual aquifers penetrated by the well are joined. Unless the aquifers are isolated by solid casing and adequate cementing of the country rock, pressures are transmitted from one aquifer to the next. This is a common situation in systems which have a series of alternating impermeable rocks and aquifers.

After temperature and pressure equilibrium has been attained, the well is discharged and the mass and heat outputs are measured over a range of discharge pressures. Measurements of mass output are made by a variety of techniques (James, 1970; Wainwright, 1970). Chemical sampling is commenced, although downhole samples are sometimes collected during the "heating-up" period (Klyen, 1973). Mass output and heat measurements determine whether a discharge originates from steam, from water at temperatures present in the most permeable zone, or from a mixed intake of steam and water. The quantity of gas discharged is assessed.

So as to monitor pressure drawdown, downhole pressure measurements are made while the well is discharging. Pressure drawdown is related to the porosity of the aquifer about the well and, in a hot water system, to the local changes in water level. After discharging, pressures are recorded again at reference depths under static conditions. This shows the pressure recovery of the aquifer, which can be related to its production capacity. Temperature measurements made immediately after the well has been closed record the movement of colder water into the production zone at various depths and the presence of two or more hot water inflows in the well. The wellhead pressure of a well immediately after closure can be used to estimate the supply water temperature (James, 1970).

After a number of wells have been drilled, static pressures at certain depths are compared and the hydraulic gradient across the field determined. Movements of hot and cold water can be traced, zones of vertical or horizontal permeability outlined, and impermeable rock barriers located. Monitoring static pressures in one well while other wells in the field are discharging supplies information on horizontal permeability and interaction between wells. These measurements enable the correct spacing between future production wells to be determined and further test the capacity of the aquifer or aquifers.

A great deal of information relating to the type of fluid flow into a well and the movement of fluids in the system can be obtained from the results of the initial wells. Considerable advances in the technology have taken place in the last five years and much of the work has been summarized by Bolgarina (1975); Coury (1975); James (1975a,b); Kavlakouglu (1975); Lowell (1975); Ramey *et al.* (1975) and Ramey and Gringarten (1975).

Geological Assessment

In the course of drilling, rock cores are recovered from representative depths. In exploration wells, cores are collected at frequent intervals (possibly every 15–16 m), and rock chips are sifted from the drilling mud. The lithology of the drilled formations is determined from visual, microscopic, and x-ray inspection of the rock material. The porosities of the formations are determined from the cores and the mass permeability of the rocks assessed. Permeable faulted or fissured zones are recognized by silicification or calcification of cores and the type of hydrothermal rock alteration present. The dip and strike of intersected faults are assessed, and the type of hydrothermal alteration present in the rocks at different depths is correlated with the temperatures measured in the well and levels of boiling of water (Browne, 1970). The grades of hydrothermal alteration of the subsurface rocks help to outline the zones of highest temperature and greatest water flow. After several wells have been drilled, a stratigraphic correlation is made of the subsurface rocks and major discontinuities are recognized and traced across the active area.

Geophysical Assessment

Temperature gradients in deep exploratory wells are compared with those in shallow wells, enabling a more detailed interpretation to be made of surface heat surveys. Temperature inversions in the wells are investigated to relate the heat transfer processes to the rock formations and movement of cold water through the hot zone.

Vertical resistivity variations are calculated from the measured rock porosities, fluid temperatures, and water compositions in the initial wells. The results are compared with the resistivity structure deduced from electrical soundings. An example of this comparison, taken from Hochstein (1971), is shown in Fig. 6.4. After the completion of several investigation wells, the lateral resistivity structure, obtained from surface soundings, is compared with the actual stratigraphy and structure. Interpretations made from surface measurements in un-drilled zones of the system are then modified.

Samples of rock cores taken from the wells are tested for density,

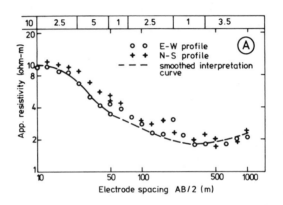

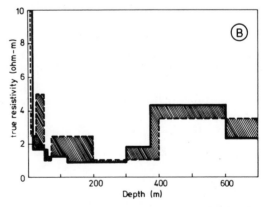

Fig. 6.4 Resistivity soundings (A) and resistivity versus depth columns (B) near well 4, El Tatio, Chile. Solid lines in B are computed from cores and temperatures and broken lines are interpreted from E-W resistivity soundings (from Hochstein, 1971).

magnetic susceptibility, and thermal conductivity, the values being compared with those used for the preliminary interpretation of results from seismic, gravity, and magnetic surveys. The model of the geothermal system deduced from the initial surveys is modified in the light of the structures revealed by drilling.

Geochemical Assessment

The chemical compositions of fluids at different depths are obtained from the exploration wells. The results are compared with the spring and fumarole chemistry and this, together with differences between the chemistry of wells, helps in selecting further well sites.

The initial discharge from a well is usually contaminated with drilling fluids and rock debris, but steam or water from a well with high mass output will usually approach a steady chemical composition within a few hours. For a well of smaller output a constant composition may not be achieved for many days. For example, Table 6.3 shows the change in chloride concentration with time following the initial opening of well 2, Broadlands, New Zealand. The discharge enthalpy remained approximately constant during the discharge period.

Where several hot aquifers occur at different levels, the discharge composition of a well may change with the rate of discharge. For example, Mercado (1969) found that for well M-20, Mexicali, there was an upward trend in Na/K ratios in discharged water as the control orifice was increased in diameter. At high discharge rates the well apparently obtained a high proportion of its supply from shallow aquifers containing water of a lower temperature than at deeper levels.

Chemical constituents are used for assessing the distance that water or steam has traveled from a source (extent of rock–water interaction) and in this respect lithium, bicarbonate, and hydrogen sulfide are of particular importance. A comparison of the ratios of unreactive constituents, such as Cl/B, Cl/As, and Cl/Cs, may enable a correlation between well discharges and particular hot spring flows.

TABLE 6.3

Chloride Concentrations in Waters Discharged from Well 2, Broadlands, during Initial Opening[a]

Time (h)	0.5	1	2	4	6	8	(7 days)
Cl⁻ (ppm)	1180	1300	1400	1430	1514	1630	1750

[a] Concentrations at atmospheric pressure and 100°.

The Na/K ratio, the concentrations of silica and magnesium, and various light element isotope ratios in the water discharged from a well are used to calculate the temperature of the water supply. This temperature is not always obvious, or obtainable, from downhole temperature measurements. The temperature estimates are compared with the equilibrium temperatures measured in the well before discharging to suggest the depth and formations from which the water originates. The presence of two different hot waters in a well discharge may be recognized from the downhole sampling results.

The temperature-dependent silica solubility reaction is rapidly reversible but in comparison, the aluminosilicate reactions controlling the Na/K ratio are much slower to adjust to new temperatures. Silica concentrations are thus directly related to water temperatures in a particular zone, whereas Na/K ratios may represent conditions existing at some distance, either laterally or vertically, from the well.

When temperatures calculated from Na/K ratios are higher than those estimated from silica concentrations, upflow of hot water to the production zone is indicated with cooling in transit. If both the silica and Na/K temperatures are higher than those measured in the well, higher-temperature water is nearby. This type of information is useful for siting further wells. Table 6.4 shows the temperatures measured and the Na/K ratio and silica temperatures in the exploration stage of the Broadlands area.

The solute concentrations in waters discharged from wells vary with time and with the position in the field due to physical processes occurring in the system during exploitation, including heat transfer between rock and water, dilution by other waters, and loss or gain of steam. Correlation of subsurface chloride concentrations and silica temperatures is useful for understanding the conditions and processes operating in a system before and after exploitation. Examples of this approach are given in Chapter 7 and in Section 6.4.

After several wells have been discharged, the degree of homogeneity of the deep water is determined and the extent of dilution and/or evaporation of water in various sectors is obtained. The amount of dilution undergone by the deep water in flowing from depth to various hot springs can also be assessed.

From accurate values of the deep chloride concentrations and hot water temperatures, a more exact estimate of the natural heat escaping from the area is made. A reconsideration of water flow patterns may be obtained from deuterium and ^{18}O concentrations in well waters. Tritium levels are used to assess the extent of dilution by young meteoric waters in various parts of the area (Banwell, 1963).

TABLE 6.4

Comparison of Measured Temperatures with Those Estimated from SiO_2
Concentrations and Na/K Ratios in Waters from Broadlands Wells[a]

Well number	Measured (°C) maximum well temp.[b]	Temperature (°C)	
		Na/K	SiO$_2$
1	278	230	236
2	287	278–292	262
3	281	277–288	252
4	273	268–280	260
5	244	197–200	—
6	143	192	—
7	279	252–260	270–280[c]
8	273	300	256
9	294	292	259–261
10	280	240–242	245
11	271	290	261
12	274	220	212
13	276	282	255
14	297	277	260
15	298	273	290[c]
16	277	234	184
17	291	298	256
18	290	288	241
19	272	292	267

[a] Silica temperatures have been corrected for enthalpy variations.
[b] Measurements made by Ministry of Works and Development, Wairakei, N.Z.
[c] Suspect results.

Well water analyses may show concentrations of elements, such as cesium, rubidium, lithium, and potassium, of potential economic value. More rarely, appreciable concentrations of heavy metals, such as silver, copper, lead, gold, zinc, and tungsten, may be present. Where appropriate, the economics of mineral extraction processes are examined, and these may make the exploitation of an area more attractive than would power production alone. As an example, Table 6.5 shows the concentrations of minor elements in waters from three wells (about 500 m deep) in the El Tatio system.

The interpretation of gas concentrations in a well discharge in a hot water area can be made only if the discharge enthalpy is known. Measurement by physical techniques may not be possible immedi-

TABLE 6.5

Trace Metals in Waters Discharged from Wells and Springs at El Tatio, Chile[a]

Constituent	Well number			Spring number					
	2	3	4	65	80	109	186	218	238
Magnesium	0.81	2.4	3.2	7.2	1.3	0.6	1.5	2.5	0.5
Strontium	4.9	3.8	4.5	1.0	1.12	1.3	1.19	1.15	1.14
Iron	0.05	0.06	0.61	0.55	0.07	0.06	0.03	0.06	1.10
Manganese	0.42	0.29	0.96	0.43	0.17	0.30	0.16	0.49	0.41
Copper	0.05	0.04	0.06	All less than 0.02					
Zinc	1.4	0.03	0.04	All less than 0.004					
Lead	0.1	0.1	0.1	All less than or equal to 0.06					
Rubidium	7.3	3.5	2.5	—	—	—	—	—	—
Cesium	13.3	10.9	13.4	10.0	10.8	10.9	10.9	11.7	12.6
Lithium	43	31	31	28	34	34	34	37	45
Barium	2.5	0.3	3.5	—	—	—	—	—	—
Silver	All less than 0.005								
Nickel	All less than 0.03								
Thallium	All less than or equal to 0.06								

[a] Concentrations in parts per million.

ately and a chemical method can be used (Mahon, 1966). The method involves measuring the gas concentrations (normally CO_2) in steam separated from the discharge at two different pressures. It is assumed that no heat loss occurs along the pipe through conduction, and that there is only minor conversion of thermal energy into kinetic energy in the two-phase fluid.

The enthalpy is determined from the relationship

$$E = (RH_1 - rH_2)/(R - r)$$

where E is the enthalpy of the discharge, R the ratio of concentrations of CO_2 in steam at separation pressures P_1 and P_2, H_1 and H_2 the heat contents of liquid water at pressures P_1 and P_2, and r the ratio of the latent heats of vaporization of water at pressures P_1 and P_2.

Under the normal conditions used for sampling (greater than 5% steam in the discharge) practically all of the carbon dioxide is in the steam phase. Most reliable enthalpy results are obtained when the two collection pressures differ by at least 3 bars. Table 6.6 shows a comparison between the enthalpies determined by the gas method and by physical methods on two wells at Wairakei.

In a hot water system, a well which initially discharges a steam-water mixture with an enthalpy equivalent to that of the underground water may, after a short period, discharge a steam–enriched mixture. The gas content of the deep water can only be accurately assessed under the first condition, so that the gas concentration in a well discharge should be measured immediately after the well is opened. It also may be possible to measure the gas concentrations in the deep water by use of a downhole sampling bottle before a well is discharged.

The gas compositions and isotope ratios in the steam phase are used for estimating underground temperatures and in revising interpretations of the composition of fumarole gases. Changes in the gas concentrations in individual well discharges and over the whole area are used to monitor physical and chemical changes underground (see Chapter 7).

For wells drawing on a liquid phase, complete discharge analyses are used to calculate the pH of the deep water and to interpret the interaction of the waters with various rock minerals (refer to Chapters 3 and 7).

Logging Investigation Wells

Various techniques are available for logging geothermal wells, but because many are expensive, private consultants may be employed to do the work. Some forms of logging are

(a) temperature versus depth;

TABLE 6.6

Comparison of Discharge Enthalpies Determined by Gas Method and by Physical Measurement

Wellhead pressure (bars)	P_1 (bars)	P_2 (bars)	R	E (cal/g) Gas	E (cal/g) Physical
12.9	11.78	2.04	2.24	235	222
10.4	8.99	2.38	2.05	221	222
13.5	12.53	1.63	2.07	249	250
12.7	11.92	1.63	1.84	258	250
13.3	12.26	1.97	2.15	245	250
10.9	10.56	1.63	1.52	277	278
14.6	14.30	1.77	1.72	280	278

(b)   pressure versus depth;

(c)   induction electrical logging, which gives an indication of the resistivity of various formations;

(d)   compensated formation density logging, which gives a record of rock porosity through determining the bulk rock density and the fluid contained in the rock pores;

(e)   gamma-ray neutron logging, a means of determining rock porosity;

(f)   sonic logging, a further method of determining rock porosity.

Similar information, possibly less detailed, can sometimes be obtained less expensively by other means. For example, porosity can be determined from core samples, and the resistivity of formations can be deduced from surface soundings and the chemistry of well discharges.

6.4   THE EL TATIO SYSTEM: AN EXAMPLE OF PREDEVELOPMENT GEOCHEMICAL INVESTIGATIONS

The El Tatio geothermal system lies at an altitude of 4250 m, 80 km west of Calama, in the Antofagasta province of northern Chile. The thermal area lies in a basin between the volcanos of the Andes and the Serrania de Tucle. A volcanic sequence of ignimbrites, tuffs, volcanic breccias, lavas, and interbedded sedimentary layers rests on shales of Cretaceous age.

Zeil (1959) and Trujillo (1968) mapped some 200 geysers, hot springs, and fumaroles in the area. The waters are highly saline and the rapid evaporation rate associated with the high altitudes has created extensive salt evaporates in the Tatio valley. Figure 6.5 from Trujillo (1973) and Mahon (1974) outlines the area. Cusicanqui *et al.* (1975) described the geochemical program that examined the natural thermal features, selected sites for exploration wells, and interpreted the well discharge compositions.

Chemistry of Surface Activity

The natural activity occurs mainly in four zones (Fig. 6.6): the wide northern part of the valley, where there are extensive salt flats with a line of geysers and hot springs on the western margin; along the

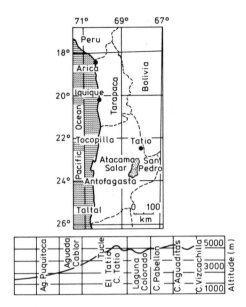

Fig. 6.5 Location of the El Tatio geothermal area, Chile, and the section along 22°65′ south latitude from longitude 67°30′ to 68°30′ west.

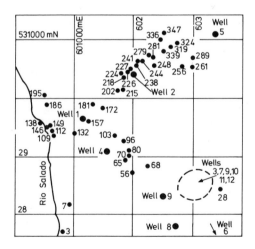

Fig. 6.6 Locations and numbers of the springs (small points) and wells (large points) sampled at El Tatio, Chile.

banks of the Rio Salado in the western valley; a zone of sinter flats and pools continuing southwest from the northern area and into the southeast valley and plateau; and about 2 km southeast of the main activity and at the base of the High Cordillera where steam-heated pools, mud pools, and fumaroles occur at the head of a stream. Most of the hot water flows drain ultimately into the Rio Salado.

The extensive sinter flats and a number of sinter cones of extinct geysers, when compared with the present-day low water output, give the impression that activity was greater in the past. However, the high evaporation rate and low boiling point (85°–86°) may be partly responsible.

There are also warm springs in a valley northeast of Copacoya and boiling springs, fumaroles, and hot water seepages occur at about 4900

TABLE 6.7

	Temp. (°C)	pH (15°)	Li	Na	K	Cs	Mg	Ca	F	Cl	SO$_4$
Spring number											
65	85.5	6.70	28	2880	145	10.0	7.2	225	—	5236	42
80	84	6.93	34	3200	165	10.8	1.3	252	—	5878	50
109	85.5	7.70	34	3300	270	10.9	0.6	245	—	5874	37
149	85	7.78	34	3430	256	11.0	1.8	256	—	6011	32
181	84.5	6.90	—	2250	230	8.0	6.0	170	—	4009	36
186	83.5	7.00	34	3380	320	10.9	1.5	313	—	6062	35
202	85.5	7.25	33	3140	336	10.2	4.2	218	—	5628	37
218	85	7.30	37	3980	475	11.7	2.5	248	—	6638	35
227	85	7.40	47	4600	520	13.0	0.3	280	2.6	8220	38
238	85	7.32	45	4340	520	12.6	0.5	272	3.1	7922	30
244	85	7.30	46	4330	525	13.2	0.4	274	—	8126	27
339	82	6.22	46	4580	525	13.0	0.5	269	2.9	8037	32
Well number											
1		7.46	30.2	4480	420	15.0	1.1	270	2.75	7943	64
2		7.36	42.0	5080	663	17.4	0.82	272	2.9	9134	49
3		7.65	31.1	3512	168	13.0	2.08	268	3.0	6160	60
4		7.45	28.0	4537	193	17.0	3.7	228	2.6	7774	62
5		7.4	44	5000	684	17.0	1.0	280	3.5	8360	45
6		8.2	17.1	1900	111	6.8	1.3	99	—	3048	177
7		7.0	45.2	4840	830	17.4	0.16	211	3.0	8790	30
10		7.05	43.9	4795	799	16.9	0.72	253	—	8986	31
11		6.98	44.9	4900	825	17.2	0.15	208	—	8716	32
12		7.56	45.1	4850	778	17.5	0.24	163	—	8450	62

[a] Concentrations in parts per million for atmospheric pressure collection.

m east of the El Tatio basin, close to the Chile–Bolivia border. The main cold water sink east of El Tatio is the Laguna Colorado, which occurs at an altitude rather lower than the El Tatio basin (see Fig. 6.5).

Extensive hydrothermal rock alteration occurs throughout the area, particularly on the El Tatio volcano. This and the deposits of salts surrounding hot water discharges outline the areas of high heat flow.

Hot Spring Chemistry

Table 6.7 gives the partial compositions of representative spring waters, while Table 6.5 contains analyses for base metals for a selection of the springs. The waters are highly mineralized (13,500 ppm NaCl) and contain unusually high concentrations of cesium (15

Chemical Compositions of Spring and Well Waters at El Tatio, Chile[a]

| | | | | | Molecular ratio | | | | | | |
| | | | | | of Na to | | | of Cl to | | | |
HCO_3	CO_2	B	SiO_2	NH_4	Li	K	Ca	B	SO_4	F	Cs
88	18	123	177	—	31	34	22.3	13.0	340	—	1960
44	5	139	174	—	28.5	19.5	22.0	13.0	320	—	2040
19	1	134	123	2.0	29.5	21	23.5	13.4	430	—	2020
12	0	136	137	—	30.5	23	23.4	13.5	510	—	2040
114	21	91	122	—	—	16.5	23.0	13.4	300	—	1880
35	4	115	142	—	32	18.0	18.8	16.0	470	—	2080
41	3	127	137	—	29	15.9	25.1	13.5	410	—	2060
40	2	149	174	2.3	32.5	14.2	28.0	13.5	510	—	2120
39	2	187	256	2.9	29.5	15.0	28.2	13.4	590	1690	2360
46	1	178	260	3.8	29	14.2	27.8	13.6	720	1370	2350
41	1	183	269	2.7	28.5	14.0	28.5	13.5	810	—	2300
36	22	182	221	3.6	30	14.8	29.6	13.4	680	1480	2310
29	1.8	173	400	1.4	44.7	18.1	28.9	13.8	335	1550	1980
53	2.0	195	452	2.2	36.5	13.6	32.6	14.3	505	1690	1965
12	—	139	238	—	34.1	35.5	22.8	13.5	280	—	1775
81	—	172	338	—	48.9	39.9	34.6	13.8	339	1600	1715
65	3.5	203	460	1.5	34.3	12.4	31.1	13.5	540	1280	1980
111	0.6	77	184	—	33.6	29.1	33.4	12.6	465	—	1620
40	5.4	203	766	2.3	32.3	9.9	40.0	13.2	800	1570	1905
40	5.5	206	740	—	32.9	10.2	33.0	13.1	786	—	1995
41	5.2	202	748	2.5	32.9	9.9	41.0	13.2	740	—	1905
69	2.2	181	655	—	32.4	10.6	51.7	14.2	370	—	1806

ppm), arsenic (45–50 ppm), and lithium (45 ppm). The frequency distribution of concentrations and of ratios for some constituents is shown in Fig. 6.7.

Apart from steam-heated and dilute waters, there is a small range of Cl/B ratios, suggesting a common source for the hot water. The constant value of 13 in the southern area suggests that the waters pass to this zone in a single flow, or else through very homogeneous rock. The scatter of ratios about a mean of 13.45 for the northern and western areas is more usual, showing minor local variations. The Cl/Cs ratio is also almost constant over the field within the possible errors in the determinations, and the Na/Li ratio also has a compact range of values. Since lithium has a slight tendency to be absorbed into secondary minerals as water cools, the asymmetrical distribution of Na/Li ratios is as expected.

Variations in chloride concentrations in the waters about the area are given in Fig. 6.8. The distribution of chloride values shows major peaks at about 8000 and 6000 ppm, corresponding with the northern geyser area and parts of the western and southern areas, respectively.

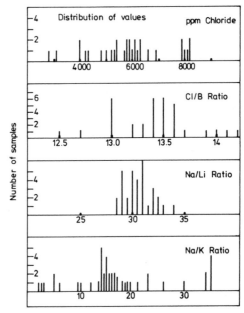

Fig. 6.7 Distribution of constituent concentrations and ratios in springs at El Tatio, Chile.

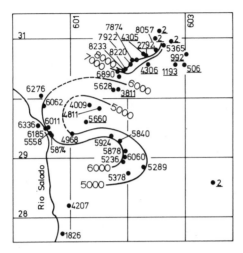

Fig. 6.8 Isochloride lines for waters of the hot spring areas of El Tatio, Chile. Underlined springs are minor and below boiling point.

The near-constant concentrations of about 8000 ppm suggest that waters in the northern area rise from depth with little dilution. Temperature indicators such as silica and Na/K ratios show that the northern springs receive the most direct supply of deep hot water and that minimum underground temperatures are 200°–220°. Rubidium concentrations follow the trends in potassium concentrations.

The major constituents of the deposits forming from the waters are silica, calcium, arsenic, and antimony, a common element suite for neutral-pH high-chloride hot spring areas. The ratios of sodium to strontium, rubidium, and cesium are about the level found in many volcanic hot spring areas. Iron and manganese are rather higher than occur in the dilute hot waters of New Zealand, but as pointed out earlier (Chapter 4), the higher salinity would account for this. Base metal concentrations are all low.

The water chemistry indicates that near-surface drainage from the east sweeps into a shallow hot aquifer underlying the western and southern areas, diluting the high-chloride water rising into it from deeper levels. The high chlorides in the low-lying western area near the Rio Salado also suggest an upflow of undiluted water. The northeast springs have the characteristics of a deep chloride water that has been considerably diluted. The eastern springs contain only minor chloride and are steam-heated surface waters.

Fumaroles

The concentrations of gases are low in the steam of the northern and western fumaroles (Table 6.8). The concentrations are approximately 10 times higher in the southern fumarole areas, but this may be to some extent due to concentration by steam condensing in the surface waters of the areas. The lower CO_2/H_2S ratios and appreciable sulfide concentrations in the large northern fumarole and at the head of the camp stream suggest that the steam originates beneath a layer of oxidation which seems to underlie parts of the system, and possibly in a high-temperature aquifer.

Isotope Results

The D/H and $^{18}O/^{16}O$ ratios in hot and cold waters were surveyed (Fig. 6.9). The hot high-chloride waters do not have δD values identical with local surface waters, but follow a trend which suggests mixing of a deep water from outside the El Tatio basin with waters of local origin. The steam-heated low-chloride waters are related to local meteoric waters concentrated in deuterium and ^{18}O by evaporation.

Water samples collected some distance east of El Tatio show a scatter of δD values and various degrees of evaporation. Although only a limited number of samples were collected, the trend in δD and $\delta^{18}O$ was toward more negative values than for cold waters at El Tatio. The isotopic composition of water discharged from wells at El Tatio (discussed later) suggest that the source water for the geothermal system could have δD and $\delta^{18}O$ values of -74 to -78 and -10.5 to

TABLE 6.8

Analysis of Fumaroles at El Tatio

Fumarole	Temp.	Millimoles of gas per 100 moles of water						Molecular ratio CO_2/H_2S
		CO_2	H_2S	N_2	O_2	CH_4	H_2	
266	86.6	75	0.6	19.8	5.2	0.02	0.0	125
233a	86.5	21	Nil	11.1	2.9	0.01	0.0	1000
140	86.4	28	0.03	7.9	2.1	0.00	0.0	1000
29	86.0	700	0.73	124	32	0.00	0.0	960
Camp stream head	86.0	360	2.4	39	11	0.00	0.0	150

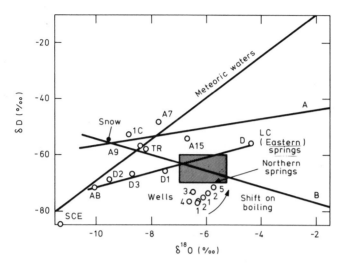

Fig. 6.9   Isotope results (δD, $\delta^{18}O$) for hot springs and geothermal wells at El Tatio, Chile. The hot springs fall within the square. Lines A and D represent evaporation; line B dilution. AB, D1, D2, D3, SCE, and LC are eastern springs; TR, 1C, A7, A9, and A15 are western springs.

−11.0, respectively. The eastern El Tatio cold water trends toward these values and it would appear that the supply water of the system originates some distance (15–20 km) to the east or southeast.

The hot waters have an appreciable ^{18}O "shift." Extensive rock-water interaction is indicated and this will be aided by the long distance of travel from the source area. Although the source water for the system may only be 15 km distant in a horizontal plane, the topography suggests that the actual distance traveled may be considerably more.

Preliminary data are available for the tritium concentrations in El Tatio waters. They indicate that the tritium content of undiluted deep Tatio water is 3.2 ± 0.2 T.U.. On the basis of a simple cyclic flow of water through the rock without dilution, the time of transit would be about 15–17 yr. However, it is possible that the positive tritium values result from some near-surface dilution by "young" waters.

Chemistry of Well Discharges

Figure 6.6 shows the positions of exploration wells (numbered 1–6) and production wells (7–13). Table 6.9 shows data relating to the wells.

TABLE 6.9

El Tatio Wells: Physical Data

Well number	R.L.[a] (m)	Depth (m)	Solid cased (m)	Max. temp. before discharge
1	4254.04	617	243	211
2	4262.41	652	295	226
3	4344.02	616	238	253
4	4263.96	733	247	229
5	4271.97	568	282	212
6	4478.99	551	401	180
7	4363.2	867	590	250
8	4335.9	1585	600	231
9	4323.3	1816	594	225
10	4343.5	1504	590	236
11	4341.4	894	575	240
12	4356.3	1410	595	247
13	4354.0	1010	501	?

[a] Relative level (to sea level).

Wells 1, 2, 4, 5, 8, 9, and 10, drilled in the western part of the area, passed through a temperature inversion at depths of 250–400 m but wells 3, 7, 11, 12, and 13 in the eastern part did not and temperatures at the base of these wells were still increasing. Well 6 in the southeast was comparatively shallow but temperatures were still increasing at the bottom (180°).

The wells penetrated nearly horizontal rock formations in most cases and only one major discontinuity was recognized, the "geyser fault" in the northwest of the field. Many of the springs and geysers in the northern part of the field occur along the surface trace of this fault. Figure 6.10 gives a typical cross section of the field.

Representative water compositions for wells are given in Table 6.7. The compositions are similar to those of spring waters and show their common origin. Initially, higher concentrations of chloride were discharged from wells 2 and 9. Trends for well 2 were followed and after some weeks the concentrations trended toward those for the northern springs.

Samples of water were taken by a downhole sampling bottle at different depths in the wells before and after discharge. A saturated brine was discovered at a depth below about 600 m in well 2 and

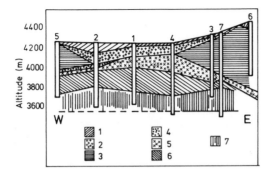

Fig. 6.10   A cross section of the El Tatio system through wells 5, 2, 1, 4, 3, 7, and 6: 1, intermontane sands and gravels; 2, Tatio ignimbrite; 3, Tucle dacite and andesite; 4, Tucle rhyolite; 5, Tucle tuffs; 6, Puripicar ignimbrite; 7, Salado ignimbrite and breccias.

evidence of this brine was also found in well 5 at a similar depth, and in well 9 and possibly 10 at deeper levels (800 m).

Figure 6.11 correlates for individual wells the deep temperatures and chloride concentrations. Temperatures were estimated by silica and Na/K geothermometers and deep chloride concentrations were calculated from surface discharges as well as from downhole sampling.

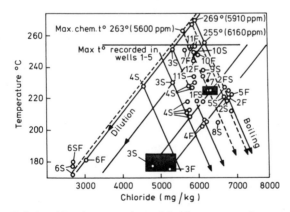

Fig. 6.11   Relationships between deep chloride concentrations and supply water temperatures in wells 1–12, El Tatio, Chile. Trend lines show the effect of boiling and dilution on temperatures and chloride concentrations of well supply waters. The lines representing boiling are straight over the temperature range in the figure. At higher temperatures the relationship between chloride concentration and temperature deviates slightly from linearity. S is a static downhole measurement; F a result from a well flow. Upper shaded area indicates probable supply water for the northern springs; lower one that for springs in the south and west.

Trend lines show the effects of boiling and dilution on the waters and the probable source of waters for the northern, southern, and western springs. The diluting water is assumed to be local surface water.

Extrapolation of the best chloride data and the maximum chemical temperatures suggested that the parent water entering the El Tatio system at about 800 m contained 5600 ppm chloride and had a temperature of 263°. Extensive testing of well 7, including the measurement of deep temperatures during discharge, subsequently showed that the water supply was 261°–263° (R. James, personal communication), although static measurements gave maximum temperatures of about 250°.

Well 7 water is reasonably representative of the parent water composition. However, the Cl/B ratio of 13.2 was rather higher than for a number of southern springs or wells 6 and 8. This may indicate a small amount of brine contaminating the water (the Cl/B ratio in the brine was 300).

The carbon dioxide and hydrogen sulfide concentrations in the water supply of well 7 were 1.8×10^{-2} and 2.1×10^{-4} molal, respectively, with a CO_2/H_2S ratio of approximately 90. The western wells in general had lower gas concentrations, except for well 4, which probably had a mixed intake of water plus steam. Ratios of carbon dioxide to hydrogen sulfide in the western wells were considerably higher (3–15 times) than for the eastern wells, suggesting fluid migration from east to west across the system. There is evidence from downhole sampling that some steam loss occurred from well 7 water at depths beneath the well and the concentration of carbon dioxide in the parent deep water has been estimated as $8 \times 10^{-2} m$ (Cusicanqui *et al.*, 1975).

Carbon dioxide makes up over 99.5% by volume of the total gases present in the deep parent water. The nonacidic gases, consisting essentially of nitrogen with small amounts of methane and hydrogen, are present to the extent of about 0.25% of the total gases.

El Tatio Brine

The occurrence of a saturated salt brine in the central part of the El Tatio area is unusual and interesting. At 25° the brine had a pH of 2.0, a density of 1.2, and a chloride concentration of $18–19 \times 10^4$ ppm. It occurred at a temperature of 180°–200° at a depth considerably below the point of boiling. Its origin is uncertain, but Table 6.10 compares its composition with that of waters from Crater Lake on the andesite

TABLE 6.10

Constituent Ratios in El Tatio Brine

Ratio	El Tatio brine	Well 7, El Tatio	Crater Lake, Ruapehu	White Island	Well IID-1, Salton Sea
Na/K	38.7	9.9	31.2	13.0	3.9
Na/Li	6100	32.3	124	—	51.0
Na/Ca	29.8	40.0	2.1	5.2	2.3
Na/Rb	12,100	2450	13,430	—	4434
Cl/B	302	13.2	244	2900	108
Cl/Cs	11,500	1900	230,000	—	34,500
SiO_2[a]	200	810	712	180	—

[a] Concentrations in parts per million.

volcano Ruapehu, New Zealand , a fumarole condensate from the active andesitic volcano on White Island, New Zealand, and water from well IID-1, Salton Sea, California. The high-density brine had a negative temperature gradient throughout its depth and it apparently diverts the hotter, less dense fluids rising from deeper levels, causing them to flow toward the eastern parts of the system.

Interpretation of El Tatio from Chemistry

The general model of the El Tatio area arising from the chemical investigations is as follows. Meteoric waters falling 15–20 km to the east or southeast of El Tatio penetrate under the Andes and flow under a hydraulic gradient toward the sea. The waters are heated within the volcanic complex and are diverted upward through permeable strata.

Hot water rising into the Tatio basin in the east separates off some steam, which reaches the surface to the east of El Tatio, creating steam-heated pools and fumaroles. The hot water entering the basin below 800–1000 m is diverted initially to the east around the brine accumulation and then to the west following the direction of rock permeability. There are two aquifers in the basin; the lower aquifer of Puripicar and Salado ignimbrites is overlain by impermeable Tucle tuffs. In the center, south, and southeast of the basin, a second aquifer, the Tucle dacite, is overlain by the impermeable Tatio ignimbrite. The Tucle dacites are not present in wells 1, 2, 4, or 5.

As the hot water moves upward toward the base of the Tucle tuffs, boiling occurs at a temperature of about 215°–230° and at a pressure of about 25 bars. Water which has cooled by boiling moves horizontally or, in zones where there is no hot water upflow, downward. The density difference between cold and hot water is appreciable, and downward movement of the cooled water may be partly responsible for the temperature inversion in the west of the basin.

Steam and gas liberated from the flowing hot water move along the top of the ignimbrite with the water, or may be released to the surface where permeability allows.

There is little dilution in the lower aquifer, particularly where the upper Tucle dacite is absent. Waters move upward through permeable zones in the Tucle tuffs and are discharged into the Tucle dacite or through the Tatio ignimbrite. The Tucle dacite contains cold water flowing from the western side of the Andes and this mixes with the rising hot water. The temperature of the mixed water in the dacite aquifer is variable. In the central basin the chloride concentration and water temperature suggest that there is twice as much hot water passing into this aquifer as there is cold water moving through it. The amount of cold water and the degree of dilution increase to the southeast at well 6.

The broad picture of the El Tatio hydrology and fluid compositions was developed from the initial survey of natural activity and formed a good base on which to recommend the siting of the prospecting wells. The complementary nature of the chemistry and isotope chemistry studies was an important feature of the work in the area.

REFERENCES

Alfano, L. (1960). *Quad. Geofis. Appl.* **21,** 3.

Arnorsson, S. (1970). *Geothermics (Spec. Issue 2)* **2** *(Pt. 1),* 547.

Arnorsson, S., Bjornsson, A., Gislason, G., and Gudmundsson, D., (1975). *Proc. U.N. Symp. Develop. Use Geothermal Resources, 2nd, San Francisco, California, May* **2,** 853.

Baba, K. (1975). *Proc. U.N. Symp. Develop. Use Geothermal Resources, 2nd, San Francisco, California, May* **2,** 856.

Baldi, P., and Ferrara, G. (1975). *U.N. Symp. Develop. Use Geothermal Resources, 2nd, San Francisco, California, May,* Abstract III-5.

Banwell, C. J. (1963a). *N.Z. J. Geol. Geophys.* **6,** 52.

Banwell, C. J. (1963b). *In* "Nuclear Geology on Geothermal Areas." (E. Tongiorgi, ed.), pp. 95–112. Consiglio Nazionale Delle Ricerche, Lab. di Geol. Nucl., Pisa.

Banwell, C. J., and Gomez Valle, R. (1970). *Geothermics (Spec. Issue 2)* **2** (Pt. 1), 27.

Banwell, C. J., and Macdonald, W. J. P. (1965). *Commonwealth Min. Metall. Congr., 8th Aust.* N.Z. Paper 213, 1.

Banwell, C. J., Cooper, E. R., Thompson, G. E. K., and McKree, K. J. (1957). "Physics of the N.Z. Thermal Area." N.Z. Dept. Sci. and Ind. Res. Bull. 123.

Battini, F., and Menut, P. (1961). *Proc. U.N. Conf. New Sources Energy, Rome* **2**, 73.

Beck, A. C., and Robertson, E. I. (1955). "Geothermal Steam for Power in New Zealand." Dept. Sci. and Ind. Res. Bull. 117.

Benseman, R. F. (1959a). *J. Geophys. Res.* **64**, 123.

Benseman, R. F. (1959b). *J. Geophys. Res.* **64**, 1057.

Bodvarsson, G. (1951). *Timarit Verkfraed Ingafelags. Isl.* **5**, 1.

Bolgerina, E. F. (1975). *U.N. Symp. Develop. Use Geothermal Resources, 2nd, San Francisco, California, May* Abstract IV-5.

Bowman, H. R., Hebert, A. J., Wollenberg, H. A., and Asaro, F. (1975). *Proc. U.N. Symp. Develop. Use Geothermal Resources, 2nd, San Francisco, California, May* **1**, 699.

Browne, P. R. L. (1970). *Geothermics (Spec. Issue 2),* **2** (Pt. 1), 564.

Cassinis, R. (1961). *Quad Geofis. Appl.* **21**, 28.

Cataldi, R., Ferrara, G. C., Stefani, G., and Tongiorgi, E. (1969). *Bull. Volcanol.* **31**, 1.

Celati, R., Ferrara, G., and Panichi, C. (1975). *U.N. Symp. Develop. Use Geothermal Resources, 2nd, San Francisco, California, May* Abstract III-11.

Combs, J., and Rotstein, Y. (1975). *Proc. U.N. Symp. Develop. Use Geothermal Resources, 2nd, San Francisco, California, May* **2**, 909.

Coury, G. E. (1975). *U.N. Symp. Develop. Use Geothermal Resources, 2nd, San Francisco, California, May* Abstract VI-11.

Cusicanqui, H., Mahon, W. A. J., and Ellis, A. J. (1975). *Proc. U.N. Symp. Develop. Use Geothermal Resources, 2nd, San Francisco, California, May* **1**, 703.

Dawson, G. B. (1964). *N.Z. J. Geol. Geophys.* **7**, 156.

Dawson, G. B., and Dickinson, D. J. (1970). *Geothermics (Spec. Issue 2),* **2** (Pt. 1), 466.

Dickinson, D. J. (1975). *Proc. U.N. Symp. Develop. Use Geothermal Resources, 2nd, San Francisco, California, May* **2**, 955.

Dominco, E., and Samilgil, E. (1970). *Geothermics. (Spec. Issue 2)* **2**, 553.

Ellis, A. J. (1970). *Geothermics (Spec. Issue 2),* **2** (Pt. 1), 516.

Ellis, A. J., and Wilson, S. H. (1955). *N.Z.J. Sci. Technol.* **36**, 622.

Fancelli, R., Nuti, S., and Noto, P. (1975). *U.N. Symp. Develop. Use Geothermal Resources, 2nd, San Francisco, California, May* Abstract III-23.

Finlayson, J. B. (1970). *Geothermics (Spec. Issue 2)* **2** (Pt. 2), 1344.

Fournier, R. O., and Truesdell, A. H. (1970). *Geothermics (Spec. Issue 2),* **2** (Pt. 1), 529.

Fournier, R. O., and Truesdell, A. H. (1973). *Geochim. Cosmochim. Acta* **37**, 1255.

Fournier, R. O., and Truesdell, A. H. (1975). *J. Res. U.S. Geol. Surv.* **2** (No. 3), 263.

Fujii, Y., and Akeno, T. (1970). *Geothermics (Spec. Issue 2)* **2** (Pt. 2), 1416.

Glover, R. B. (1970). *Geothermics (Spec. Issue 2)* **2** (Pt. 2), 1355.

Gupta, M. L., Saxena, V. K., and Sukhija, B. S., (1975). *Proc. U.N. Symp. Develop. Use Geothermal Resources, 2nd, San Francisco, California, May* **1**, 741.

Gutsalo, L. K. (1975). *Proc. U.N. Symp. Develop. Use Geothermal Resources, 2nd, San Francisco, California, May* **1**, 745.

Hamilton, R. M., Smith, B. E., and Knapp, F. (1972). U.N.D.P. Programme Kenya. U.N.D.P. Rep, New York.

Hayakawa, M. (1970). *Geothermics (Spec. Issue 2),* **2** (Pt. 1), 459.

Hayakawa, M., and Mori, K. (1962). *Bull. Geol. Surv. Jpn.* **13**, 643.

Hayakawa, M., Takaki, S., and Baba, K. (1967). *Bull. Geol. Surv. Jpn.* **18**, 73.

Healy, J. (1970). *Geothermics (Spec. Issue 2),* **2** (Pt. 1), 571.

Hermance, J. F., Thayer, R. E., and Bjornsson. (1975). *Proc. U.N. Symp. Develop. Use Geothermal Resources, 2nd, San Francisco, California, May* **2**, 1037.

Hochstein, M. P. (1971). Survey for Geothermal Development in Northern Chile. U.N.D.P. Rep, U.N., New York.

Hochstein, M. P. (1975). *Proc. U.N. Symp. Develop. Use Geothermal Resources, 2nd, San Francisco, California, May* **2,** 1049.

Hochstein, M. P., and Dickinson, D. J. (1970). *Geothermics (Spec. Issue 2),* **2** (Pt. 1), 420.

Hochstein, M. P., and Hunt, T. M. (1970). *Geothermics (Spec. Issue 2),* **2** (Pt. 1), 333.

Hodder, D. T. (1975). *U.N. Symp. Develop. Use Geothermal Resources, 2nd, San Francisco, California, May* Abstract III-41.

Hoover, D. B., and Long, C. L. (1975). *Proc. U.N. Symp. Develop. Use Geothermal Resources, 2nd, San Francisco, California, May* **2,** 1059.

Hunt, T. M. (1970). *Geothermics (Spec. Issue 2),* **2** (Pt. 1), 487.

Ingersoll, L. R., Zobel, O. J., and Ingersoll, A. C. (1955). "Heat Conduction," p. 325 Thames and Hudson, London.

Isherwood, W. F. (1975). *Proc. U.N. Symp. Develop. Use Geothermal Resources, 2nd, San Francisco, California, May* **2,** 1065.

Iyer, H. M., and Hitchcock, T. (1975). *Proc. U.N. Symp. Develop. Use Geothermal Resources, 2nd, San Francisco, California, May* **2,** 1075.

James, R. (1970). *Geothermics (Spec. Issue 2)* **1,** 91.

James, R. (1975a). *Proc. U.N. Symp. Develop. Use Geothermal Resources, 2nd, San Francisco, California, May* **3,** 1685.

James, R. (1975b). *Proc. U.N. Symp. Develop. Use Geothermal Resources, 2nd, San Francisco, California, May* 1693.

Kartokusumo, W., Mahon, W. A. J., and Seal, K. E. (1975). *Proc. U.N. Symp. Develop. Use Geothermal Resources, 2nd, San Francisco, California, May* **1,** 757.

Kavlakoglu, S. (1975). *Proc. U.N. Symp. Develop. Use Geothermal Resources, 2nd, San Francisco, California, May* **3,** 1713.

Keller, G. V., (1970). *Geothermics (Spec. Issue 2),* **2** (Pt. 1), 318.

Keller, G. V., and Frischknect, F. C. (1966). Electrical Methods in Geophysical Prospecting." Pergamon, Oxford.

Klyen, L. E. (1973). *Geothermics* **2** (No. 2), 57.

Koga, A. (1970). *Geothermics (Spec. Issue 2)* **2** (Pt. 2), 1422.

Koga, A., and Noda, T. (1975). *Proc. U.N. Symp. Develop. Use Geothermal Resources, 2nd, San Francisco, California, May* **1,** 761.

Lloyd, E. F. (1959). *N.Z. J. Geol. Geophys.* **2,** 141.

Lowell, R. P. (1975). *Proc. U.N. Symp. Develop. Use Geothermal Resources, 2nd, San Francisco, California, May* **3,** 1733.

Lumb, J. T., and Macdonald, W. J. P. (1970). *Geothermics (Spec. Issue 2),* **2** (Pt. 1), 311.

Maasha, N. (1975). *Proc. U.N. Symp. Develop. Use Geothermal Resources, 2nd, San Francisco, California, May* **2,** 1103.

Mahon, W. A. J. (1966). *N.Z. J. Sci.* **9,** 791.

Mahon, W. A. J. (1970). *Geothermics (Spec. Issue 2)* **2** (Pt. 2), 1310.

Mahon, W. A. J. (1974). Survey for Geothermal Development in Northern Chile. U.N.D.P. Rep., New York.

Mahon, W. A. J. (1975). *Proc. U.N. Symp. Develop. Use Geothermal Resources, 2nd, San Francisco, California, May* **1,** 775.

Marsh, S. E., Honey, F., and Lyon, R. J. P. (1975).*Proc. U.N. Symp. Develop. Use Geothermal Resources, 2nd, San Francisco, California, May* **2,** 1135.

Mazor, E. (1975). *Proc. U.N. Symp. Develop. Use Geothermal Resources, 2nd, San Francisco, California, May* **1,** 793.

Mazzoni, A., and Breusse, J. J. (1954). *Proc. Int. Geol. Congr., 19th, Algiers* Section 15, 17, 161.

McEuen, R. B. (1970). *Geothermics (Spec. Issue 2),* **2** (Pt. 1), 295.

McKenzie, W. F., and Truesdell, A. H. (1975). *U.N. Symp. Develop. Use Geothermal Resources, 2nd, San Francisco, California, May* Abstract III-65.

McKenzie, W. F., Thompson, J. M., and Truesdell, A. H. (1975). *U.N. Symp. Develop. Use Geothermal Resources, 2nd, San Francisco, California, May* Abstract III-66.

Meidav, T. (1970). *Geothermics (Spec. Issue 2),* **2** (Pt. 1), 303.

Mercado, S. (1969). *Geol. Soc. Am. Bull.* **80,** 2623.

Mercado, S. (1970). *Geothermics (Spec. Issue 2)* **2** (Pt. 2), 1367.

Modriniak, N., and Studt, F. E. (1959). *N.Z. J. Geol. Geophys.* **2,** 654.

Mouton, J. (1969). *Bull. Volcanol.* **33,** 165.

Noguchi, T. (1966). *Bull. Volcanol.* **23,** 520.

Nolan, T. B., and Anderson, G. H. (1934). *Am. J. Sci.* **27,** 215.

O'Neil, J. R., Truesdell, A. H., and McKenzie, W. F. (1975). *U.N. Symp. Develop. Use Geothermal Resources, 2nd, San Francisco, California, May* Abstract III-71.

Palmason, G., Friedman, J. D., Williams Jr., R. S., Jonsson, J., Saemundsson, K. (1970). *Geothermics (Spec. Issue 2),* **2** (Pt. 1), 399.

Panichi, C., and Tongiorgi, E. (1975). *Proc. U.N. Symp. Develop. Use Geothermal Resources, 2nd, San Francisco, California, May* **1,** 815.

Ramey, H. J., and Gringarten, A. C. (1975). *Proc. U.N. Symp. Develop. Use Geothermal Resources, 2nd, San Francisco, California, May* **3,** 1759.

Ramey, H. J., Kruger, P., London, A. L., and Brigham, E. W. (1975). *Proc. U.N. Symp. Develop. Use Geothermal Resources, 2nd, San Francisco, California, May* **3,** 1763.

Risk, G. F. (1975). *Proc. U.N. Symp. Develop. Use Geothermal Resources, 2nd, San Francisco, California, May* **2,** 1185.

Risk, G. F., Macdonald, W. J. P., and Dawson, G. B. (1970). *Geothermics (Spec. Issue 2),* **2** (Pt. 1), 287.

Sigvaldasson, G. E., and Cueller, G. (1970). *Geothermics (Spec. Issue 2)* **2** (Pt. 2), 1392.

Strangeway, D. W. (1970). *Geothermics (Spec. Issue 2)* **2** (Pt. 2), 1231.

Studt, F. E. (1958). *N.Z. J. Geol. Geophys.* **1,** 219.

Studt, F. E. (1959). *N.Z. J. Geol. Geophys.* **2,** 746.

Thompson, G. E. K. (1975). Geophysics Div., D.S.I.R., N.Z. Circ. G.E.K.T., 21 September.

Thompson, G. E. K., Banwell, C. J., Dawson, G. B., and Dickinson, D. J. (1961). *Proc. U.N. Conf. New Sources Energy, Rome* **2,** 386.

Tonani, F. (1970). *Geothermics (Spec. Issue 2),* **2** (Pt. 1), 492.

Truesdell, A. H., and Fournier, R. O. (1975). *Proc. U.N. Symp. Devel. Use Geothermal Resources, 2nd, San Francisco, California, May* **1,** 837.

Truesdell, A. H., Fournier, R. O., McKenzie, W. F., and Nathenson, M. (1975). *U.N. Symp. Develop. Use Geothermal Resources, 2nd, San Francisco, California, May* Abstract III-87.

Trujillo, P. (1968). Survey for Geothermal Development in Northern Chile. Corfo Project Rep., Santiago, Chile.

Trujillo, P. (1973). Geotermico de Chile, Seminario de Conyeitsobre Energia, Santiago, Chile.

Turner, W. J. (1969). U.N.D.P. Survey for Geothermal Resources in El Salvador. U.N.D.P. Rep., United Nations, New York.

Vecchia, O. (1961). *Quad. Geofis. Appl.* **21,** 18.

Wainwright, D. (1970). *Geothermics (Spec. Issue 2),* **2** (Pt. 1), 764.

Wait, J. R. (1962). "Electromagnetic Waves in Stratified Media." Pergamon, Oxford.

White, D. E. (1970). *Geothermics (Spec. Issue 2)* **1,** 58.

Whiteford, P. C. (1970). *Geothermics (Spec. Issue 2),* **2** (Pt. 1), 478.

Whiteford, P. C. (1975). *Proc. U.N. Symp. Develop. Use Geothermal Resources, 2nd, San Francisco, California, May* **2,** 1255.

Wilt, M., and Combs, J. (1975). *Proc. U.N. Symp. Develop. Use Geothermal Resources, 2nd, San Francisco, California, May* **2,** 917.

Yuhara, K. Sekioka, M., and Ijichi, S. (1975). *Proc. U.N. Symp. Develop. Use Geothermal Resources, 2nd, San Francisco, California, May* **2,** 1293.

Zeil, W. (1959). *Abhandl. Bayer Akad. Wiss. Math-Naturwiss. Kl. No. 96* 1–14.

Zohdy, A. A. R., Anderson, L. A., and Muffler, L. J. P. (1973). *Geophysics* **38,** 1130.

Chapter 7

DATA PROCESSING AND CHANGES WITH TIME

7.1 WELL DISCHARGES AND DEEP FLUID CHEMISTRY

Necessity for Recalculation of Results

For geothermal areas producing steam only, as at Larderello or The Geysers, the chemistry of a steam sample from a well can be related with little difficulty to the conditions at the production depth.

For high-temperature water systems the chemical composition of the fluids collected from a well differs from that of the deep water. During the passage of fluid up the well, steam separates from the water with consequent cooling and concentration of its solutes, while the accompanying loss of acidic gases from the water increase its pH. The nature of the steam and water phases varies according to the separation pressure. Separate collections of water, steam, and gas samples are usually made at either atmospheric pressure or a controlled higher pressure, and the separate analytical results on these samples have to be combined to give information on the conditions at the intake level of the well.

Several types of calculations are necessary to interpret the chemical results from hot water areas. The degree of water concentration resulting from steam formation must be calculated (e.g., estimation of the aquifer temperature requires the silica concentrations in the deep water). The concentrations and molecular proportions of various gases in the total discharge are calculated to help understand steam–water separation processes in the aquifer. Their partial pressures are estimated for use in chemical equilibrium calculations and for the contribution they make to total vapor pressures. This type of calculation requires the integration of water and steam phase compositions in the correct proportions determined by the enthalpy of the well discharge.

The ratios between constituents dissolved in the water (e.g., Na/K or Na/Rb ratios) are required to check on its origin or on the homogeneity of the aquifer, as well as to estimate the deep-water temperature. The quality of the water analysis should be examined by an anion/cation balance.

The pH of the underground water in conjunction with ion concentrations is required to estimate its approach to equilibrium with various mineral phases (see Chapter 4). The calculation of the water pH requires the consideration of many interacting acid–base equilibria, the distribution of acidic gases between steam and water, and the effect of temperature on the equilibria.

Most of these calculations are not complex but are time consuming. Manual successive approximation methods as used by Ellis (1967, 1970) are too slow when the results must be interpreted from many wells in a large field such as Wairakei.

Most of the computations necessary for practical assessment of fluid compositions and temperature and for interpreting the processes within an aquifer were incorporated into a computer program written by Truesdell and Singers (1971). The input to the program includes the enthalpy of the fluid discharged from the well, steam–water separation pressures, and the chemical analyses of water and steam samples. A variation of the program also interprets the analytical results obtained from a downhole sampling device.

Example of Calculation for Hot Water System

The types of calculations incorporated in the program of Truesdell and Singers can be described by working through an example.

Assumptions in Calculations

Separate pieces of analytical information must be integrated to obtain the composition of the deep hot water supplying a well. A summation is required of the individual products of an analytical concentration of a constituent in a phase multiplied by the proportion of the phase in the discharge, to give the bulk composition of the total discharge. The application of chemical equilibrium equations and mass and charge balances for the system then enables a calculation of the chemical entities present at equilibrium in the original high-temperature fluid.

Several assumptions are made: (1) the fluid from a well is produced from a single aquifer; (2) the fluid does not gain or lose significant heat or material during its transit through the well and surface piping; and (3) chemical equilibrium exists both in the samples analyzed and in the original high-temperature fluid.

As pointed out by Truesdell and Singers, the justification for these assumptions must be investigated for individual wells. If a well produces from a single liquid phase, its discharge enthalpy and the silica concentration in the water should indicate the same aquifer temperature. Contributions to a well discharge from more than one aquifer may be recognized through anomalous water compositions or gas concentrations in the discharge (Glover, 1970; Mahon, 1970).

Estimates show that heat losses through the pipeline are negligible in comparison with the heat delivered to the surface in the fluid. Losses of even traces of solute from the discharge would soon be recognized through the buildup of scale in the pipeline. The second assumption may therefore be easily checked.

The assumption of internal equilibrium is considered reasonable. In the high-temperature aquifer fluids, saturation with quartz is maintained exactly (Chapter 4) and it would be expected that most reactions within the aquifer water would be more rapid than those requiring equilibrium with solid phases. Although in general equilibrium is expected in the solutions stored at ambient temperature for analysis, dissolved silica is present in a metastable form, but this has little effect on the calculations.

When collecting and analyzing a steam–water discharge from a geothermal well, it is usually assumed from solubility data (Chapter 4) that carbon dioxide is concentrated entirely into the steam phase when it is present to the extent of at least several percent. However, a study of wells at Wairakei with two successive stages of steam separation (Glover, 1970) revealed that due to the rapid transit of discharge through the production separators there was insufficient time for complete exsolution of carbon dioxide. It appeared that up to 2% of the total carbon dioxide in the discharge could be retained in the water phase, but the significance of this result is somewhat obscured by the progression of the following type of reactions as pressures are reduced:

$$H_3BO_3 + HCO_3^- = CO_2 + H_2BO_3^- + H_2O \qquad (7.1)$$

$$H_3BO_3 + HS^- = H_2S + H_2BO_3^- + H_2O \qquad (7.2)$$

The small uncertainties introduced will not have a significant effect on the calculation of the aquifer water pH or on the calculated underground gas pressures, within the possible errors of the collection and analysis methods.

Input Data for Well Discharge Processing

The input data consist of physical data and chemical analyses of the well discharge. For the Truesdell and Singers program the following are entered:

(1)   The enthalpy of the well discharge measured at or near the time of water sampling.

(2)   The first or higher pressure at which water was separated from steam. If no separator was used and water collection was at atmospheric pressure, this figure is zero.

(3)   The second or lower separation pressure for wells having two stages of steam separation. This figure is zero for unseparated or single separator wells.

(4)   The collection pressure for the water samples. This is usually atmospheric pressure absolute.

(5)   The CO_2 content in the steam at the gas plus steam separation pressure, in millimoles per 100 moles of water. A CO_2 analysis is ordinarily necessary for the calculation of the aquifer pH.

(6)   The H_2S content of steam. The conditions and units are as for CO_2.

(7)   The pressure for the gas plus steam collection. Frequently this is equal to the higher water separation pressure, with the steam sample being taken from a large production separator, but if a sampling minicyclone was used, its operating pressure is entered.

(8)   The temperature of the pH measurement.

(9)   The pH of the water measured in the laboratory.

(10)   The analytical concentration, in milligrams per kilogram, of constituents in the water sample. These are entered in the order Li, Na, K, Rb, Cs, Ca, Mg, Ca + Mg, F, Cl, Br, I, SO_4, HBO_2, SiO_2, As, NH_3, HCO_3, CO_3, H_2S. For minor or noncritical elements in the calculation, zero is written if no analysis is available. The Ca + Mg entry is required in New Zealand for early analyses in which low Mg concentrations could not be distinguished by titration procedures.

For the example that follows, the input sheet is shown in Table 7.1. The unfortunate split between English and metric units results from separate records being kept by engineers and scientists in the New Zealand project.

Output from Program

Tables 7.2a and 7.2b are printouts from the program using the preceding set of analyses for a Wairakei well. Referring to Table 7.2a,

TABLE 7.1

Input Sheet of Physical Data and Chemical Analyses of Well Discharge[a]

Wairakei Well XY; 31 January 1965; W.A.J. Mahon

Discharge enthalpy (Btu/lb)	Water sep press. 1 (psig)	Water sep. press. 2 (psig)	Water col. press. (psia)	CO_2 (in steam) (mmole/100 moles)	H_2S (in steam) (mmole/100 moles)	Gas sep. press. (psig)	Water analysis temp.
490	180	0	14.2	200	6.5	180	20
pH (lab) 8.40	Li (ppm) 14.2	Na 1320	K 225	Rb 2.8	Cs 2.5	Ca 17.0	
Mg 0.03	Ca + Mg 0	F 8.3	Cl 2260	Br 6.0	I 0.2	SO_4 36	
HBO_2 117.0	SiO_2 660	As 4.8	NH_3 0.2	HCO_3 25.0	CO_3 1.0	H_2S 1.0	

[a] The program will not run if enthalpy, collection pressure, CO_2, gas separation pressure, water analysis temperature, pH, SiO_2, and HCO_3 are not filled in. If data are not available for other quantities, estimate or fill in (0) zero.

TABLE 7.2a

Printout from Program Using Input Data of Table 7.1

```
WAIRAKEI WELL XY = 31/1/45 = W.A.J. MAHON

   3 ITERATIONS       2 ITERATIONS

TOTAL ENTHALPY              490.000 BTU/-3= 272.222 CAL/GM      SILICA TEMPERATURE   246.841 DEG C
EXCESS ENTHALPY             29.9375 BTU/-3= 16.6320 CAL/GM      SILICA PRESSURE       37.6524 BAR ABS= 531.763 PSIG
WATER COLLECTION PRESSURE    .979090 BAR ABS= 14.1968 PSIA
WATER SEPARATION PRESSURES  13.3901 BAR ABS= 179.960 PSIG       .979090 BAR ABS= .000000 PSIG

PERCENT STEAM AT HIGHEST WATER SEPARATION PRESSURE  14.1972  BY WEIGHT     96.1221 BY VOLUME
PERCENT STEAM IN AQUIFER                             4.01948  BY WEIGHT     54.0659 BY VOLUME


ANALYTICAL CONCENTRATIONS IN MG/KG AT 20.0000 DEGREES C
PHM      LI       NA       K        RB       CS       CA       MG       CA+MG    F        CL       B        I        SO4      H3O2
8.40000  14.2000  132.000  225.000  2.60000  2.55000  17.7000  .030000  .000000  8.30000  2260.00  5.00000  .200000  36.0000  117.000

SIO2     AS       NH3      HCO3     CO3      H2S      H3SIO4   B02      HS       HF       HSO4     H4SIO4   NH4      NACL
660.000  4.80000  .20000   25.0000  1.00000  1.00000


ORIGINAL CONCENTRATION IN AQUIFER WATER PHASE IN MG/KG
PHM      LI       NA       K        RB       CS       CA       MG       CA+MG    F        CL       B        I        SO4      H3O2
8.40000  10.1725  245.425  151.152  2.00544  1.79057  12.1759  .021497  .000000  5.94471  1615.58  4.29733  .143246  25.7843  83.7989

SIO2     AS       NH3      HCO3     CO3      H2S      H3SIO4   B02      HS       HF       HSO4     H4SIO4   NH4      NACL
472.712  3.43770  .143246  17.9057  .715230  .715230
```

MOLAR RATIOS

NA/K = 9.77333   CL/F   = 145.920   -LOG(NA/K)=   .998839   LOG(NA/K) TEMP=
NA/LI= 28.0545   CL/SO4   = 170.603   -LOG(NA/K)+1/3_LOG(0.5*CA/NA)=   .874637   LOG(NA/K)+1/3_LOG(0.5*CA/NA) TEMP= 255.839
NA/CA= 155.339   CL/B   = 23.8555   -LOG(NA/K)+4/3_LOG(0.5*CA/NA)=   .502029   LOG(NA/K)+4/3_LOG(0.5*CA/NA) TEMP= 255.644
NA/MG= 1769.00   CL/AS   = 993.459   TEMP= 327.500
CA/MG= 343.627   CL/HCO3= 153.578
                 CL/BR   = 347.500

RECIPROCAL SILICA TEMPERATURE (DEGREES ABS*1000)= 1.92307

GAS SEPARATION PRESSURE   180.010 PSIG= 13.3203   BARS ABS
PERCENT STEAM AT GAS SEPARATION PRESSURE   16.1967 BY WEIGHT   70.1217 BY VOLUME

GAS CONCENTRATIONS IN MMOLES/100 MOLES H2O
CONCENTRATION CO2   200.000   CONC IN TOTAL DISCH CO2   32.6707   PRESSURE CO2 (BAR ABS)   2.63743e-01
CONCENTRATION H2S   6.50000   CONC IN TOTAL DISCH H2S   1.08776   PRESSURE H2S (BAR ABS)   6.94792e-03
CO2 CONC IN WATER   4.90648   RATIO CONC CO2/CONC H2S   30.7592
PCO2 IF NO EXCESS STEAM   1.75574
TOTAL PRESSURE   37.9231   BARS ABS

MEQ CATIONS = 66.4092   MEQ ANIONS = 65.7741

CONCENTRATION AT ANALYSIS TEMPERATURE AND AQUIFER TEMPERATURE IN MMOLES PER KG
PHH   LI   NA   K   RB   CS   H3   CA   MG   CA+MG   F   I   BR   SO4   HBO2
8.40000 2.05502 57.5579 5.77116 .032914 .019900 .120219 .001227 .420219 .001227 .050177 .439894 64.0515 .075447 .001583 .324901 2.30560
7.27159 1.47053 39.6072 4.02544 .023343 .013518 .284798 .050177   .307826 44.3318 .053966 .001133 .082418 1.89870

SIO2   AS   NH3   HCO3   CO3   H2S   H3SIO4   BO2   H2SIO4   HS   HF   MG124   NH4   MGOH   NACL
11.0517 .064370 .030862 .419717 .009568 .000928 .621219 .379159 .000000 .023562 .000002 10.4301 .010944 .000000 .056857
7.90505 .046042 .038877 .364929 .000230 .126630 .051961 .020777 .097339 .006096 .007935 7.85195 .000168 .000007 1.39982

KCL   MGSO4   CASO4   KSO4   NASO4   CACO3   MGOH   H2CO3   M2SIO4
.005772 .000012 .024991 .000597 .046691 .000000 .000001 .003406 .000340
.093448 .000461 .015515 .155702 .006096 .000249 .003406 2.18348 .000380

LOG (I/H) = 4.29507
LOG CA/H*2 = 10.4335
LOG NA/H = 5.71194
LOG MG/H*2 = 7.28182
LOG K /H = 4.77772

CALCIUM BICARBONATE SOLUBILITY PRODUCT (MOLES 3 /(KG 3 BARS)) = 1.03583e-10

255

TABLE 7.2b

Printout from Program Using Input Data of Table 7.1

WAIRAKEI WELL XV - 8/21/65 - W.A.J. MAHON
CALCULATED WITHOUT EXCESS ENTHALPY

0 ITERATIONS   0 ITERATIONS

TOTAL ENTHALPY	460.001 BTU/LB= 255.500 CAL/GM	SILICA TEMPERATURE	246.893 DEG C
EXCESS ENTHALPY	-.006395 BTU/LB=-.003553 CAL/GM	SILICA PRESSURE	37.6600 BAR ABS= 531.873 PSIG
WATER COLLECTION PRESSURE	.579090 BAR ABS= 14.1948 PSIA		
WATER SEPARATION PRESSURES	13.3901 BAR ABS= 179.940 PSIG	.979090 BAR ABS= .000000 PSIG

PERCENT STEAM AT HIGHEST WATER SEPARATION PRESSURE 12.6602 BY WEIGHT   94.8953 BY VOLUME
PERCENT STEAM IN AQUIFER   .000000 BY WEIGHT   .000000 BY VOLUME

ANALYTICAL CONCENTRATIONS IN MG/KG AT 20.000 DEGREES C

PPM	LI	NA	K	RB	CS	CA
8.41000	14.2000	1327.00	225.000	2.80000	2.50000	17.0000

MG	CA+MG	F	CL	SO4	I	HBO2
.030000	.000000	3.39000	2260.00	5.00000	.200000	36.3000	117.000

SIO2	AS	NH3	HCO3	CO3	H2S	H3SIO4
660.000	4.65000	.200000	25.0000	1.00000	1.00000	

BO2	HS	HSO4	HCL	NACL

ORIGINAL CONCENTRATION IN AQUIFER WATER PHASE IN MG/KG

PPM	LI	NA	K	RB	CS	CA
8.41000	10.1737	945.721	161.202	2.00607	1.79114	12.1797

MG	CA+MG	F	CL	SO4	I	HBO2
.021494	.000000	5.94658	1619.19	4.29573	.143291	25.7924	83.9253

SIO2	AS	NH3	HCO3	CO3	H2S	H3SIO4
472.860	3.43598	.143291	17.9114	.716455	.716455	

BO2	HS	HSO4	HCL	NACL

MOLAR RATIOS
NA/K = 9.97333   CL/F = 145.920   LOG(NA/K)=   .998839   LOG(NA/K) TEMP=   255.839
NA/LI = 28.0346   CL/SO4 = 170.002   -LOG(NA/K)+1/3LOG(0.5+CA/NA)= .574614   LOG(NA/K)+1/3LOG(0.5+CA/NA) TEMP= 255.647
NA/CA= 135.339   CL/B = 23.8556   -LOG(NA/K)+4/3LOG(0.5+CA/NA)= .501935   LOG(NA/K)+4/3LOG(0.5+CA/NA) TEMP= 327.520
NA/RB= 1749.00   CL/AS = 993.458
CA/MG= 343.627   CL/HCO3= 155.578
                     CL/BR = 547.500

RECIPROCAL SILICA TEMPERATURE (DEGREES ABS*1000)= 1.72303

GAS SEPARATION PRESSURE   150.020 PSIG= 13.3915   BARS ABS
PERCENT STEAM AT GAS SEPARATION PRESSURE   12.6593   BY WEIGHT   24.5943   BY VOLUME

GAS CONCENTRATIONS IN MMOLES/100 MOLES H2O
CONCENTRATION CO2   200.300   CONC IN TOTAL DISCH CO2   25.0249   PRESSURE CO2 (BAR ABS)   1.37747=00
CONCENTRATION H2S   6.50000   CONC IN TOTAL DISCH H2S   .859857   PRESSURE H2S (BAR ABS)   1.63257=-02
CO2 CONC IN WATER   25.6248   RATIO CONC CO2/CONC H2S   30.7592
PCO2 IF NO EXCESS STEAM 1.37712
TOTAL PRESSURE   39.0538   BARS ABS

MEQ CATIONS = 66.4032   MEQ ANIONS = 65.7741

CONCENTRATION AT ANALYSIS TEMPERATURE AND AQUIFER TEMPERATURE IN MMOLES PER KG
PH%   LI   NA   K   F   CL   SO4   I   804   H802
8.40000 2.05602 57.5578 5.77416 4.02775 .018900 .420249 .287240 64.0515 .077447 .001393 .324901 2.39560
6.55895 1.47110 39.7101   .013523 .023350   .006230   44.5445 .053593 .001133 .061983 1.91627

GAS CONCENTRATIONS
CO2   H3SIO4   H2S   B02   HS   MG504   N4SI04   S04   I   NACL
         .620249 .001227 .379198 .029962 .000000 10.4301 .324901 .000000 .056857
.439894 .283657 .006230 .003551 .062162 .001441 7.68742 .061983 .000037 1.39074

SI02   AS   NH3   HCO3   CO3   H2S   CA   MG   H2SIO4   MGOH   H2C03   H2SI04
31.0517 .064370 .000062 .419717 .008558 .000928 .420249 .000000 .621219 .000001 .003406 .000104
7.90735 .045957 .007614 .970721 .000049 .136151 .287240 .000000 .009672 .000000 13.3579 .000013

KCL   MG04   CASO4   NA504   KS04   HCO3   CO3   CASO3
.005772 .000012 .004901 .045981 .004591 .970936 .003936 .000000
.093523 .000587 .017003 .155923 .019441 .000666

MOLAL RATIOS
LOG LI/H = 3.56254
LOG CA/H+1 = 5.99152
LOG NA/H = 4.97741
LOG MG/H+1 = 5.93099
LOG K/H = 3.97519

CALCIUM BICARBONATE SOLUBILITY PRODUCT (MOLES 3 /(KG 3 BARS)) = 2.63994=-11

257

we see that it includes the temperature of the aquifer fluid, estimated from silica concentrations. In doing this estimation, the silica content in the deep water must be calculated from the water analysis, allowing for the concentration caused by steam separation to maintain the liquid–vapor equilibrium at particular pressures. If, as in the present case, the well discharges excess steam, this has an additional effect on the concentration of solutes in the water phase during the reduction of pressure. Since the calculation of silica concentrations (and other solutes) in the deep water then requires enthalpy values for saturated steam and water at the aquifer temperature, an iteration procedure is necessary.

In addition to the silica temperature, the pressure of water vapor at this temperature and the excess enthalpy (excess steam) in the discharge are calculated (but, as discussed later, there are uncertainties in the measured enthalpy).

The program calculates the weight and volume percent of steam at the highest water separation pressure, as well as the percentage of steam taken into the well from the aquifer. Taking into account the effect of steam boiling from the water during the reduction of pressure (and that due to the presence of excess steam), we calculate the concentration of solutes in the aquifer water; following this, the molal ratios between various solutes are estimated and printed. From the Na/K ratio and Na–Ca–K geothermometer, further estimates of the aquifer temperature are made (see Chapter 4).

In calculating the pH of the aquifer water and the concentration of various chemical species in solution, the program reconstructs the chemical conditions at the inflow point to the well by recombining the separate pieces of analytical and sampling information.

First, a chemical model of the water as analyzed is produced; it takes into account the analytical concentrations, known dissociation constants, mass balances, and the water pH measured at a temperature. The concentrations of weak acid/weak base species and of ion pairs are calculated. For each element having more than one solution species, a series of mass balance and equilibrium constant equations are solved, considering also the activity coefficients for various ionic species. The activity coefficients are calculated from the ionic strength by an iterative procedure. The total concentration of potentially available hydrogen ions in the solution is then calculated, including not only the free ions but those in combination in weak acids. The total "ionizable" hydrogen concentration is calculated by summing the concentrations of hydrogen ions and all of the undissociated and partly dissociated weak acids—HCO_3^-, HBO_2, H_2S, NH_4^+, HSO_4^-, HF,

HCl, $H_3SiO_4^-$, H_4SiO_4 (two times), and H_2CO_3 (two times); then the total ionizable hydrogen concentration in the aquifer water is estimated by correction for steam formation (and excess enthalpy concentration effects, if any). The H_2S and CO_2 concentrations in the aquifer water are then added (the CO_2 two times as H_2CO_3) to the ionizable hydrogen concentration. The H_2S and CO_2 concentrations in the aquifer water are derived from the total discharge analysis in the case of wells without excess enthalpy; for wells with excess enthalpy, they must be calculated for the original downhole conditions, taking into account the fractions of steam and water at the well intake level and the solubility coefficients of the gases.

Second, the aquifer water chemical model is calculated, using the water analysis corrected for steam separation and excess enthalpy effects, the CO_2 and H_2S concentrations in the aquifer water, the total ionizable hydrogen concentration, and values of dissociation and ionization constants at the aquifer temperature and saturated water vapor pressure. Using a series of mass and charge balance equations, the pH of the aquifer water is calculated by an iterative process which terminates when the mass balances for particular acid–base systems agree to within 0.1% of the total content. The pH and the concentrations of the various species in the aquifer water are then presented.

Ratios such as a_{Na^+}/a_{H^+} and $a_{Ca^{2+}}/a_{H^+}^2$ are given and a check on calcite saturation is made through the ratio $a_{Ca^{2+}} a_{HCO_3^-}/P_{CO_2}$. Other information printed includes the concentration of CO_2 and H_2S in the total discharge and their molecular ratio, as well as the CO_2 concentrations in the aquifer water. The pressures of CO_2 and H_2S in the aquifer are calculated (for two-phase conditions if the well has excess enthalpy, or for a single phase if there is no excess enthalpy). However, on the assumption that the enthalpy measurement could be wrong, the program always gives in addition the partial pressure of CO_2 that would result from a single-phase aquifer condition. The total pressure of water vapor plus gas pressure is printed.

It should be noted that although the concentrations of nonvolatile elements in the aquifer water can be calculated accurately by making an allowance for excess steam in the discharge, calculations of acid–base equilibria, of the pH of the aquifer water, and of gas pressures may be less meaningful. In many cases it is difficult to be certain that the excess steam entering the well was in equilibrium with the aquifer water. Derived quantities, such as the Na^+/H^+ and K^+/H^+ ratios, also should be treated with reserve for excess enthalpy discharges.

As noted in the foregoing, there are usually appreciable errors possible in measured enthalpy values. Accordingly, when the enthalpies are within ±28 cal/g (±50 Btu/lb) of saturation values for the aquifer temperature, a recalculation of all equilibria is made, assuming that there is no excess enthalpy (i.e., a single liquid phase at the well intake).

Table 7.2b shows this recalculation. Note that the silica temperature is slightly higher as a result of the higher silica concentrations calculated for the aquifer water. There are also higher concentrations and partial pressures of CO_2 and H_2S in the aquifer fluid, while the calculated aquifer water pH is lower. The CO_2 concentration in the aquifer water is equal to that in the total discharge. Derived quantities are also effected.

7.2 STORAGE AND RETRIEVAL OF DATA

Requirements

At the beginning of a geochemical program a decision is required on the way in which data will be recorded, tabulated, and stored. Preliminary spring and fumarole sampling in a new geothermal area can produce more than 1000 separate constituent results to be collated and interpreted in relation to the temperatures, flows, hydrology, etc. Chemical monitoring of production wells also produces vast quantities of data. During 20 years of geochemical work at Wairakei, where the discharges of between 60 and 70 wells have been chemically monitored six to eight times a year, over 100,000 separate analyses have been made.

More often than not, the facilities for storing data are limited in the early investigational stages of a project, and the results are tabulated on standardized report forms. These are placed on files which grow at an alarming rate, until storage, and more particularly easy recovery and reference, becomes difficult and time consuming. Faster sorting and recovery is obtained by entering results directly onto manual sorting stack cards. However, entering the data is slow and there is still the problem of storage space and of interpreting the results.

The greatest efficiency in storage, recovery, and application of results is obtained through the use of a large computer and by placing the results on magnetic tape or disc files. The transference of analytical results, physical information, and collection conditions into computer storage is still time consuming but when completed the results are highly accessible, can be recalculated for standard conditions of collection or comparison, and can be manipulated through suitable programs to greatly aid interpretation.

Example of Computer System

Palmer (1977) developed a computer program for storing, processing, and interrelating the large accumulated amount of chemical and physical data for monitoring Wairakei and Broadlands wells. The program has three main functions:

(1)   To store sets of analyses and related sampling data for steam and water samples, and to present them systematically on demand.

(2)   To process the results from steam and water samples to produce downhole concentrations of constituents, ratios, gas partial pressures, temperatures from chemical geothermometers, and various equilibrium calculations. This aspect incorporates the program of Truesdell and Singers (1971) described earlier, and any of the variables derived by this program are available for examination.

(3)   To display the stored data in the form of tables and graphs which can interrelate most of the variables concerning chemistry, temperature, well position, and time. This may be done by interactive processing using a visual display unit, or by batch job processing.

On command, the user can interrelate any of the chemical variables with position, time, or values of another variable. The variables are any of 56 physical and chemical quantities or the sum, difference, product, or ratio of two of these quantities. For example, a chemical variable may be plotted versus

(a)   time, for a well or wells;

(b)   depth, for a period of time and series of wells;

(c)   grid coordinates for a given time (i.e., in map preparation);

(d)   another variable for a series of wells (seeking general correlations);

(e)   another variable for a well over a given period.

Two examples are given from the use of this powerful processing procedure. For two Wairakei wells, Fig. 7.1 gives the variation with time in the partial pressure of carbon dioxide associated with the deep inflow water. Figure 7.2 compares, for three Wairakei wells, the temperatures obtained by the silica and sodium–potassium–calcium geothermometers during the decade 1960–1970.

A few hours of interactive processing of geothermal data by this

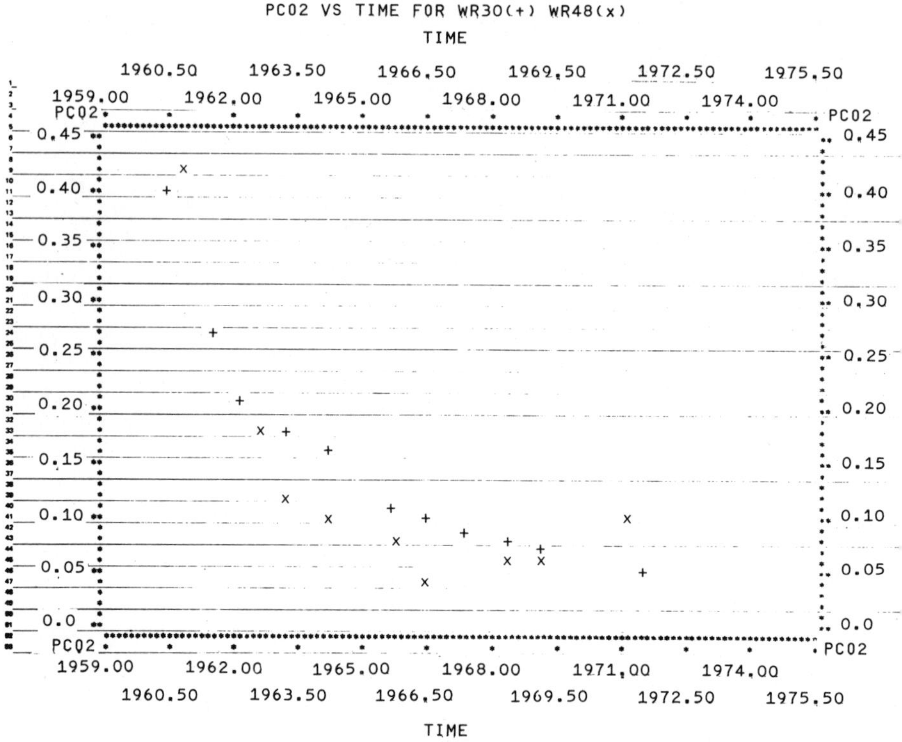

PCO2 VS TIME FOR WR30(+) WR48(x)

TIME

Fig. 7.1 Computer printout of the changes with time in the partial pressure of carbon dioxide in the aquifers supplying Wairakei wells 30 and 48.

means can quickly examine correlations between variables, changes with time, and variations within a production field.

7.3 CHANGES IN DISCHARGE CHARACTERISTICS

Chemical analysis of fluids from wells and from natural activity is used extensively in following the changes that occur within a geothermal field during exploitation. There are only a few geothermal fields in which chemical records extend back more than a few years, and probably the most extensive records are for Larderello and Wairakei.

Large quantities of fluid are produced from geothermal production wells. For example, the approximate rate of fluid production from Larderello is 2.5×10^7 and from Wairakei 5×10^7 tons/yr. During 17 years of production at Wairakei, 1 km³ of liquid water has been removed from the system. Following major fluid withdrawal, steam may be created to fill a void in the aquifer, or there may be vertical or lateral inflow to the production levels. The new fluid may or may not have the same composition as the original. Aquifer pressures may decrease and cause boiling of water to occur at increasingly deeper levels.

The removal of fluid may cause the ground level to fall if the near-surface rocks are of low strength. In the central part of the Wairakei field, the rate of fall of ground level has exceeded 0.3 m/yr (Bolton, 1973), but this tapers off to zero on the perimeter. However, the volume involved in the decrease in ground level is less than 1% of the fluid volume removed from the system over a given period (R. B. Glover, personal communication). Gravity surveys (Hunt, 1970)

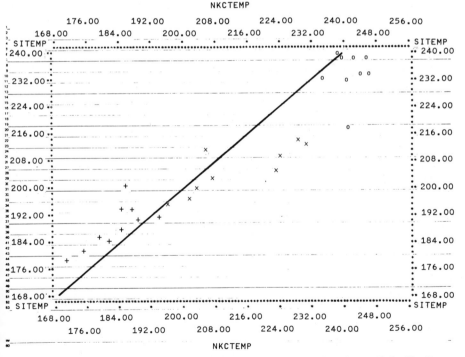

Fig. 7.2 Computer printout comparing temperatures from the silica and the Na–K–Ca geothermometers for Wairakei wells 8, 40, and 58 during the decade 1960–1970. Line added by hand.

showed that in the early stages of the Wairakei field, only a portion of the water removed by wells was replaced by inflow, but in recent years the mass inflow to the aquifer has approximately equaled that drawn off.

Within an area there may be extensive changes in the natural activity following fluid removal from the system by wells. Monitoring the physical and chemical nature of the natural activity can show the lateral underground spread of the effects from production and may also be important from a tourist or public safety viewpoint.

The rate at which physical and chemical changes occur in well characteristics varies widely between geothermal areas. It depends largely on whether the inflow to the well is from a fissure or an aquifer, the rock permeability, the volume of stored hot fluid, and in a hot water system, whether boiling is initiated at the well intake level.

The gathering of adequate geochemical records for each well in production is an important part of geothermal field management. From the time that a new well is discharged, samples of steam and water should be collected, at first hourly, then daily, until stable production conditions are attained. The subsequent frequency of collection depends on the rate of change observed. For example, at Wairakei (a very stable production field) water and steam collections are made at approximately 4- to 6-month intervals after 17 years of running, whereas in the initial trials of the Broadlands field, where appreciable changes were still occurring after several years of running, collections were made every 3–4 weeks.

Analytical information must be compared on a common basis. For steam fields this is no problem, but in hot water fields solute concentrations in waters and gas concentrations in steam vary with the conditions of collection. The results need to be recalculated to a standard collection condition, or to refer to the conditions existing at the inflow point to the well.

Changes in Water Chemistry

In fields where separate aquifers occur at different levels, well water compositions may change when the pressure in an aquifer is reduced to the extent that water flows in preference to the well from another level. For several years some shallow wells (150–300 m) in the Wairakei area produced a water with 0.7–0.8 times the salinity of the main deep-water supply (over 600 m), but gradually the more saline deep water predominated (Ellis, 1961).

Variations in underground water temperatures during production can be monitored by chemical geothermometers. With the silica method, temperatures may be obtained to within about ±2°C. The chemical methods have the advantage of measuring the temperature of the water supplying a well. Downhole measurements may require production to be ceased, and the temperatures measured under static conditions are often of questionable relevance to production.

The differing rates of reequilibration of the silica solubility and Na–K exchange reactions with changing temperatures may be used as a check on the mechanism causing the changes. For example, a temperature decrease caused by either rapid boiling or inflow of cold water may not immediately affect the Na/K ratio, whereas a mixing of two hot waters previously equilibrated at different temperatures would affect both silica and Na/K ratios.

Figure 7.3 gives as an example the trends with time in the temperatures obtained by the Na–K–Ca geothermometer for various production wells at Wairakei.

The solute concentrations in water discharged from a well are affected by many processes within the system, including dilution by other waters, loss or gain of steam, and heat conducted from rock. For wells obtaining fluid from low-permeability rocks, drawdown of water levels may cause the water to boil and solutes to be concentrated by rock-to-water heat transfer. A well which discharges water with constant solute concentrations for a long period (months, years) almost certainly obtains its supply from major fissures or a highly permeable

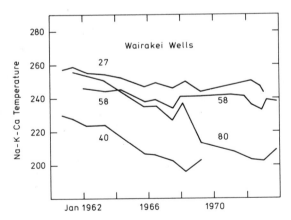

Fig. 7.3 Changes with time in Na–K–Ca temperatures for some Wairakei production wells.

TABLE 7.3

Average Chloride Concentration in Wairakei Well Water[a]

Year	1960	1961	1962	1963	1964	1965	1966	1967	1968	1969	1970
Cl^- (ppm)	2251	2258	2243	2244	2225	2230	2225	2235	2222	2223	2220

[a] From 19 production wells in the western Wairakei field. (Concentrations for separation at atmospheric pressure.)

aquifer. This is the case for the western Wairakei field, for which the chloride concentrations in waters have remained almost constant (±1%) for major production wells (see Table 7.3). In contrast, some examples are given in Fig. 7.4 of more rapid changes with time that were observed for some wells at Broadlands, where permeabilities are much lower (Mahon and Finlayson, 1972).

Wilson *et al.* (1967) evaluated carefully the chloride changes with time for wells in the Wairakei field. Their map for the year 1965 is reproduced as Fig. 7.5, showing contours of change of chloride with time. The maximum changes were on the peripheral part of the field and in the relatively impermeable area separating the western and eastern production zones. There was a zone of dilution appearing in the northeastern part of the western field.

Trends in water solute concentrations within a field and changes with time can be examined on a graph of chloride (as the most accurately determined soluble element) in the deep water supplying the well versus the water temperature obtained by the silica method.

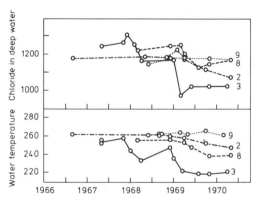

Fig. 7.4 Changes with time in chloride concentrations for some Broadlands wells (from Mahon and Finlayson, 1972).

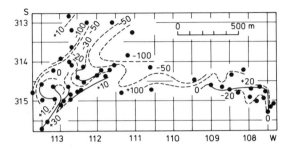

Fig. 7.5 Wairakei production area contoured for changes in chloride concentrations in waters discharged by wells during 1965. Contour figures are changes in parts per million per year (adapted from Wilson *et al.*, 1967).

This type of interpretation can add considerably to the understanding of water flow patterns in a field during production. Figure 7.6 gives the trend lines which would be expected for various processes which may occur in the aquifer.

As an example Fig. 7.7 gives the changes in water supply chloride concentration and silica temperature for three Wairakei wells during the period 1960–1972. For well 27 the changes are small and variable and seem to be dominated by a complex interplay of boiling, condensation, and conductive processes. Well 80 shows progressive dilution by a cold water during the period of observation. Well 40 had a steam plus water discharge which increased in enthalpy and eventually

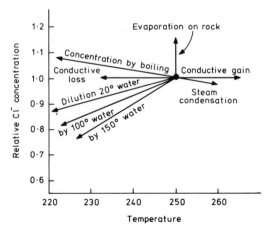

Fig. 7.6 Trend lines relating changes in deep-water chloride concentrations to various physical processes.

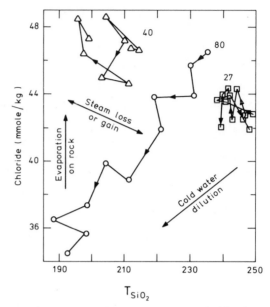

Fig. 7.7 Examples of changes in silica temperatures and chloride concentrations in deep waters supplying Wairakei wells during 1960–1972. Arrows show time direction.

produced steam alone; the changes are due mainly to loss or gain of steam together with some dilution.

Finlayson and Mahon (1976) outlined the variation in solution concentrations in the Broadlands field with position and time. The initial chloride concentrations in the deep waters of the northern part of the field were originally 1150 ppm, but in the southern sector concentrations were lower (approximately 800–1000 ppm in the deep waters). Chloride concentrations decreased rapidly at the boundary of the system. The Broadlands waters could be derived from a hypothetical deeper water at about 300°–310°, containing between 950 and 1050 ppm chloride. Figure 7.8 shows that the northern well waters could be formed from the deeper high-temperature water by boiling as it rose to the shallower levels of the wells. The southern water could be explained by a variable amount of dilution of the deep water.

Following exploitation there were changes in the temperatures and composition of waters tapped by individual wells. For many wells the chloride concentrations and supply water temperatures decreased during the first 1–3 years of discharge but the rate of decrease then in general tended to zero, with individual wells showing variable characteristics. In the northern zone, for example, wells 3 and 11 showed

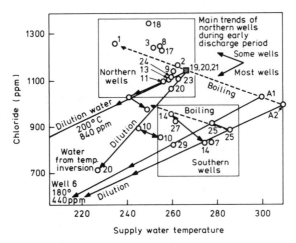

Fig. 7.8 Chloride concentrations and silica temperatures in deep waters about the Broadlands field. Mechanisms by which various waters could be derived from high-temperature original waters A1 or A2 much below the level of drilling are shown, along with some trends with time.

signs of further dilution; in contrast, well 8 increased its chloride content at constant temperature, suggesting evaporation of water on rock surfaces. The dilution waters appeared to be derived from a shallower aquifer, possibly that originally supplying the Ohaki pool.

Well 7 in the southern part of the field is in impermeable formations and has been discharged for approximately 6 years. Figure 7.9 illus-

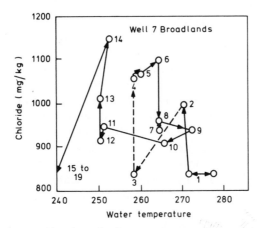

Fig. 7.9 Changes in chloride and silica temperatures for water supplying well 7, Broadlands. Numbers are a time sequence.

trates the trends with time in water temperatures and chloride concentrations for this well (obtained from both well discharges and downhole samples). Soon after opening, the water increased in chloride concentration at constant temperature, which suggests that in the vicinity of the well it evaporated on rock surfaces. The chloride and temperature subsequently decreased (possibly by dilution by water from a shallower level), and then increased again at a constant temperature lower than that initially present. Some steam heating and dilution by condensate then appears to have occurred, causing chloride to decrease again, accompanied by complex temperature changes (points 6–10). There was then a period of boiling of water (points 10–11) as pressures decreased in the well vicinity; this continued to a stage where the heat of vaporization was extracted mainly from the rock (points 12–14). Later observations showed marked decreases in chloride and temperature caused predominantly by dilution.

In an area where there are lateral variations in water compositions due to differing rock types, changes with time in ratios of "soluble" constituents, such as Cl, Br, B, and Cs, can give evidence of migration of fluid from one part of the field to another. For example, cesium concentrations are typical of a particular rhyolite pumice aquifer in the El Tatio, Chile, geothermal field. High ammonia concentrations may identify waters contacting organic-rich sediments.

In the Broadlands area the Cl/B ratio varies across the field, being lower in the east, due to greater contact of deep water with greywacke rock. For well Br-7 the Cl/B ratio fell from about 6.7 to 4.5 during several years of discharge, showing an increasing contribution of water from the eastern part of the field.

Because of the convective heating cycle involved in most geothermal fields, inevitably there is concern during utilization that increased water output will cause the system to "wash through," and that the high-temperature mineralized water may become depleted and replaced by a front of cooler and less mineralized water flowing up through the system. Whereas most of the heat in a geothermal system is stored within the rock, there is a selective extraction of many soluble chemical constituents from the rock into the hot water during hydrothermal alteration (e.g., chloride, boron, cesium). For elements of limited availability in the rocks, such as boron or cesium, changes in their concentrations in well discharge waters could give an early indication of the system's becoming depleted. Table 7.4 gives the concentrations of boron and cesium in waters from some Wairakei wells over an extended period. The changes are minor and variable

TABLE 7.4

Cesium, Boron, and Chloride Concentrations in Waters Discharged from Some Wairakei Wells at Two Widely Separated Periods[a]

Well	1956–1958			1974–1975		
	Cs	B	Cl	Cs	B	Cl
18	2.0	25.0	2150	2.0	25.5	2033
27	2.9	27.0	2180	2.4	25.4	2136
41	2.4	26.8	2115	2.1	22.9	1829
46	2.2	26.5	2220	2.7	26.1	2200
48	3.0	27.3	2280	2.6	25.2	2152
58	2.4	27.8	2115	2.5	25.3	2152

[a] Concentrations in parts per million.

(some localized dilution) and the figures give an interesting commentary on the size and general homogeneity of this hot water system.

Changes in isotopic ratios of hydrogen and of oxygen in the discharge waters can provide a useful check on a changing origin of water. Also, decreasing $^{18}O/^{16}O$ ratios would reflect a decreasing rock–water interaction. An increasing tritium concentration in the water would be viewed with suspicion, as indicating a short transit time for water in the geothermal system or increasing dilution by meteoric water.

The tritium concentration in a series of samples collected from Wairakei wells in 1960 ranged from 0.04 to 0.44 T.U. (Wilson, 1963). In 1972 the concentrations for a series of wells (not all identical with the 1960 ones) ranged from 0.14 to 0.5 T.U. (Hulston, 1975). This change is negligible, considering the errors involved in collection and processing. This finding again reemphasizes the stability of the Wairakei situation.

The changes in pH of the aquifer water due to steam and carbon dioxide separation are discussed in the next paragraphs, but in geothermal areas in regions of active volcanism (e.g., in Taiwan, where the rock systems are high in sulfur), the acidity of water produced from wells may change very considerably during production. From wells in rocks of low permeability, water of slight acidity or neutral pH may be produced first, due to neutralization by contact with rocks under stagnant conditions. Later production may draw water more directly from an aquifer containing acidic water.

With decreasing water levels in a hot water field, boiling may occur within the country either about an individual well or throughout the production zone. Carbon dioxide, hydrogen sulfide, and other gases are lost during boiling, and the water pH rises due to the loss of the acidic gases.

In theory this increased water pH should not persist, since pH-dependent mineral–solution equilibria should gradually restore it to its original value for equilibrium with the mineral assemblage. In practice, however, these mineral reactions are slow and the pH of the residual water is controlled largely by the changing solution composition. For example, Table 7.5 shows the change in the calculated average pH of the deep hot waters supplying the western Wairakei wells, from 1958 to 1970. The table includes the pH value (pH_c) at which calcite saturation occurs for the solution composition and the value of $pH_c - pH$ (i.e., the difference between the pH at which the water becomes saturated with calcite at the temperature and the actual pH of the water).

Over this period the pH of the deep Wairakei waters increased by almost 1 unit. Whereas in the early stages of development the waters were undersaturated with calcite, the waters later approached saturation because of the increasing pH. Although no problem has arisen yet with calcite deposition at Wairakei, this type of chemical interpretation gives a warning of the potential danger of mineral deposition which can be brought about by rapid drawdown of pressures.

The effect of high production from geothermal wells is usually to reduce the temperature and flows of natural hot springs within the area. At Wairakei, the flow of mineralized thermal water to the surface has become very small. Springs which had an appreciable output of high-chloride water at temperatures near boiling point have been replaced by mud pots or fumaroles. Alternatively, some water flows persist but the characteristics are those of steam-heated groundwater.

Figure 7.10 gives the changes with time in the chloride concentra-

TABLE 7.5

Average pH of Hot Waters Supplying Western Wairakei Wells[a]

	1958	1962	1964	1968	1970
pH	6.3	6.6	6.9	7.0	7.1
$pH_c - pH$	0.5	0.2	0.1	0.0	0.0

[a] pH_c is the point of saturation with calcite.

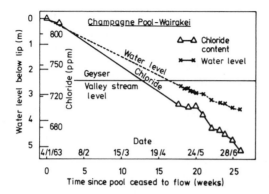

Fig. 7.10 Changes with time in the chloride concentration and water level in the Champagne Pool, Geyser Valley, Wairakei (R. B. Glover, personal communication).

tions in the water of Champagne Pool, Wairakei, which was originally a large spring of major flow, but now has a water level much below the surface.

Variations in Steam Chemistry

In a hot water geothermal area the concentrations of gas in the steam discharged from wells or fumaroles give valuable information on the conditions at deep levels. They can also give information on the direction of migration of steam or water within a field and on changes in the underground aquifer. Trends in the average enthalpy for all the well discharges in the field give complementary information on whether or not the supply is based on upward flows from very deep levels through fissures or on a closed aquifer system.

The gas concentrations and ratios from different wells can be used to identify fluids from particular aquifers or flow zones within a geothermal field. This has been done in the New Zealand fields as well as in the Italian steam region where gas/steam ratios have been found to group themselves into families of wells in an area. It has been possible to deduce the position of faults feeding steam into particular parts of the Larderello field. The gas content of steam increases outward from the Larderello area (e.g., from 27 l of gas per kilogram of steam at Larderello to 500 l of gas per kilogram of steam near Monte Amiata; Cataldi *et al.*, 1969). High proportions of hydrogen in gas from steam wells were correlated with permeable Mesozoic rocks in the south-western parts of the boriciferous steam region.

In steam reservoir systems such as Larderello or The Geysers, if the fields originally consisted of steam at shallow levels coexisting with water at a very considerable depth, it could be expected that the concentration of gases in steam would eventually decrease with time as the outflow from the fields was accelerated by wells. Also, as suggested by White (1970), the proportions of the gases in the steam should remain nearly constant if it was being obtained from storage in the reservoir, but should change when the steam is produced predominantly by boiling of the deep water. The more volatile constituents would be expected to decrease at the most rapid rate (i.e., in the order N_2, H_2, CH_4, H_2S, and NH_3). This trend was confirmed at The Geysers, where there has been an increasing proportion of the more soluble gases (CO_2, H_2S, and NH_3) and a lower proportion of the insoluble gases, such as H_2 and N_2 (Truesdell and White, 1973).

At both Larderello and The Geysers, the total gas content of the steam decreased with time from a few percent down to about 0.5% and a similar trend was observed at Matsukawa (White 1970; Celati *et al.*, 1973), where temperatures determined by the $\Delta^{13}C(CH_4 - CO_2)$ isotope exchange reaction changed from about 310°–320° in 1964–1967 to the present values of about 240°–260° (Nakai, 1976). In the Monte Amiata, Italy, and Ngawha, New Zealand, fields the well discharges were of a mixed steam–water type, but the steam initially contained a very high proportion of carbon dioxide. This decreased rapidly with time to a few percent of the steam (White, 1970). At Monte Amiata the well outputs have decreased by about 30% during 5 years of operation, with some cooling by cold water inflow.

Celati *et al.* (1973) reviewed the changing hydrological condition in the Larderello district as revealed by ^{18}O, deuterium, and tritium measurements. The boundaries to the Larderello field that have been well defined by drilling to the south and southwest show a gradual change from superheated steam wells on the inside of the field, trending into two-phase mixture zones, to cold aquifers on the outside. Steam from the inner Larderello zone has a high $\delta^{18}O$ shift of about 7‰, but toward the outer areas the oxygen shift becomes small (e.g., at Castelnuovo it is about 1.5‰). Tritium concentrations are high in wells in the southeast and trend to zero in the north and northwest production zone. In general there was a good correlation between low tritium concentrations and high $\delta^{18}O$ shift.

Production of steam has lowered the pressures in the central steam area, upsetting the balance between central steam zones and the surrounding cold water. Where there is a permeable flow route, as in the southeast, cold water is now entering the steam zone, with a consequent lowering of $\delta^{18}O$ values and increasing tritium concentrations. Where there is limited permeability to the outside, however, there has been a widening of the central zone where wells produce steam from a deep origin.

Figure 7.11, from Celati *et al.* (1973), shows the complex hydrological balance in the Larderello field. The heavy arrows pointing inward indicate zones where $\delta^{18}O$ values and tritium concentrations show an increasing fraction of young water in the steam produced. Arrows pointing outward are in zones where there is an increasing fraction of "deep" steam in the fluid delivered by wells. Reinjection of condensate has complicated the pattern and, for example, has tended to produce a wider steam zone low in tritium concentrations.

Other changes have been noted in the chemistry of Larderello steam. D'Amore (1975) discussed the concentrations of ammonia, chloride, and boron in the steam. Although the total alkali present in the condensates was less than 1 ppm, for 116 wells the chloride reached a maximum of 60 ppm and the NH_3/Cl ratio was about 5. The presence of chloride is of concern due to its deleterious effect on metals. High boric acid concentrations in the steam were in general accompanied by high chloride, whereas ammonia showed an inverse relationship to boron and chloride. There was a rather constant $CH_4/$

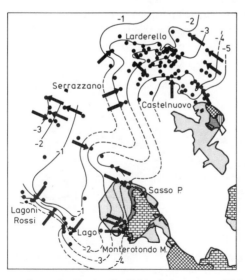

Fig. 7.11 Contours about the Larderello field of changes in $\delta^{18}O$ values with time during the period 1960–1970. See text for explanation of arrows (adapted from Celati *et al.*, 1973).

NH₃ ratio. The increasing chloride in steam from some parts of the field may be due to the drying out of rock which initially contained an adsorbed liquid phase. Ammonium chloride is volatile as molecular NH_4Cl in a dry steam situation. There was no sign of chloride being present as free HCl as the pH of condensates ranged from 5.5 to 7.0.

The differences between areas were considered to arise from a balance between steam derived from deep levels and that produced from flashing of meteoric water entering the steam zone. There was an inverse relationship between boric acid concentrations and $\delta^{18}O$, with steam from the deep source showing low boric acid and a large $\delta^{18}O$ shift. The boric acid concentration in steam from the central field has decreased over the years from about 700 ppm to about 100 ppm. The original high value was considered due to the accumulation of boric acid in the higher parts of the reservoir (Celati *et al.*, 1975). After several years of production, concentrations trended to a stable boron level for a period and then decreased further as a greater contribution of steam appeared from the flashing of meteoric water, at least in the peripheral wells.

With continuing production from a hot water geothermal field, the water level about a well may decrease so that over a period of time the well successively draws upon liquid water, a mixed steam–water inflow, and finally a steam phase. A regular program of measuring the gas concentrations of discharges can show the rate and direction of change of water pressures in the field at the level of well inflow.

Figure 7.12 shows an example for well 27 in western Wairakei field.

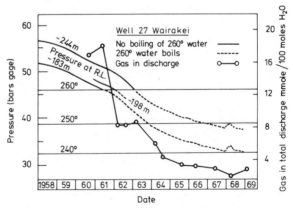

Fig. 7.12 Changes with time in aquifer pressure at various levels and in gas content of the total discharge from well 27, Wairakei. Saturated steam pressures at 240°, 250°, and 260° are also shown.

This well is supplied through a fissure at a level relative to sea level (R.L.) of −198 m. The change in aquifer pressure at −183, −198, and −244 m over the years is shown, together with the changing carbon dioxide concentration in the total well discharge. The horizontal lines show the saturated water vapor pressures at various temperatures. The rapid decrease in gas concentration in water supplying the well in 1961 corresponds to the time at which 260° water would begin to boil at about R.L. −198 m. With the continued downward trend in aquifer pressures, boiling increased at the level of the well inflow and the gas concentrations also decreased in a parallel manner. By 1969, there was a tendency for both aquifer pressures and gas concentrations in the total discharge to stabilize, and boiling water temperatures at the −198-m level would be somewhat over 240° (silica temperatures for waters in mid-1969 averaged 243°).

In areas of low rock permeability there may be a rapid local drawdown of water levels over a period of weeks or months, following the discharge of wells. At constant wellhead pressure (WHP) the discharges will consist of increasing proportions of steam, and the total gas concentration and the molecular ratio CO_2/H_2S in the total discharge will increase. This occurred in parts of the Broadlands area (an example is shown in Table 7.6). At a later stage the gas concentrations in the water phase of the aquifer may become depleted, so that the gases in the total discharge show a maximum and then decline.

Well discharges in hot water areas may change in enthalpy either with time or with different discharge conditions (e.g., varying wellhead pressure). A change in discharge enthalpy with WHP is characteristic of wells in hot water areas within low-permeability rocks.

TABLE 7.6

Changes with Time in Enthalpy and Gas Concentrations for Total Discharge for Well 11, Broadlands[a]

	Date				
	9/17/68	10/15/68	11/27/68	1/21/69	3/21/69
Enthalpy (cal/g)	296	317	333	374	356
CO_2 (mmole/100 moles)	223	370	480	560	855
CO_2/H_2S ratio	43	44	49	50	59

[a] At a wellhead pressure of 13.8 bars gage.

Consideration of the related changes in the gas concentration of the total discharge may show whether the well takes its supply from a single inflow zone or from two levels. For example, Fig. 7.13 shows results for two Broadlands wells over a series of output tests at different wellhead pressures. Well 2 has a constant supply source and both enthalpy and gas concentrations change in proportion, but well 9 changes in gas concentration at a particular enthalpy (and WHP) when fluid of a different composition enters the well at a second depth.

Chapter 4 gave data on the solubility of gases commonly found in geothermal discharges. From this solubility data, the equilibrium distribution of various gases between steam and water at various separation pressures (or temperatures) can be shown. Figures 7.14 and 7.15, from Ellis (1962), show the equilibrium concentrations in steam and water phases of carbon dioxide, hydrogen sulfide, and ammonia for increasing percentages of steam formation from 260° water. After several percent steam formation, most of the hydrogen sulfide and carbon dioxide is in the steam phase, so that after considerable steam

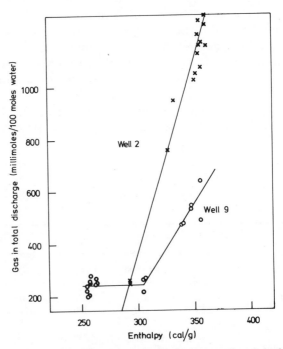

Fig. 7.13 Relationship between gas content and enthalpy of total discharges for two Broadlands wells with varying wellhead pressure.

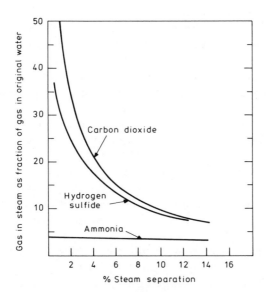

Fig. 7.14   Variation in concentrations of CO_2, H_2S, and NH_3 in steam with increasing percentage of steam formation, as multiples of the gas concentrations in original water (from Ellis, 1962).

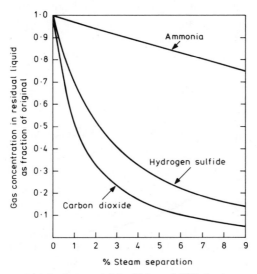

Fig. 7.15   The gradual depletion of CO_2, H_2S, and NH_3 in the water phase with the equilibrium separation of increasing percentages of steam (from Ellis, 1962).

separation, the concentration of these gases in the steam is inversely proportional to the percentage of steam formed. Insoluble gases such as nitrogen, methane, and hydrogen are released from water even more rapidly than carbon dioxide. Ammonia is concentrated slightly into the steam phase, but the water retains an appreciable amount even after extensive steam formation.

The proportions of the different gases in steam and water phases change with increasing steam formation. The concentrations and interrelationships between CO_2, H_2S, NH_3, and the insoluble gases (nitrogen, methane, hydrogen) are used to judge the degree of steam separation in processes relating to a well inflow in different parts of a field or at various stages of production (Ellis, 1962). Figure 7.16 shows, for example, the CO_2/H_2S ratio in steam and water for equilibrium steam separation.

Table 7.7 gives examples of gas concentrations in the discharges from Wairakei wells which tapped various types of inflows. At the date of sampling, well 27 tapped the original 260° water before boiling had occurred at the intake level for this well (original CO_2, H_2S, and NH_3 concentrations in the waters). Well 43 had a water supply from which steam had separated and gases were lost approximately in the amounts that would be expected for an equilibrium single-stage separation. The water was low in all gases and had low CO_2/H_2S and CO_2/NH_3 ratios. Well 42 tapped water which was cooled by steam loss but also received an inflow of steam at a higher intake level, which returned the enthalpy almost back to the same level as that for the original 260° water. A distinction from the well 27 type of inflow could, however, be made on the basis of the high CO_2/H_2S and $CO_2/$

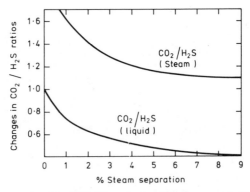

Fig. 7.16 Changes in the CO_2/H_2S ratio in the steam and liquid phases with increasing equilibrium steam separations (from Ellis, 1962).

TABLE 7.7

Carbon Dioxide, Hydrogen Sulfide, and Ammonia in Wairakei Wells[a]

Well	S.P.[b] (bars)	E (cal/g)	CO_2[c] (ppm)	H_2S[c] (ppm)	NH_3[c] (ppm)	NH_3[d] (ppm)	Total discharge				
							CO_2 (ppm)	H_2S (ppm)	NH_3 (ppm)	Molecular CO_2/H_2S	Molecular CO_2/NH_3
27	14.5	271	2910	76	4.8	0.7	426	11.1	1.3	29.7	125
43	13.8	239	930	47	4.9	0.8	76	47	1.15	15.1	25
42	6.7	269	3270	74	4.4	0.65	660	14.9	1.4	34.3	180
40	15.4	461	4600	128	5.1	0.8	2560	71	3.2	27.8	300
203	20.7	667	7850	183	5.1	no water	7850	183	5.1	33.2	600

[a] With different types of inflow (about 1960).
[b] S.P., separating pressure (gage).
[c] Concentration in steam at separating pressure.
[d] Concentration in water at separating pressure.

NH₃ ratios. Well 40 was supplied mainly by a steam phase, and the steam evidently resulted from extensive boiling of the original water (the CO_2/H_2S ratio is similar to that of the original water). Some ammonia was lost from the steam, presumably through absorption on moist rock surfaces or into secondary clay minerals. Well 203 was supplied entirely by separated steam formed by a few percent of steam boiling from the original water. This well was away from the main production area, and compared to other gases, a considerable loss of ammonia had occurred.

As noted earlier, a well is most likely to tap unboiled water during the first stages of discharging. The gas concentrations in the original high-temperature water are important, since deductions about later processes in the field are based on this value. Also, this gas concentration gives a good indication of whether or not the geothermal field is likely to be one which gives trouble with calcite deposition (see Chapter 3).

In the early stages of the Wairakei field, the gas concentrations in steam and residual liquid water tapped by wells gave reasonable agreement for equilibrium distribution of gases between two phases for a single-stage steam separation (Ellis, 1962). Glover (1970) showed that after there was appreciable lowering of the aquifer pressures in the field, the gas concentrations in well discharges of various types tended to give a better fit with a multistage model for steam separation. It is apparent that much of the water now supplying the wells has lost steam continuously while traveling some distance through the country.

The steam flows from natural fumaroles in a field under exploitation may also change in composition and magnitude. For example, the Karapiti fumarole area to the south of the Wairakei production field has enlarged considerably and has more extensive steam escape. There have also been several major hydrothermal blowouts in the area. The trends in fumarole steam compositions parallel those for steam from the Wairakei wells and indicate a steam supply obtained by increased boiling of an underground water system.

The gas concentration in steam being supplied to a power station should be monitored so that the effectiveness of a gas extraction plant in the station may be assessed and to enable a chemical mass balance for the station to be made. Frequent checks should also be made on the effectiveness of steam–water separators by examining the quality of the steam (its chloride and silica concentrations) entering the power station. These measurements are simple but of considerable practical importance to maintain the efficiency of the engineering operations.

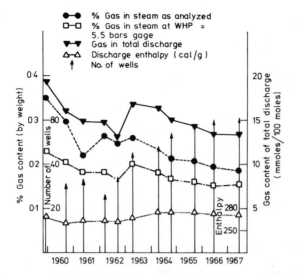

Fig. 7.17 Changes with time in the gas concentration of steam entering the Wairakei power station. The changes in the integrated enthalpy and gas concentrations for all discharges are also given (adapted from Glover, 1970).

Figure 7.17 gives the changes with time in the characteristics of steam entering the Wairakei power station over a period of several years (Glover, 1970). It also gives the gas concentration ($CO_2 + H_2S$) in the integrated discharges of steam and water for the field. The gas concentrations in the steam must be compared at a constant separation pressure if changes with time are to be relevant (in this case, a pressure of 5.5 bars gage). The gas concentration in the integrated steam supply from the Wairakei field has decreased only slightly with time and now has a tendency to stabilize. The trends in gas compositions, gas ratios, and integrated enthalpy for the field agree with an initial and rather rapid drawdown of water levels in the field, combined with increased boiling and steam formation. More recently, pressures at production levels have stabilized at values where the water upflow approximately equals the outflow through wells.

REFERENCES

Bolton, R. S. (1973). *In* "Geothermal Energy" (H. C. H. Armstead, ed.), pp. 175–184. UNESCO Earth Sci. Ser., No. 12, Paris.

Cataldi, R., Ferrara, G. C., Stefani, G., and Tongiorgi, E. (1969). *Bull. Volcanol.* **33,** 1.

Celati, R., Noto, P., Panichi, C., Squarci, P., and Taffi, L. (1973). *Geothermics* **2,** 174.

Celati, R., Ferrara, G., and Panichi, C. (1975). *U.N. Symp. Develop. Use Geothermal Resources, San Francisco, California, May,* Abstract III-13.

D'Amore, F. (1975). *Proc. Int. At. Energy Agency Advisory Group Meeting Appl. Nucl. Tech. Geothermal Stud., Pisa, Italy, Sept.* (in press).

Ellis, A. J. (1961). *Proc. U.N. Conf. New Sources Energy, Rome* **2,** 208.

Ellis, A. J. (1962). *N.Z. J. Sci.* **5,** 434.

Ellis, A. J. (1967). *In* "Geochemistry of Hydrothermal Ore Deposits" (H. L. Barnes, ed.), pp. 465–514. Holt, New York.

Ellis, A. J. (1970). *Geothermics (Spec. Issue 2),* **2***(Pt 1),* 516.

Finlayson, J. B., and Mahon, W. A. J. (1976). *Conf. Volcanolog. Geothermal Stud., February.* Univ. of Auckland, New Zealand.

Glover, R. B. (1970). *Geothermics (Spec. Issue 2),* **2***(Pt 2),* 1355.

Hulston, J. R. (1975). *Proc. Int. At. Energy Agency Advisory Group Meeting Appl. Nucl. Tech. Geothermal Stud., Pisa, Italy, Sept.* (in press).

Hunt, T. M. (1970). *Geothermics (Spec. Issue 2),* **2***(Pt 1),* 487.

Mahon, W. A. J. (1970). *Geothermics (Spec. Issue 2),* **2***(Pt 2),* 1310.

Mahon, W. A. J., and Finlayson, J. B. (1972). *Am. J. Sci.* **272,** 48.

Nakai, N. (1976). *Int. Conf. Stable Isotopes, Lower Hutt, New Zealand, August* (Abstracts, p. 46).

Palmer, R. A. (1977). *Geothermics* (in press).

Truesdell, A. H., and Singers, W. (1971). Computer Calculation of Down-hole Chemistry in Geothermal Areas. Rep. CD. 2136, Chem. Div., DSIR, New Zealand.

Truesdell, A. H., and White, D. E. (1973). *Geothermics* **2,** 154.

White, D. E. (1970). *Geothermics (Spec. Issue 2)* **1,** 58.

Wilson, S. H. J. (1963). *In* "Nuclear Geology on Geothermal Areas" (E. Tongiorgi, ed.), pp. 173–184. Consiglio Nazionale delle Ricerche, Lab. di Geol. Nucl., Pisa.

Wilson, S. H. J., Mahon, W. A. J., and Aldous, K. J. (1967). *N.Z. J. Sci.* **10,** 843.

Chapter 8

SOME SPECIFIC CHEMICAL TOPICS

8.1 VOLATILITY OF CHEMICALS IN STEAM

In planning utilization schemes, the concentrations of various impurities of geothermal steam must be considered, together with their effects on materials of construction. The main impurities are usually ammonia, boric acid, carbon dioxide, hydrogen sulfide, and other gases (nitrogen, hydrogen, methane). For mixed steam–water discharges, if the separation of steam from water is not efficient, droplets of water spray with salts and silica may also be carried in the steam.

Many constituents in geothermal waters are slightly volatile in steam at high temperatures and pressures. The distribution of a solute between steam and water phases can generally be expressed to a good approximation by the relationship

$$K_d = C^v/C^l = (\rho^v/\rho^l)^n$$

where C^v and C^l are the mass concentrations in vapor and liquid phases, ρ is the density of the phases, and n is a constant over a wide range of temperatures (Styrikovich *et al.*, 1955).

Experimental data on the volatility of various substances from solution were obtained by Styrikovich *et al.* (1955, 1960). These included boric acid, silica, alkali chlorides, and sulfates. Further results on silica distribution between steam and water were given by Martynova (1972) and for sodium chloride by Sourirajan and Kennedy (1962). Experimental results on boric acid distribution were reported by Wilson (1974) and some values were also obtained at Wairakei by the analysis of coexisting phases in discharge separators (R. B. Glover, personal communication). The results for boric acid are shown in Fig. 8.1.

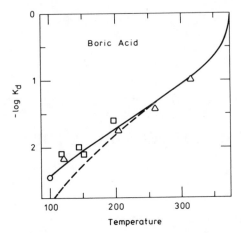

Fig. 8.1 The distribution of boric acid between steam and water expressed as K_d. The circle is from Von Stackelberg *et al.* (1937), triangles from Wilson (1974), squares from Glover, personal communication, and the dashed line expresses the results of Styrikovich *et al.* (1960).

The variation in volatility with temperature has important implications for steam-heated aquifers within a geothermal field. Surface or near-surface waters kept at boiling point by the passage of high-temperature (250°–300°) steam originating from a deep aquifer can have up to about ten times the boric acid concentration of the deep water (Tonani, 1970).

Table 8.1 gives estimated concentrations of sodium chloride, silica, and boric acid in steam separated at various temperatures and pressures from a water containing 1000 ppm NaCl, 100 ppm H_3BO_3, and 500

TABLE 8.1

Concentrations in Steam Phase in Equilibrium with Water[a]

Equilibrium temp. (°C)	Equilibrium gage pressure (bars)	Concentrations in steam (ppm)		
		NaCl	H_3BO_3	SiO_2
150	3.7	10^{-7}	0.6	0.03
200	14.5	10^{-5}	1.5	0.2
250	39	0.0005	3.8	1.0
300	85	0.02	8.9	5.0

[a] Water contains 1000 ppm NaCl, 100 ppm H_3BO_3, and 500 ppm SiO_2.

ppm SiO_2. At the usual separation pressures for geothermal steam utilization (less than about 15 bars), contamination of steam through the equilibrium volatility of these substances is not of major concern.

The preferential distribution of halides, silica, and boric acid into the water phase even at high temperatures means that they can be washed from a steam flow by scrubbing it with a small proportion of water. This practice was used in the Larderello field for recovering boric acid from the steam.

The distribution of ammonia between steam and water phases at various temperatures was obtained experimentally by Jones (1963). His results are summarized in Table 8.2, which also includes pK_a values for the dissociation reaction $NH_4^+ = H^+ + NH_3$ (see Chapter 4).

With decreasing temperature, ammonia is increasingly concentrated into the steam phase. Only a proportion of total ammonia in geothermal waters is present as free NH_3, although for near-neutral-pH geothermal waters NH_3 should predominate over NH_4^+ under separation conditions between 100°–200°. For example, some results for separation of steam and water at atmospheric pressure from Wairakei wells are shown in Table 8.3. The average pH of the Wairakei waters during separation at 100°C was about 8.0, so that about 80% of the total ammonia in the waters was present as NH_3. Comparison of Tables 8.2 and 8.3 shows that under the Wairakei separation conditions slightly less ammonia was found in the condensate than expected for equilibrium distribution.

The distribution of HF between steam and water phases at different temperatures has been obtained from a limited number of laboratory experiments at Chemistry Division (Table 8.4). For near-neutral-pH geothermal waters total fluoride is present mainly as the fluoride ion (pK_a values for HF ionization are included in Table 8.4). For example, during steam–water separation from Wairakei discharges at 100°, only

TABLE 8.2

Distribution of NH_3 between Steam and Dilute Water Solutions at Various Temperatures[a]

$t°C$	100	150	200	250	300
K_d	13.5	9.8	7.1	5.0	3.4
pK_a	7.4	6.5	5.8	5.2	4.6

[a] From Jones (1963). K_d = ppm steam/ppm water; also pK_a for NH_4^+ dissociation in water.

TABLE 8.3

Distribution of Total Ammonia between Water and Steam Phases from Wairakei Wells[a]

Well number	4/2	8a	9	11	12
Total ammonia (ppm)					
Condensate	2.8	3.05	16	2.3	5.1
Water	0.5	1.05	2.8	0.4	0.75
Ratio of conc.	5.6	2.9	5.7	5.7	6.8

[a] Separated at atmospheric pressure.

about 1 in 10^4 of total fluoride is present as HF. Table 8.5 gives the distribution of total fluoride between steam and water for some Wairakei well discharges, confirming that in practice the concentrations of total fluoride in the condensate are extremely low.

In geothermal areas with acid waters, a high proportion of the total fluoride may be present as HF, and in this case steam may also contain appreciable concentrations of HF. Where wells tap acid waters containing chloride there is also the possibility that free HCl may be present in the discharge and that during separation traces of this may be volatilized into the steam phase. Hydrochloric acid becomes a weak acid at high temperatures (Appendix Table A.3). For molecular HCl the coefficient of solubility in water is likely to be similar in magnitude to that for H_2S, that is, $m_{HCl}/P_{HCl} \sim 0.1$ mole kg^{-1} bar^{-1}. Therefore, approximate values of P_{HCl} can be estimated.

$$\log P_{HCl} \simeq 1 + \log m_{Cl} - pH - \log K_a$$

Table 8.6 gives the concentration of HCl expected in steam separated

TABLE 8.4

Distribution of HF between Steam and Water at Various Temperatures[a]

$t°C$	100	150	200	250	300
K_d	0.25	0.19	0.25	0.35	0.5
pK_a	3.85	4.3	4.9	5.8	6.8

[a] K_d = ppm steam/ppm water; also pK_a for HF ionization in water.

TABLE 8.5

Distribution of Total Fluoride between Water and Steam Phases from Wairakei Wells[a]

Well number	1	4/2	10	11	14
Total fluoride (ppm)					
Condensate	0.025	0.035	0.006	0.04	0.01
Water	7.5	7.4	3.7	9.1	11.8
Ratio of conc.	0.003	0.005	0.006	0.004	0.001

[a] Separated at atmospheric pressure.

at various temperatures from a molal solution of NaCl of different pHs. For example, steam separated at 100° and pH 3 would contain HCl at a partial pressure of 10^{-5} bar. The steam would contain about 5 × 10^{-4} m HCl, which would be fully ionized in the cold condensate (pH about 3.3).

High chloride in waters, low pH, and high temperatures of steam separation may together produce an acidic condensate containing free HCl. The only known test of this calculation arises from a series of results from wells in the Matsao, Taiwan, field, which has an aquifer containing high-temperature acidic water with 0.5 m chloride. Steam separated at atmospheric pressure from a mixed steam–water discharge (water pH 3.3) formed a condensate which had a pH of 4–5 when cooled and a chloride concentration of less than 10 ppm. A nearby well producing an almost dry, high-pressure steam flow had a condensate slightly contaminated with water, with pHs down to 1.0 and containing much (~0.1 m) free HCL.

The equation above may therefore overestimate the concentration of HCl in steam separated from water at atmospheric pressure, but steam separated from acidic waters at higher pressures and temperatures

TABLE 8.6

Calculated m_{HCl} and pH (Cold) for Condensate of Steam[a]

$t°$ (steam separation)	100	100	200	200
pH water ($t°$)	3	7	3	7
m_{HCl} (condensate)	5 × 10^{-4}	5 × 10^{-8}	0.5	4 × 10^{-5}
pH (cold condensate)	3.3	7.3	0.3	4.3

[a] Separated at 100° or 200° from water containing 1 m NaCl at pH 3 or 7.

may produce a corrosive condensate containing appreciable hydro-chloric acid.

In the early stages of the Wairakei project, an experiment was set up to detect any constituents in the steam not found by the usual analysis methods. Two fractionating columns were used. Steam was fed into one at the bottom to meet a condenser at the top. This was a refining column to detect compounds with boiling points less than that of water. In the other, a stripping column, condensate was fed in at the top and trickled down to a reboiler which sent vapor up again. This collected samples with boiling points higher than the condensate feed. A week of operation was allowed to obtain equilibrium. The only components identified in the low boiling point fraction were small amounts of silica and straight-chain hydrocarbons. Except for constit-uents which would be carried by traces of water droplets in the steam, only traces of hydrocarbons were found in the higher boiling point fraction. At the top of the refining column, crystalline deposits of impure ammonium bicarbonate were formed, plus traces of sulfides. The experiments gave no evidence of any unusual volatility of ele-ments contained in the Wairakei geothermal water.

8.2 CORROSION CHEMISTRY

Corrosion control in geothermal plants was reviewed extensively by Marshall and Braithwaite (1973) with particular reference to New Zealand experience. Only a brief outline is given here, with an emphasis on the chemical aspects.

Geothermal waters vary widely in composition, from waters strongly acidic with sulfur and halogen acids which actively corrode most common alloys, to the more usual neutral-pH or slightly alkaline pH waters, which may lay down protective scales of metal oxides, silica, or calcium carbonate. The effect of geothermal fluids on metals varies widely with the alloy and the type of application, and depends on whether or not air is present.

Geothermal steam that is either obtained directly from wells in a natural steam field or formed by separation from steam–water dis-charges contains carbon dioxide, hydrogen sulfide, ammonia, traces of boric acid, and other minor constituents, all of which limit the use of certain construction materials in geothermal power plants.

Water-soluble constituents which are carried as volatiles or in droplets in saturated steam are enriched considerably in small quan-

TABLE 8.7

Concentrations in Steam Flow and in Condensate in a Pipeline at Lago

	Geothermal steamflow (ppm)	Condensate (ppm)		
		1st drain	2nd drain	3rd drain
H_3BO_3	119	2750	5690	7850
Cl^-	103	273	550	2020
Fe^{2+}	Present	Present	Abundant	Abundant
pH	—	3.7	3	3

tities of condensate that may collect in a steam line, and may cause local accelerated corrosion. Allegrini and Benbenuti (1970) gave the results in Table 8.7 for a well at Lago. Note the low pH of the condensate.

In a new geothermal area, the corrosion characteristics of geothermal fluids should be examined in exposure tests, using the discharge of a typical investigation well. Metals are exposed to steam, steam condensate, or waters, with the test pieces assembled in appropriate conditions of stress. Tests are carried out also with steam containing a minor proportion of water to simulate incomplete or inefficient steam separation. The effects of aeration and fluid velocity must be included in the tests. Early corrosion test work at Wairakei was summarized by Marshall (1958) and for Icelandic areas by Einarsson (1961). More recent corrosion work at Cerro Prieto, Mexico, was described by Tolivia (1970, 1975), in Iceland by Hermannsson (1970), in Italian geothermal plants by Allegrini and Benbenuti (1970), and at The Geysers by Dodd *et al.* (1975) and Hanck and Nekoksa (1975).

Results from the high-temperature water area of Cerro Prieto were in general similar to those encountered at Wairakei except that the higher temperatures and more saline waters at Cerro Prieto led to slightly higher corrosion rates, particularly when water was present in the test fluids.

Surface Corrosion of Metals

As summarized by Marshall and Braithwaite, the surface corrosion rates for common engineering alloys are usually higher in air-free geothermal steam or water than in clean steam or water under comparable utilization conditions. The rates, however, are usually acceptable for practical uses, except for copper-based alloys. From test programs and plant experience it is known that contamination of geothermal fluids with oxygen (e.g., by aeration) drastically accelerates the surface corrosion of most engineering alloys other than austenitic stainless steels, titanium, chromium plating, and in some cases aluminum alloys. However, acidic sulfate conditions will also cause pitting attack on stainless steels.

In external uses, carbon steel and galvanized or cadmium-plated steel are severely attacked by the saline spray from a geothermal plant, and copper-based alloys are rapidly corroded by sulfide. Austenitic stainless steels and aluminum alloys can be satisfactory (Dodd et al., 1975).

Mercado (1975) noted that corrosion of outer well casings was an important problem at Cerro Prieto. Outside casing corrosion caused one well to get out of control, and control was regained only after considerable effort. Double casing is now utilized as a practical solution, and wells with two casings had no failures in nine years' service.

Troublesome tarnishing of silver and copper electrical and switch-gear equipment in power stations can occur through the presence of hydrogen sulfide in the atmosphere. It may be necessary to house electrical equipment in clean-air cabinets using activated carbon or manganese-impregnated sawdust air filters (Sewell, 1973). The use of nontarnishing platinum contacts may be necessary in critical control instruments.

Stress Corrosion

Stress corrosion of alloys in the presence of chloride- or sulfide-containing geothermal media is of particular importance in designing a geothermal plant. Stress corrosion of austenitic stainless steel is likely to occur under aerated conditions, both from wet chloride-containing steam and from contact with high-temperature water. This limits the application of this corrosion-resistant alloy to air-free situations.

Medium- and high-strength carbon and alloy steels are susceptible to sulfide stress corrosion; the effect is more critical in cooled (50°) geothermal waters than at high temperatures (250°). The corrosion of steels by sulfide is accompanied by hydrogen infusion into the metal with consequent blistering or embrittlement, and measurable effects were reported after exposure of steel to geothermal condensate saturated with hydrogen sulfide (Marshall and Braithwaite, 1973). The inverse temperature effect is probably due to the formation of stable and protective magnetite coatings on steel in high-temperature geothermal waters. Hydrogen embrittlement of steels by hydrogen sulfide solutions has not yet been investigated exhaustively and its significance in the design and performance of a geothermal plant requires more detailed study.

Low-strength carbon and alloy steels resist sulfide stress corrosion in most geothermal fluids, although in all cases experimental exposure tests should be undertaken for each new geothermal field. However, the general resistance of low-strength steels to stress corrosion means that they can be used satisfactorily for pipelines transmitting geothermal fluids and for well casings (Allegrini and Benbenuti, 1970; Marshall and Braithwaite, 1973).

Stress corrosion effects usually limit the choice for rotor and turbine blade construction to low-strength carbon and alloy steels, and to protect these from surface corrosion all possible steps must be taken to eliminate air from turbine housings. Dodd *et al.* (1975) reported failures of 12% chromium stainless steel turbine blades by fatigue in The Geysers steam. The heat treatment of the alloy was critical.

Condensers and Cooling Towers

Severe corrosion conditions exist within the condenser–cooling tower systems of a geothermal power station due to the presence of oxygen, hydrogen sulfide, ammonia, and carbon dioxide. However, the relatively low temperatures allow the use of coated steels (lead, or coal-tar epoxy resin coatings), aluminum, plastics, and concrete. Titanium also should have suitable resistance to geothermal condensates (Tolivia, *et al.*, 1975). Wood has been used extensively for the construction of geothermal cooling towers (Finney, 1973).

Allegrini and Benbenuti (1970) noted that although the aerated water in the condensation units at Larderello was of near-neutral pH (6.5–7), common carbon steel corroded rapidly, forming ferrous hydroxides, oxides, and sulfides. Cast iron was not appreciably corroded but formed sulfide scales. Type 316 stainless steel gave good corrosion resistance, but type 304 underwent intergranular and electrochemical corrosion. The major corrosion problems within the plant came from the oxidation of hydrogen sulfide by air present in the condensation and cooling tower systems. At The Geysers severe pitting of steel pumps was noted in the cooling systems (Hanck and Nekoksa, 1975).

Corrosion in cooling towers is to a major extent related to the ratio of ammonia to hydrogen sulfide and carbon dioxide in the steam condensate. Depending on this, the circulating water may range from slightly alkaline when the ammonia and acidic gases are in comparable proportions, to highly acidic when sulfur is in considerable excess of ammonia.

Severe etching of concrete and brick materials by aerated sulfurous gases was observed at Larderello in the barometric wells, cooling towers, and tunnels. Attack below the water line was less serious (Allegrini and Benbenuti, 1970).

Standby Corrosion

Corrosion and pitting of metals under standby conditions can be severe due to the oxidation of hydrogen sulfide to sulfuric acid. It is usual to maintain a slight flow of steam through lines and plant to avoid this problem.

Erosion

Severe erosion of wellhead pipes, caused by debris entrained in the fluids, may occur in the first stages of discharging a well. Complete penetration of a T-joint has sometimes occurred, and wells are usually discharged vertically first to avoid this situation.

Carry-over of sediment, sand, and iron corrosion materials in the steam flows has been troublesome at Matsukawa, causing erosion of equipment and incorporation of the materials in scales (Ozawa and Fujii, 1970). Steam flows at The Geysers carry both coarse and fine particulate matter. The former is handled by T-piece "dirt legs" in the pipeline, and fine particulate matter is 99% removed by centrifugal separators (Budd, 1973).

Nonmetallic Materials

There is little systematic information on the performance of nonmetallic construction materials in geothermal fluids, particularly for plastics and protective coating materials. An accelerated aging of plastics through contact with geothermal gases and waters has been observed in many fields.

Geothermal waters do not usually have an adverse effect on concrete unless the waters have been aerated, with subsequent acid sulfate attack. In some situations epoxy resin coatings on concrete have provided a satisfactory protection, although the persistence of such coatings under hot and appreciably acid conditions is questionable. Concrete impregnated with organic resins is also expected to find geothermal applications.

Many paints have a short lifetime on geothermal equipment exposed to sulfurous and saline atmospheres. The most satisfactory results have been obtained with epoxy paints after careful sand blasting or scraping of surfaces, and silicone paints have given promising results (Allegrini and Benbenuti, 1970). In general, oil paints, acrylic, vinyl, phenolic, chlororubber, polyester, and polyurethane paints are unsatisfactory. Lead-containing paints blacken rapidly and must be avoided.

Hot Water Reticulation

Hermansson (1970) reviewed experiences with the corrosion of metals in the reticulated hot water (85°–130°) heating service for Reykjavik, Iceland. Early wells producing waters about 87° and low in sulfide gave little corrosion and laid down a film of silica and metal oxides. The installation of airlifts to the reticulation system allowed oxygen into the waters and caused severe internal corrosion of pipes. However, this was cured by the addition of 10 ppm Na_2SO_3 for every 1 ppm of oxygen present. The addition to the reticulation system of hot waters containing hydrogen sulfide from another geothermal area tended to eliminate oxygen through the formation of sulfur but created a problem through corrosion of copper alloys.

Higher-temperature waters (130°) were later obtained from wells within Reykjavik city and these formed a protective magnetite film within pipes. The mixed waters from several sources now reticulated contain a trace of sulfide and 0.1–0.5 ppm oxygen. The water forms a protective white coating on iron, consisting mainly of iron oxide and silica with a variable amount of magnesia. Copper–zinc and copper–tin alloys do not give good service, but a protective film is laid down on copper. First copper sulfide is formed, which is replaced by a white film of cuprosilicate containing magnesium. Zinc-coated steel gives satisfactory service, although a thin and sporadic white silicate coating is formed.

Acid Sulfate Waters

In recent volcanic areas, extremely corrosive high-temperature geothermal waters may occur in sulfur-rich areas. For example, in the Tatun geothermal zone of Taiwan, acid sulfate–chloride waters were encountered by wells in both the Tahuangtsui and Matsao areas (Chen, 1970). The pH of the deep waters ranged from 1.5 to 5 but was frequently in the range of 2–3.5. Under these conditions extreme corrosion of steel well casing occurred, leading to its destruction. In some wells, liquid sulfur was encountered along with hot water, adding to the corrosive nature of the environment. Similar acidic conditions were encountered at Hakone, Japan (Noguchi *et al.*, 1970). In these situations there is no material of construction available at present that would enable the economic production and separation of geothermal steam.

At Matsukawa, Japan, the sulfate-rich and slightly acidic waters produced by several wells had a pH ranging from 3.7 to 5.5. Some corrosion of steel casing occurred due to acid and sulfide attack. Fortunately, many of the wells, after running for a period, changed to dry steam discharges. The Matsukawa wells in general now produce high enthalpy discharges and there is a delicate balance between corrosion and the deposition of silica and sulfate scales. Ozawa and Fijii (1970) and Fujii and Akeno (1970) reviewed the complexities of corrosion and scale formation in the Matsukawa plant.

8.3 MINERAL DEPOSITION

Introduction

Extensive deposits of siliceous sinter or calcium carbonate (travertine) in or around springs are features of many hot spring areas. Deep drilling has shown that hydrothermally formed calcite and quartz also occur extensively in rock fissures. Unfortunately, in some hot water fields mineral deposition may also occur along the artificial flow channels (pipes, plant drains) associated with geothermal utilization schemes.

High-temperature steam transports small concentrations of volatile silica, boric acid, sulfide, and halides, and in steam fields minor scale formation can occur in pipes or plant through either vapor phase deposition or the reaction of steel with steam constituents. Where wells tap steam carrying a minor amount of entrained water, scales of complex composition may form through the partial dehydration of water droplets as steam flows down a pressure gradient (e.g., the complex metal sulfate–silica scales at Matsukawa discussed later).

Natural hot waters are saturated with silica in equilibrium with quartz, and are frequently close to saturation with calcite, calcium sulfate, and calcium fluoride (Chapter 4). Some highly mineralized or acid waters of high temperature contain appreciable and near-saturation concentrations of heavy metals, and with changes in temperature and pressure are capable of depositing sulfides in the country rock, in piping, and in surface drainage channels.

Table 8.8 shows the general types of deposits found in several geothermal areas. The most common deposits formed in hot water systems are calcium carbonate (mainly calcite but sometimes aragonite in lower-temperature systems) and silica in its various forms. Very high-temperature, or very acid, water systems may also form sulfide deposits. Steam-producing wells may carry over rock dust and silica to form a major depositional problem in pipes and equipment.

When utilizing high-temperature water systems, detailed consideration must be given to the possibility of scale or deposit formation. With adequate design the deposition may be induced to occur at noncritical parts in the geothermal plants, and preferably at the water disposal side of the operation. The factors governing the formation of various types of scales are now examined.

Several physical and chemical processes occur as high-temperature water is brought to the surface through wells and by pipeline into surface equipment. A steam phase forms in increasing proportion as pressures decrease along the pipeline and it contains a large proportion of the gases originally present in the deep water. The cooled water is depleted in acidic gases, while nonvolatile constituents undergo concentration. Mineral deposition may be initiated by these changes.

A small change in the concentration of a constituent can represent a sizable deposition rate in a geothermal well of high output. For example, a loss of 1 ppm of calcite or silica from solution in a 20-cm diameter well producing 100 tons of water per hour would give a deposit about 2 mm thick per day over a 1-m length of pipe.

Deposition rates can be much greater than this. For example, a well in the Tongonan, Leyte, Philippines, area developed aragonite scale at the rate of 1 mm/h (R. B. Glover, personal communication). Complete blockages of wells have occurred in a matter of days in severe scaling situations.

TABLE 8.8

Examples of Deposits Found in Various Geothermal Utilization Schemes

Scheme	Deposits
New Zealand	
Wairakei	Minor calcite (pipes), silica (drains)
Broadlands	Calcite (pipes), silica (silencers, drains)
Kawerau	Calcite (pipes), silica (drains)
Turkey	
Kizildere	Calcite, aragonite (pipes and plant)
Chile	
El Tatio	Calcite (pipes)
Japan	
Matsukawa	FeO, FeS, $FeSO_4$, silica, sediments
Otake	Calcite, silica (drain pipes)
Taiwan	
Matsao	Silica, PbS, FeS, Pb (pipes)
Iceland	
Reykjavik	Fe oxides, MgO, silica (pipes, plant)
Namafjall	Amorphous silica, chalcedony
Italy	
Larderello steam fields	Silica, FeS, borates, silicates
Soviet Union	
Bolshe-Bannoe	Calcite (pipes)
Philippines	
Tiwi	Calcite
Tongonan, Leyte	Aragonite
United States	
Steamboat Springs	$CaCO_3$, probably aragonite (pipes)
Casa Diablo	$CaCO_3$, probably aragonite (pipes)
Salton Sea	Silica, hydrated Fe oxides, metal sulfides (pipes and plant)
The Geysers	Siliceous material, rock dust (pipes and plant)
Mexico	
Cerro Prieto	Calcite, silica, Fe oxides (pipes, plant)

Calcium Carbonate

Most high-temperature geothermal waters are close to saturation with calcite (Ellis, 1970). As reviewed in Chapter 4, the solubility of calcite decreases with increasing water temperature. At constant concentrations of carbon dioxide and other constituents in a water, calcite cannot be deposited from solution by simple cooling. Therefore, if the formation of steam is avoided in a utilization scheme (e.g., a pressurized binary cycle system), no problem with calcium carbonate scale would be anticipated.

The changes in water chemistry which occur when steam boils from high-temperature water can be summarized in the following equations. A water which was originally close to saturation with calcite rapidly becomes supersaturated following boiling, loss of carbon dioxide, and rising pH.

$$CO_2(aq) = CO_2(vap) \tag{8.1}$$

$$HCO_3^- + H^+ = CO_2(aq) + H_2O \tag{8.2}$$

$$Ca^{2+} + 2HCO_3^- = CaCO_3 + CO_2(aq) \tag{8.3}$$

$$HCO_3^- + H_3BO_3 = H_2BO_3^- + CO_2(aq) + H_2O \tag{8.4}$$

$$HCO_3^- + H_4SiO_4 = H_3SiO_4^- + CO_2(aq) + H_2O \tag{8.5}$$

Most high-temperature geothermal waters contain appreciable concentrations of boric and silicic acids, and during the later stages of carbon dioxide loss there is a tendency for reactions of the type (8.4) and (8.5) to occur. Bicarbonate ions are destroyed and the degree of supersaturation with calcite may be reduced; in some situations possibly to the extent that the water becomes undersaturated again.

Figure 8.2 shows semiquantitatively how the pH and the product $a_{Ca} a_{HCO_3}^2 / P_{CO_2}$ change in a Wairakei water from an original one-phase 250° condition during the equilibrium formation of various percent-

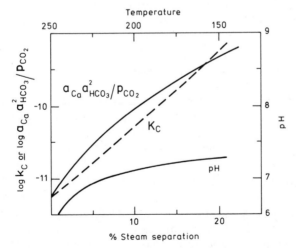

Fig. 8.2 The trend in pH and in the product $a_{Ca} a_{HCO_3} / P_{CO_2}$ in Wairakei water with various percentages of steam formation. Values of K_C for calcite saturation are given for comparison.

ages of steam. With the first formation of steam from the water there is a rapid rise in pH and in $a_{Ca}a_{HCO_3}^2/P_{CO_2}$, but as the percentage of steam increases the rate of change becomes less. In contrast, the value of $a_{Ca}a_{HCO_3}^2/P_{CO_2} = K_C$ for saturation with calcite changes with decreasing temperatures in such a way that the water first becomes supersaturated with calcite, and then at a later stage of steam separation (between 15% and 20%) becomes undersaturated again. The trends for individual waters will vary according to their original composition, particularly the carbon dioxide concentration and the relative concentrations of weak acid anions.

In practice the most troublesome calcite deposition usually occurs in the well casing at the level of first boiling, with a heavy band of calcite being deposited over a short length (a few meters). Calcite deposition may taper off to become minor or zero from this point upward. Wells should be run at a wellhead pressure sufficient to ensure that calcite deposition, if it is to occur, is within the pipe and not in the rock fissures outside the well.

Kryukov and Larianov (1970) described experiments to determine the level of calcite saturation of waters at various depths in wells during production. Water sampling probes and deposition tests were used to demonstrate that calcite deposition was concentrated at the depth of first boiling in the wells of the Bolshe-Bannoe area. In this area waters were calculated to be saturated with calcite before steam formation, but at Pauzhetsk, where waters were undersaturated before steam formation, there was no calciting in the wells.

Calcite deposits formed in a large number of Broadlands wells but the rate of calcite deposition was variable and was highest for wells with discharge enthalpies much higher than that of the deep water.

The tendency to form calcite deposits is greatest in geothermal areas with a high carbon dioxide content in the waters. For example, little or no calcite deposition occurred in wells at Wairakei ($m_{CO_2} \sim 0.01$) but it was troublesome in the Broadlands and Kawerau areas ($m_{CO_2} \sim 0.1$). From mineralogical observations Ellis (1970) suggested that for geothermal waters in the temperature range 230°–300° the critical concentration of CO_2 in the deep water may be about 0.1 m, with calcite scaling likely to be troublesome at higher CO_2 levels.

In the Kizildere field in Turkey, the carbon dioxide concentration of the deep water was several times this level, and bicarbonate concentrations were high (2500 ppm) (Dominco and Samilgil, 1970). The comparatively low temperature (200°) allowed the product of calcium and carbonate in the water to be high. Not only did calcite form in wells at the level of first boiling and to a lesser extent higher up the casing, but with the extensive degree of supersaturation caused by steam loss a heavy scale of aragonite was formed in the near-surface and surface pipes and equipment at temperatures approaching 100°.

Following extensive withdrawal of water from a high-temperature water aquifer a general depletion of carbon dioxide in solution may occur through boiling. The reaction of the water with rock minerals is not sufficiently rapid to buffer out the consequent pH rise, and calcite may then form in the aquifer rock. This is possibly of little importance if it is disseminated homogeneously, but if it is deposited in a major flow fissure at a specific point, an irreparable decay in water flows may occur.

Silica

Silica Chemistry

The solubilities of various forms of silica in water were summarized in Fig. 4.18. In geothermal waters at temperatures in excess of about 150° the solubility of quartz controls the concentrations of silica. When discharged at atmospheric pressure, the cooled and concentrated geothermal waters are usually supersaturated with respect to amorphous silica solubility. This may eventually be precipitated in various forms, including colloidal silica, gelatinous silica, and fibrous or opaline silica sinter. The chemical processes leading to deposition are complex.

The molecular form of silica in waters can be determined by its rate of reaction with ammonium molybdate reagent (O'Connor, 1961). At ambient temperatures, monomeric silicic acid reacts to form a yellow complex within two minutes and noncyclic silicic acids react within five minutes, whereas cyclic polysilicic acids require up to several hours for reaction.

The silica in solutions in equilibrium with quartz or with amorphous silica at high temperatures is in a reactive monomeric form, possibly as H_4SiO_4. Studies of high-temperature hot spring waters (White *et al.*, 1956; Fournier, 1973) and of geothermal well waters (Ellis, 1960) showed that in freshly discharged waters silicic acid was present in a simple monomeric form.

Cooled geothermal waters show concentrations of monomeric silica which decrease with time to approach a value which is approximately equal to the steady-state concentration for equilibrium with amorphous silica at the holding temperature (Krauskopf, 1956). The silica present in solution in excess of amorphous silica solubility polymerizes through a series of reactions involving linear, cyclic, and three-dimensional polymeric silicic acids.

The rate of polymerization of a solution initially supersaturated with respect to amorphous silica depends on various factors (White *et al.*, 1956):

(a) the initial degree of supersaturation (related to the probability of nuclei formation);

(b) pH of the solution;

(c) temperature;

(d) presence of colloidal or particulate siliceous material;

(e) ion concentrations.

Numerous studies have been made of the rate of polymerization of silica in supersaturated solutions. Figure 8.3 shows the characteristic form of the reaction curve. There is usually an induction period which is inversely related to the concentration of silica in excess of amorphous silica saturation, and which is longer at low pHs (Baumann, 1959). This induction period can be reduced or eliminated by adding some aged solution which already contains polymerized silica (White *et al.*, 1956). A relatively rapid polymerization reaction follows, and in individual runs the rate can be expressed in terms of Eq. (8.6) (Kitahara, 1960), where C is the concentration of monomeric silica at time t and C_e is the equilibrium "solubility" of amorphous silica at the temperature. Finally, the monomeric silica concentration trends to C_e.

$$-dC/dt = k(C - C_e)^n \qquad (8.6)$$

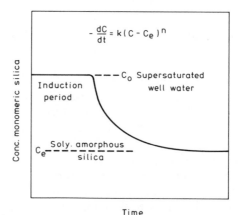

Time

Fig. 8.3 Characteristic form of the silica polymerization reaction curve at constant temperature and pH.

The great effect of pH on the polymerization rate of silica in natural thermal waters is shown in Fig. 8.4 (examples from Kitahara, 1960). The kinetics of the polymerization reaction are complex, and n in Eq. (8.6) varies widely both with the degree of supersaturation and with temperature. The reactions are assumed to be autocatalytic chain reactions catalyzed in particular by the OH^- ion. The ions Na^+ and K^+ and SO_4^{2-} have little effect on polymerization rates, but both Cl^- and Br^- catalyze the reaction (Baumann, 1959).

Yanagase *et al.* (1970a,b) used a light-scattering photometer to show that the growth of silica colloids in a high-temperature geothermal water began almost as soon as the water emerged from the well. When the temperature was kept near boiling point, large relatively uniform polymers (colloidal particles of about 0.3 μm) were formed over a period of approximately 1 hour, and when the temperature was allowed to drop slowly to 50°, rather similar results were obtained. However, when the water was rapidly cooled to 58° and then allowed to polymerize, light scattering could not detect many large polymers. High temperatures of polymerization apparently produce higher polymers than are formed at low temperature. Growth of silica colloids was also influenced by stirring, filtering, mixing of hot waters, and mixing with cold water.

Yanagase *et al.* (1970a,b) showed that the tendency for silica deposits to grow on surfaces contacting the water was greatest during the period of maximum polymerization rate. H. P. Rothbaum (personal

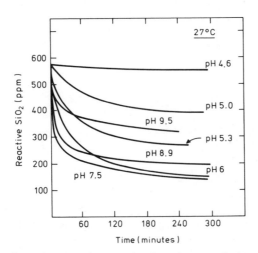

Fig. 8.4 The effect of pH on the rate of polymerization of silica in water (adapted from Kitahara, 1960).

communication) has, however, not been able to confirm this with Wairakei or Broadlands waters.

For temperatures of 52°, 92°, and 120°, Rothbaum (1974) studied the polymerization of silica in geothermal well waters containing different concentrations of silica, but initially all in the monomeric form. He confirmed that the induction period depended on the degree of silica supersaturation and that there was no consistent reaction process operating through the temperature range (the apparent reaction order changed with both temperature and initial silica concentration). After the polymerization reaction ceased, the concentration of monomeric silica approached that for amorphous silica solubility. An example from his study is given in Fig. 8.5, showing the induction period and polymerization rate for three high-temperature geothermal waters cooled to 92° after discharging, and initially containing 1020, 650, and 570 ppm monomeric silica, all at pH 7.0–7.1.

The implications of diluting a geothermal water with fresh cold water should be noted. As well as cooling the solution, the term $(C - C_e)$ in Equation (8.6) is reduced, which lowers the polymerization reaction rate and increases the induction period (see the effect of diluting a Wairakei well water, in Fig. 8.5). However, induction periods could be eliminated by the presence of particulate material, or polymerized silica in solid or solution form.

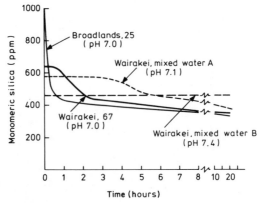

Fig. 8.5 The silica polymerization rate for a Broadlands well water and three Wairakei waters (mixed water A consists of water from wells 26A, 26B, 76, 107, 108, and 48; mixed water B is the same water diluted 20%). All reactions at 92° (from Rothbaum, 1974).

Practical Implications

Research on silica deposition is highly relevant to the disposal of geothermal waters, since silica blockage of surface pipes or flumes or of natural channelways during reinjection could cause great difficulties and expense.

High-temperature geothermal waters are in general cooled in one of two ways:

(1) Steam is allowed to flash from a well discharge at a controlled pressure and is subsequently used in turbines or as process steam. The solutes are concentrated by steam loss, and Fig. 8.6 gives trend lines for the silica concentrations in waters initially saturated with quartz and cooled by adiabatic steam loss from 200°, 250°, and 300°, respectively. The concentrations of silica in the particular waters exceed that for amorphous silica solubility when they have cooled to approximately 95°, 150°, and 200°, respectively. Below these temperatures polymerization of the excess silicic acid occurs, and there is a tendency to deposit amorphous silica on surfaces. The tendency increases as the water cools further.

(2) If a high-temperature geothermal water is utilized through heat exchangers operating at high temperature and pressure, no steam

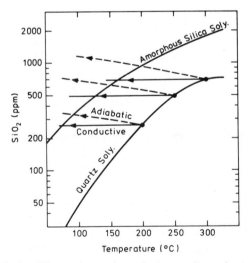

Fig. 8.6 Trends in silica concentrations during cooling of geothermal waters by steam loss and by heat exchange. Lines for quartz and amorphous silica solubility are shown.

phase is formed and the silica concentration remains constant during cooling. In this situation amorphous silica solubility would not be exceeded until the water temperatures fell below 75°, 125°, or 160°, respectively, for the three samples.

To be certain that silica deposition does not occur, steam–water separation, or heat exchange, in utilization schemes should not be planned to occur much below the temperature at which the water becomes saturated with amorphous silica; that is, for a 250° water, silica scale deposition could occur following cooling by steam separation below 150°, or by conduction below 125°. Because polymerization and precipitation does not occur at a significant rate until a reasonable degree of supersaturation is achieved, trouble is not experienced in practice until temperatures rather lower than these theoretical points are reached. The actual point will depend on the nature of the water (composition; presence of nuclei).

At Wairakei, where the deep-water temperatures were in the range of 230°–250°, there were few cases of amorphous silica deposition in bypass pipes, pressure reducing valves, or silencers when water outputs were high (200–400 tons/hr). However, for some wells where water outputs decreased with time, amorphous silica deposits were found in the last section of the bypass pipe and in the silencers. Deposits on the floors of some silencers in 1975 were 15 cm thick and on the walls 2–3 cm. The borderline nature of the situation is emphasized by the fact that at Broadlands, where the water temperatures were on average 20° higher than the original Wairakei temperatures, there was significant deposition of amorphous silica as small beads forming cone-shaped mounds on the floors of the twin-tower silencers.

There are few situations where silica has been detected within the casing of wells that have discharged continuously, and as would be expected from Fig. 8.6 the examples are in the highest-temperature fields. In the Salton Sea field (water temperatures up to 350°) major deposition of amorphous silica occurred within well casings (Skinner *et al.*, 1967). Silica deposition is also troublesome with wells at Mexicali tapping waters of similar temperature (S. Mercado, personal communication).

Silica deposition was measured at Wairakei over periods of 2–4 weeks, using metal test rods suspended in pressure tanks through which water was passed after it had been allowed to cool by steam flashing down to temperatures of (a) 149° and (b) 105°. Deposition rates after a holdup time of 10 min in the water flow were very low

(0.1-0.2 mm/month) at both pressures, and the type of metal surface had little effect on the rate or type of deposition. The growth rate on steel surfaces appeared to decrease with time, following an exponential curve.

Experimental work was also done on silica deposition from Wairakei waters under conditions which would exist in heat exchangers (G. D. McDowell, personal communication). Runs were done in a closed system with water of inlet temperature 166°C (cooled to this temperature by steam loss) passing through a 30-m pipe heat exchanger to cool to 70° (transit time 9 min), and another with water at a temperature of 108° at the inlet and of 83° at the outlet (transit time 1 min). No deposition of silica occurred. The tendency of a water to deposit silica may be increased when it is exposed to the air, presumably because of the formation of minor amounts of iron or manganese hydroxides which act as silica nucleation sites. Pumping oxygen into the water flows induced silica scaling in the pipes.

The immediate deposition of silica on steel from water emerging from Wairakei wells at 98° is slight (50–60 mg cm^{-2} yr^{-1}). After exposure to the atmosphere and cooling to 90°, the immediate rate of deposition increases 15 times.

Deposition in Field Drains and Pipes

Buildup of silica deposits in drains and pipes carrying waste geothermal waters from a field has been a particular problem. The open concrete drains at Wairakei have to be continually cleared of silica by mechanical means at a considerable cost, and similar problems occur at Otake (Yanagase *et al.* 1970a,b) and Ahuachapan (Cuellar, 1975).

Many attempts were made in the early stages of the Wairakei project to prevent silica deposition in effluent channels, including the introduction of seed pellets of silica to initiate deposition away from channel walls, adjusting the pH of the water, and the addition of traces of various metallic ions. These were either ineffective or uneconomic.

At Cerro Prieto the problem was avoided by transmitting high-pressure separated water in long pipelines to allow it to discharge directly into the surface of a large lake of effluent. Silica deposition occurs in the lake.

At Otake, the problem of silica deposition in the disposal pipeline was solved by inserting a stirred holding tank (1-h holdup) in which most of the silica polymerized and some precipitation occurred, and ensuring that the flow in the outlet pipe was laminar (Yanagase et al., 1970b). Following this treatment, deposition rates in the disposal pipe were reduced to one tenth. A similar approach has been successful at Ahuachapan where deposition in drains was reduced by a factor of about four (Cuellar, 1975).

Similar experiments were made at Wairakei (water holdup time 1 h; laminar outflow from the tank), using water directly from a discharge silencer. At Wairakei however, little polymerization of silica occurred in the tank, and rates of silica deposition on test plates in, and following, the tank were only slightly less than those for untreated waters. The effect of a 1:1 dilution of the geothermal water with cold water was marked. The silica deposition rate dropped by a factor of 16 and the silica deposited was friable and more easily removed.

The different behavior of Otake and Wairakei waters, although they were of similar initial temperature, salinity, and silica concentration, emphasizes that in each field an individual investigation of silica deposition characteristics is necessary.

When a problem of silica deposition from geothermal water effluent arises, the simplest approaches are dilution with cold water, or "ponding," either with a short holdup time until the maximum precipitation tendency of the hot water has passed (as at Otake) or for long periods in an effluent lake to allow silica precipitation to equilibrate at ambient temperatures (as at Cerro Prieto). However, as discussed in Section 8.4, chemical treatment of hot waters to remove silica can also be undertaken.

Complex Scales

In the Matsukawa area of Japan many wells at first produced a mixture of steam and slightly acidic sulfate-rich water. Production later changed to saturated or dry steam. During the stage of wet steam production a complex scale of silica plus iron, potassium, and calcium sulfates was produced in wellheads, separators, and station equipment (Ozawa and Fujii, 1970). The scale composition was dependent on the water content of the discharge, water composition, and pH. Dehydration of droplets of sulfate-containing water in nearly dry steam flows was a major cause of scale formation, and carry-over of fine rock debris into the plant by steam flows also occurred. Corrosion of the steel pipes added to the complexity of the scale compositions.

Wells in the Salton Sea geothermal area produced a high-temperature ($320°-350°$) water which deposited a hard black siliceous scale in their casings (Skinner et al., 1967). The scales were layered and contained high concentrations of iron, copper, and silver sulfides, as well as metallic silver. The highly saline waters contained about 5 ppm copper, 1–2 ppm silver, 2000 ppm iron, 80 ppm lead, and 500–800 ppm zinc.

Although this is an extreme case, siliceous scales in most geothermal areas have a tendency to have elevated concentrations of trace metals. For example, Wairakei and Broadlands well waters contain only parts per billion concentrations or less of metals such as silver, lead, beryllium, thallium, and gold. Yet these appeared at the tens or hundreds of parts per million level in siliceous deposits formed from 99° waters (Table 8.9). Hot spring sinters in the area also contain appreciable concentrations of silver and gold (Ellis, 1969).

Some Broadlands geothermal wells deposited in surface channelways red precipitates rich in antimony and arsenic sulfides and containing appreciable concentrations of thallium, mercury, gold, silver, and lead as well (Weissberg, 1969). Similar precipitates were precipitated from hot spring waters of this and other areas (e.g., Steamboat Springs, Nevada; White, 1967). This type of deposit is not found within well casings.

In the Matsao geothermal field a well producing highly acid (pH 2.5) and saline water (14,000 ppm chloride) gave a minor deposit of lead sulfide and lead in surface piping. The water, originally at about 240°, contained 0.5–1 ppm lead, 11–13 ppm zinc, 40–50 ppm manganese, and about 200 ppm iron.

Hot brines (105°; up to 160,000 ppm chloride) on the Cheleken Peninsula of the eastern Caspian Sea deposit lead, sphalerite, pyrite, barite, and calcite in drill pipes and at the surface (Lebedev, 1972; Lebedev and Nikitina, 1968).

Removal of Deposits

Silica deposits in pipes in New Zealand geothermal areas have been almost pure amorphous silica. The deposits are bound tenaciously to

TABLE 8.9

Concentrations of Trace Metals in Silica Deposited on Test Pieces in Drainage Channel from Wairakei Well 76

Element	Sb	Mn	Cu	Ni	Ga	Ti	Ag	Zn	Pb	Au	Tl	Be
ppm	250	500	>500	>500	>500	1000	>100	100	100	10	15	>100

most types of surfaces and attempts at removal using strong sodium hydroxide solutions, both hot and cold, have met with little success. At Matsukawa, Japan, some success was obtained by injecting strong solutions of sodium hydroxide (10 m) under pressure into a well in which scaling had occurred. After 30 min of reaction most of the scale was removed from the pipe (Ozawa and Fujii, 1970). However, the scale was only 90% silica and it is uncertain whether the alkali dissolved the scale or weakened its structure by dissolving other constituents.

In several fields attempts have been made to dissolve calcite deposits in wells by pumping in a solution of hydrochloric acid plus a corrosion inhibitor. Unfortunately, New Zealand experience has been that at the high temperatures of production wells, the action of inhibitors is inefficient, and the well casing as well as the calcite is attacked by the acid. However, the use of an acid plus inhibitor solution for removing calcite from cooler surface equipment has been reasonably satisfactory.

For both calcite deposits and pure silica deposits in well casings, the usual removal procedure is to ream out with standard drilling equipment.

8.4   DISPOSAL OF WELL DISCHARGES

Fluid and Chemical Outflows

The rate of fluid production from wells necessary to operate a 100-MW geothermal power station in a dry steam field is about 10^7 tons/yr, while for a station of similar size in a hot water field the rate is greater by a factor ranging from about 3 to 10, depending on underground water temperatures. From the compositions of representative steam and waters given in Chapter 3, the approximate mass output of chemicals from various fields can be calculated. Table 8.10 gives the annual chemical outputs for power production at a 100-MW rate from the hot water fields of Cerro Prieto, Wairakei, and Broadlands and from the steam fields at Larderello and The Geysers. Additional information is from Reed and Campbell (1975), and from B. G. Weissberg (personal communication). It is evident that geothermal fluids bring to the surface large quantities of dissolved solids and gases. Axtmann (1975) gave an extensive review of the thermal, physical, and chemical effects on the environment from operating the Wairakei power station.

TABLE 8.10

Approximate Quantities of Chemicals Emitted per 100-MW power Generated from Five Sites

	Quantity (tons/yr)				
Constituent	Cerro Prieto	Wairakei	Broadlands	Larderello	The Geysers
Li	320	350	300	—	—
Na	1.5×10^5	3.5×10^4	3×10^4	—	—
K	4×10^4	5500	5000	—	—
Ca	1×10^4	500	80	—	—
F	40	200	200	—	—
Cl	3×10^5	6×10^4	5×10^4	—	—
Br	500	150	150	—	—
SO_4	170	700	200	—	—
NH_4	800	50	500	1300	1700
B	400	750	1100	200	200
SiO_2	4×10^4	2×10^4	2×10^4	—	—
As	30	100	100	—	—
Hg	—	0.025	0.035	—	0.04
CO_2	1.5×10^5	8000	4×10^5	4×10^5	3×10^4
H_2S	4000	300	6000	5000	2000

In hot water fields over two thirds of the total fluid mass brought to the surface is hot water, often of high salinity, and containing appreciable concentrations of elements such as arsenic, boron, and fluoride, which may affect human, animal, fish, or plant life. In a field producing dry steam, the steam condensate may contain high concentrations of boric acid, ammonia, sulfide, and other constituents. Except in favorable geographic situations, disposal of geothermal effluents into surface waters or the creation of ponds will probably not be permitted.

Gaseous Constituents

Table 8.10 shows that a geothermal power station of 100-MW capacity liberates into the atmosphere each year between 10^4 and 10^5 tons of carbon dioxide and between several hundred and several thousand tons of sulfur (as hydrogen sulfide). Although the gas outputs vary widely between fields, it is interesting to make a comparison with a coal-fired generating plant of the same capacity. When a coal containing 2% sulfur is used, such a plant would emit annually approximately 700,000 tons of carbon dioxide and 6000 tons of sulfur (emitted as SO_2), together with nitrogen oxides and smoke.

The emission of radon and other radioactive volatiles in geothermal steam also requires monitoring but does not appear to be an important environmental problem at The Geysers or at Broadlands where careful measurements have been made (Stoker and Kruger, 1975; N. E. Whitehead, personal communication).

In populated areas, hydrogen sulfide from geothermal power stations may be an embarrassment and its removal or recovery from effluents a necessity. It may also cause difficulties with electrical equipment. For example, if uncontrolled, The Geysers power stations could emit 28 tons/day of hydrogen sulfide (Reed and Campbell, 1975). The manner in which H_2S is emitted is important in designing abatement processes. At The Geysers, without treatment, about one third of the sulfide in the steam is emitted as a gas from the gas removal pump unit, about one third volatilized from condensate in the cooling towers, and the remaining third is in condensed steam which can be reinjected into the reservoir (Bowen, 1973).

Allen and McCluer (1975) and Lazlo (1976) reviewed several approaches to sulfide emission control at The Geysers. Burning of the gases from the gas extractor plant and then scrubbing them with condensate was tried, but the degree of extraction was not sufficiently high (50%). Ferric sulfate was added to the cooling water in a second trial. Ferric ions oxidized H_2S to sulfur, and the ferrous ions formed were reoxidized when the solution was aerated in the cooling towers. The condenser vent gases were passed into the cooling tower to be scrubbed by the solution. Although the degree of abatement was high (90-92%) the process caused corrosion and plugging of piping and equipment.

A third method being tried involves indirect condensation in stainless steel surface condensers. The geothermal gases undiluted by air are extracted by a chemical solution, e.g., the Stretford process using a solution of sodium carbonate, sodium ammonium polyvanadate, and anthroquinone disulphonic acid. The reduced vanadate solution in this process is reoxidized by air and high quality sulfur is recovered.

At Wairakei over 90% of the sulfide present in the original deep water is in the steam phase following steam-water separation. Of the hydrogen sulfide in the steam entering the power station, about 30% exits from the gas exhaust chimney, and the remainder enters the river water through steam condensate and drainage from the gas ejector plant. The total H_2S leaving the station is now (1977) of the order of 300 tons/yr, but Wairakei fluids have a low gas content. The Broadlands field has much higher H_2S concentrations in the deep water, and the output from a proposed 100-MW station may be sufficient to require sulfur recovery from the stack gases.

Water Constituents

The conventional operation of a power station in a hot water geothermal field faces a serious water disposal problem, since every ton of steam used requires the disposal of 3–10 tons of saline water. The manner in which this problem is solved varies with the local geographical situation and the availability of water supplies. There is the possibility in the future that binary cycle utilization schemes, at present in the pilot-plant stage of development, may eliminate this problem of water disposal. In these schemes water is removed from the aquifer, passed through a heat exchanger, and returned to the aquifer without coming into contact with the atmosphere. However, in many fields, problems with scale formation in heat exchange pipes and possibly in the reservoir remain to be solved.

In some countries where extra water supplies are needed (e.g., in northern Chile or southern California), an outflow of geothermal waters could be an advantage, and integrated power production plus water desalination would be appropriate.

In high rainfall areas, geothermal water must be considered as another industrial effluent, and its disposal must comply with local water purity and environmental regulations. With this in view it is appropriate to examine the maximum permissible concentration levels of constituents in waters intended for various purposes.

Table 8.11 gives data for elements which are likely to be critical in the disposal of geothermal effluents into waters which are later to be used for (1) drinking water, (2) freshwater fish, and (3) irrigation. For the latter two uses the figures are only a rough guide, since the quality criteria vary widely with particular fish, crops, and soils. The end uses likely to be most affected by particular elements have been italicized in the table. Water for animals should have characteristics similar to domestic water supplies, except that a higher total salinity is permissible, particularly for cattle and dry sheep (McKee and Wolf, 1971).

Table 8.12 summarizes the typical concentration ranges for critical contaminating constituents in geothermal waters discharged at the surface.

Dilution of geothermal waters at least 20–100 times by a freshwater is necessary if the subsequent mixture is to be used for domestic water supplies. The most critical factors are likely to be arsenic and salinity. Hydrogen sulfide is unlikely to be troublesome, since it is readily

TABLE 8.11

Maximum Permissible Concentration Levels (in g m⁻³) in Waters for Various Uses[a]

	Domestic water supply[b]	Freshwater fish[c]	Irrigation[c,d]
Total dissolved solids	1500	*7000*[e]	*200–1000*
Chloride	600	*250*	*100–200*
Sulfate	400	*200*	*200*
pH	6.5–9.2	*6.5–8.8*	—
F	*0.8–1.7*[f]	1	10
B	30	500	*0.5–1.0*
As	*0.05*	1	1
H₂S	*0.05*[g]	0.3	—
CO₂	—	3 cm³/1	—
Hg	0.001	—	—

[a] The end use most affected by particular constituents is italicized.
[b] W.H.O. International Standards for Drinking Water (1971).
[c] McKee and Wolf (1971).
[d] Sensitive crops, poor soil drainage.
[e] As NaCl.
[f] Depending on temperature.
[g] W.H.O. European Standards for Drinking Water (1970).

TABLE 8.12

Common Compositions of Geothermal Well Waters and Steam Condensates[a]

Constituent	Waters	Condensate
Salinity (as NaCl)	500–200000	0–1
pH	2.5–9 (usually 6–9)	5–8
SO₄	0–200	—
F	1–20	—
B	1–1000 (often 20–50)	0.01–1
	—	20[b]–200[b]
SiO₂	500–1500	—
NH₃	1–100 (often 1–10)	1–50
	—	100[b]–1000[b]
As	1–50 (often 2–10)	0.002–0.05[c]
Hg	0.0001–0.001	0.001–0.05
Cd	10⁻⁵–1 (often 10⁻⁵–10⁻³)	—

[a] Concentrations in parts per million.
[b] The Geysers and Larderello.
[c] The Geysers.

removed by oxidation. For freshwater fish and for irrigation water, similar dilution is required, but in these cases the critical constituents are, respectively, chloride, sulfide, fluoride, pH, carbon dioxide; and salinity and boron. The effects of adding more silica to lake or river water could include an increased growth of diatoms, but there is no record at present of this proving troublesome.

The same order of dilution is required to avoid thermal pollution of a receiving water by geothermal waters at 90°–99°. A temperature rise of a few degrees Celsius can be disastrous for some fish. For example, trout live most satisfactorily at 13°–16° and cannot survive above 23.5°. Their eggs will not hatch above 14.4° (McKee and Wolf, 1971). If, as at Wairakei, river water is also used for steam condensation in the power station, additional heat is liberated into the water, but in situations where this is not acceptable, cooling towers can be used to dissipate the heat of condensation into the atmosphere.

In some areas there may be problems with specific elements. For example, the waters of the geothermal field of El Tatio, Chile, contain 30–40 ppm arsenic and therefore have an unusually high potential for contaminating local drinking waters. The Ahuachapan field in El Salvador is in an area of coffee plantations (a low boron tolerance crop) and even minor concentrations of boric acid are not permissible in the local waterways used for irrigation. The Imperial Valley of southern California contains geothermal systems with exceptionally high water salinity. It is also an area of intensive agriculture, and the liberation of saline fluids at the surface is prohibited. The results of the disposal of geothermal waters into the Waikato River, New Zealand, are given as examples. Figure 8.7 shows the concentrations of chloride and arsenic at various points in the river upstream and downstream of the Wairakei power station. The dilution factor at the station is approximately 100 for average river flows. Immediately following the station the arsenic concentrations are slightly above the W.H.O. permissible level for drinking water (0.05 ppm), but with further dilution downstream acceptable levels are attained.

It is not appropriate to apply simple dilution factor reasoning to all geothermal discharges. Some geothermal waters give rise to precipitates of antimony or arsenic sulfides which concentrate within them metals such as mercury, thallium, gold, and silver, which were present in the deep geothermal waters at parts per billion concentrations or less (Weissberg, 1969). In particular, the concentrations of mercury and thallium in the precipitates may be in the range of 100–1000 ppm. The heavy precipitates may accumulate on the bottoms of local rivers

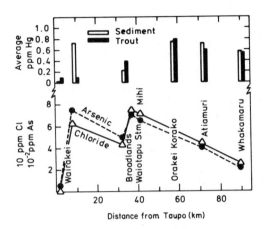

Fig. 8.7   Concentrations of chloride and arsenic in the Waikato River, showing the effects of the Wairakei and Broadlands geothermal areas. The concentrations of mercury in river sediments and in trout caught at various places are also shown.

or lakes, and in some situations high concentrations of objectionable heavy elements may be reached in bottom sediments.

Weissberg and Zobel (1973) reported high mercury levels in sediments of the Waikato river, resulting from natural geothermal areas and from geothermal wells at Wairakei and Broadlands (Fig. 8.7). The maximum mercury concentrations (present as methyl mercury) in trout caught in the river were above the permissible level for human consumption (0.5 ppm).

Disposal of Water

There are several approaches to the environmental problems arising from the disposal of large quantities of geothermal water effluents:

(1)   Reinjection of water back into the field.

(2)   Treatment of the water to remove objectionable constituents, with the possible recovery of valuable by-products, such as pure water, silica, or salts.

(3)   Piping the water away from the area to a more favorable disposal situation, for example, an isolated valley or the sea.

The disposal of aqueous effluents from geothermal power stations in dry steam fields is relatively simple, since only about 25% of the condensed steam needs to be disposed of, the remainder being evaporated by the station cooling towers. Reinjection of dilute steam condensate back into the peripheral part of The Geysers steam field is easily accomplished according to Budd (1973), and reinjection wells accept up to 300 tons/h of condensate, due to the high permeability in the formations.

Reinjection of steam condensate is also successfully being accomplished at Larderello, with careful monitoring of the effects on the production zones (Tongiorgi, 1975).

In a hot water geothermal field, reinjection faces potential difficulties. First, the volume of water to be returned is similar in magnitude to that brought to the surface by production wells unless a desalination/concentration process is operated (a 100-MW station generating from a hot water aquifer at 230°, discharges over 1 ton/sec of waste water). The water may have to be pumped back into the formations. The reinjection wells must be in zones of high permeability to accept the large quantities of fluid, and a knowledge of the temperature and permeability distribution at deeper levels in the field is required, along with the hydrology of the whole area. Recycling of cooled water back into the hot aquifer must be avoided. Bodvarsson (1972) and Drummond and McNabb (1972) reviewed the theoretical aspects of reinjection.

Most waste geothermal waters to be returned to the field will be supersaturated with silica if there is to be optimum utilization of heat. Although amorphous silica may not deposit readily from water flowing rapidly in an enclosed steel pipe, separator, or heat exchanger system, it is known to do so on concrete or rock surfaces. With time, it could therefore reduce the rate of reinjection by blocking fissures in the rock formations, unless the chemical conditions are carefully controlled.

Reinjection of hot water has been successfully operated at the Otake station since 1972 (Ejima, 1975) and at Ahuachapan since early 1976 (G. Cuellar, personal communication). Experimental reinjection is running at Broadlands. At Ahuachapan the reinjected water is at 155°, and in 1976 about 30% of the total water discharge was being reinjected. No losses in permeability were noted. Einarsson (1975) estimated the cost of reinjection to be 1 U.S. mil/kWh of power produced.

Treatment of Waste Waters

Problems with the surface disposal of waste hot water center around purifying the water of objectionable elements or the concentration of solutes into a small volume, With reinjection to the field, high permeability and flow rates are all important. The presence of silica in the water can influence both means of disposal.

Rothbaum and Anderton (1975) showed that geothermal waters treated with a minor quantity of slaked lime produce a flocculant precipitate which consists essentially of hydrated calcium silicates. The precipitates, which settle rapidly, can easily be separated from the waters by decantation or filtration, and can be air-dried to amorphous calcium silicates of low moisture content. When ignited these precipitates give a pure and fine wollastonite. It was discovered that arsenic in the waters was also removed into the calcium silicate precipitate, particularly if the arsenic was first oxidized to arsenate by a small concentration of chlorine. No mercury was removed, and to remove any boron required unrealistically high additions of lime.

Extensive work was done at Wairakei and Broadlands on the lime treatment of geothermal waters. It was found that the precipitation of silica by lime was considerably more efficient if the waters were aged to allow polymerization of silica. For Wairakei well waters and a holding temperature of 85°, silica polymerization was almost complete within 2–3 h; for Broadlands waters the time was approximately 30 min.

Table 8.13 shows the effect of adding various concentrations of lime to hot effluent waters from Wairakei and Broadlands, and Table 8.14 shows the compositions of calcium silicate materials produced. The silica concentrations in the waters were lowered to a level where deposition of amorphous silica could not occur.

A continuous-flow pilot plant was operated at Wairakei and at Broadlands to treat 2000–5000 l/h of well water, and a larger plant to take the discharge of 2–3 wells was built at Broadlands. The optimum SiO_2/CaO ratio in the product was found to be 1.7, when the gel produced had a high solid content and was easily filterable on a rotary drum vacuum filter. Of the calcium silicate formed, 93% was recovered, and the treated water had a stable residual silica concentration of about 100 ppm and a much reduced arsenic concentration.

The calicum silicate material is being evaluated for use as an insulant, an industrial filler, and a paint extender, as well as in low-density building blocks and for high-temperature ceramics.

Alternative means of removing polluting elements from waters are being tried in New Zealand, such as filtering through clay minerals and flocculation with iron hydroxides. The latter is particularly effective for arsenic removal.

TABLE 8.13

The Effects of Adding Freshly Slaked Quicklime to "Aged" Well Waters

CaO addition (ppm)	Chlorine[a] (ppm)	Total silica remaining (ppm)	Monomeric silica (ppm)	Arsenic (ppm)	pH (20°)
Wairakei well; original 99° water "aged" at an average 85°C for 2½ h					
0	0	560	390	4.3	8.0
350	0	140	150	2.5	11.2
350	10	130	140	0.45	11.3
600	0	35	35	0.5	11.6
1000	0	10	10	· 0.12	12.0
Broadlands well 22; original 99° water "aged" at an average 87°C for 30 min					
0	0	910	500	3.8	8.6
460	0	260	270	2.9	10.3
460	20	230	240	1.4	10.4
690	0	95	100	1.4	11.4
690	20	115	120	0.9	11.3
1100	0	7	7	0.1	12.0

[a] Added as NaOCl.

Recovery of Chemicals

The recovery of chemicals from geothermal fluids has been proposed for many fields, but has been in operation on an appreciable scale only at the Larderello plant, where for many years boric acid, ammonia, carbon dioxide, and hydrogen sulfide were extracted from steam flows. However, chemical recovery now is not an important part of the Larderello operations.

Salt production through the use of geothermal heat was initiated in Japan in 1940, but following the 1950s the industry was largely abandoned because of the high production costs. Komagata *et al.* (1970) noted a small salt-making operation at Shikabe, Hokkaido, and that an expansion was proposed in the plant capacity.

Lindal (1970, 1975) outlined a scheme for the production of 250,000 tons/yr of salt, potassium chloride, and calcium chloride from the high-temperature brines from wells on the Reykjanes Peninsula, Iceland. A small pilot plant has operated. The wells produce an altered seawater enriched in potassium and calcium, and depleted in sodium,

TABLE 8.14

Calcium Silicate Material Recovered from Geothermal Waters by CaO Addition[a]

CaO added to water (ppm)	Solid content of gel (by drying at 110°) (%)	Weight loss at 900°	Material dried at 110°			
			SiO_2 (%)	CaO (%)	As (%)	SiO_2/CaO ratio
Wairakei water						
350[b]	10	21.6	47.9	26.6	0.36	1.80
600	7.3	21.9	40.1	30.6	—	1.31
1000	6.7	24.6	31.7	36.1		0.87
Broadlands water						
460	15.1[c]	20.0	52.5	24.5	—	2.14
690[b]	9.2	21.8	47.0	28.0	0.17	1.68
1100[b]	4.0	21.8	29.6	40.1	0.12	0.74

[a] See Table 8.13.
[b] With added chlorine.
[c] Difficult to filter.

magnesium, and sulfate. Magnesium production from seawater was also considered.

There have been several proposals to recover chemicals from the waste discharge waters at Wairakei. Various schemes for recovering lithium were proposed in the 1950s (electrodialysis, ion exchange, and precipitation), but none proved to be economic. The proposals also incorporated the recovery of salt and of potassium, rubidium, and cesium.

In arid regions water can be the most important by-product from geothermal energy production. The production of desalinated water was proposed for the Cerro Prieto area (De Anda *et al.*, 1970). At El Tatio, Chile, an integrated scheme for producing electricity, desalinated water, and various salts has been taken to a pilot plant scale based on 250° waters which contain 1.3% salts, including appreciable potassium, cesium, lithium, and boron.

Werner (1970) reviewed the potential for recovery of salts and heavy metals from the highly saline geothermal well discharges in the Imperial Valley, California. He calculated that there was $260,000 worth of heavy metals such as Au, Ag, Zn, Pb, Ti, available each year for a brine discharge at the rate of 100 ton/h. A scheme for the recovery of common salt, potassium chloride, bromine, rubidium, and cesium was outlined, but because of the intense pipe scaling problem and problems with brine disposal in this area, no development has taken place. At the East Mesa geothermal area of the Imperial Valley a successful pilot plant experiment on the desalination of 150°–220° mineralized waters has been operated (Fernelius, 1975).

REFERENCES

Allegrini, G., and Benvenuti, G. (1970). *Geothermics (Spec. Issue 2)* **2** *(Pt 1)*, 865.

Allen, G. W., and McCluer, H. K. (1975). *Proc. U.N. Symp. Develop. Use Geothermal Resources, 2nd, San Francisco, California, May* **2**, 1313.

Axtmann, R. C. (1975). *Science* **187**, 795.

Baumann, H. (1959). *Kolloid Z.* **162**, 28.

Bodvarsson, G. (1972). *Geothermics* **1**, 63.

Bowen, R. G. (1973). *In* "Geothermal Energy" (P. Kruger and C. Otte, eds.), pp. 197–216. Stanford Univ. Press, Stanford, California.

Budd, C. F. (1973). *In* "Geothermal Energy" (P. Kruger and C. Otte, eds.), pp. 129–144. Stanford Univ. Press, Stanford, California.

Chen, C. H. (1970). *Geothermics (Spec. Issue 2)* **2** *(Pt 2)*, 1134.

Cuellar, G. (1975). *Proc. U.N. Symp. Develop. Use Geothermal Resources, 2nd, San Francisco, California, May* **2**, 1343.

De Anda, F. L., Reyes, S., and Tolivia, M. (1970). *Geothermics (Spec. Issue 2)* **2** *(Pt 2)*, 1632.

Dodd, F. J., Johnson, A. E., and Ham, W. C. (1975). *Proc. U.N. Symp. Develop. Use Geothermal Resources, 2nd, San Francisco, California, May* **3**, 1959.

Dominco, E., and Samilgil, E. (1970). *Geothermics (Spec. Issue 2)* **2** *(Pt 1)*, 553.

Drummond, J. E., and McNabb, A. (1972). *N.Z. J. Sci.* **15**, 665.

Einarsson, S. S. (1961). *Proc. U.N. Conf. New Sources Energy, Rome* **3**, 354.

Einarsson, S. S. (1975). *U.N. Symp. Develop. Use Geothermal Resources, 2nd, San Francisco, California May*, Abstract IV-6.

Ejima, Y. (1975). *U.N. Symp. Develop. Use Geothermal Resources, 2nd, San Francisco, California, May*, Abstract IV-7.

Ellis, A. J. (1960). Some Observations on Silica in Geothermal Waters. Open-file rep. 118/20. Chemistry Div., DSIR, New Zealand.

Ellis, A. J. (1969). *Proc. Commonwealth Min. Metall. Congr., 9th, London* **2**, 1.

Ellis, A. J. (1970). *Geothermics (Spec. Issue 2)* **2** *(Pt 1)*, 516.

Fernelius, E. A. (1975). *Proc. U.N. Symp. Develop. Use Geothermal Resources, 2nd, San Francisco, California, May* **3**, 2201.

Finney, J. P. (1973). *In* "Geothermal Energy" (P. Kruger and C. Otte, eds.), pp. 145–162. Stanford Univ. Press, Stanford, California.

Fournier, R. O. (1973). *In Proc. Symp. Hydrogeochem. Biogeochem., Tokyo, 1970,* **1**, 122–139. Clark, Washington, D.C.

Fujii, Y., and Akeno, T. (1970). *Geothermics (Spec. Issue 2)* **2** *(Pt 2)*, 1416.

Hanck, J. A., and Nekoksa, G. (1975). *Proc. U.N. Symp. Develop. Use Geothermal Resources, 2nd, San Francisco, California, May* **3**, 1979.

Hermannsson, S. (1970). *Geothermics (Spec. Issue 2)* **2** *(Pt 2)*, 1602.

Jones, M. E. (1963). *J. Phys. Chem.* **67**, 1113.

Komokata, S., Iga, H., Nakamura, H., and Minohara, Y. (1970). *Geothermics (Spec. Issue 2) 2(Pt 1)*, 185.

Kitahara, S. (1960). *Rev. Phys. Chem. JPN.* **30**, 131.

Krauskopf, K. B. (1956). *Geochim. Cosmochim. Acta* **10**, 1.

Kryukov, P. A., and Larianov, E. G. (1970). *Geothermics (Spec. Issue 2) 2 (Pt 2)*, 1624.

Laslo, J. (1976). Application of the Stretford Process for H_2S Abatement of The Geysers Geothermal Power Plant. Intersociety Energy Conv. Conf. (A.I.Ch.E., HT, and EC Div.), 1976.

Lebedev, L. M. (1972). *Geochem. Int.* **9**, 485.

Lebedev, L. M., and Nikitina, I. B. (1968). *Dokl. Akad. Nauk SSSR* **183**, 439.

Lindal, B. (1970). *Geothermics (Spec. Issue 2) 2 (Pt 1)*, 910.

Lindal, B. (1975). *Proc. U.N. Symp. Develop. Use Geothermal Resources, 2nd, San Francisco, California, May* **2**, 2223.

McKee, J. E., and Wolf, H. W. (1971). Water Quality Criteria. California State Water Resources Control Board, Publ. No. 3-A.

Marshall, T. (1958). *Corrosion* **14**, 159t.

Marshall, T., and Braithwaite, W. R. (1973). In "Geothermal Energy" (H. C. H. Amstead, ed.), pp. 151–160. UNESCO Earth Sci. Ser. No. 12, Paris.

Martynova, O. I. (1972). *Teploenergetika* **12**, 51.

Mercado, S. (1975). *Proc. U.N. Symp. Develop. Use Geothermal Resources, 2nd, San Francisco, California, May* **2**, 1394.

Noguchi, K., Goto, T., Ueno, S., and Imahashi, M. (1970). *Geothermics (Spec. Issue 2) 2 (Pt 1)*, 561.

O'Connor, T. L. (1961). *J. Phys. Chem.* **65**, 1.

Ozawa, T., and Fujii, Y. (1970). *Geothermics (Spec. Issue 2) 2 (Pt 2)*, 1613.

Reed, M. J., and Campbell, G. E. (1975). *Proc. U.N. Symp. Develop. Use Geothermal Resources, 2nd, San Francisco, California, May* **2**, 1399.

Rothbaum, H. P. (1974). Preliminary Report on Pilot-Plant Desilication of Geothermal Waters Broadlands. New Zealand Open-file rep. 118/16/1. Chemistry Div., DSIR, New Zealand.

Rothbaum, H. P., and Anderton, B. H. (1975). *Proc. U.N. Symp. Develop. Use Geothermal Resources, 2nd, San Francisco, California, May* **2**, 1417.

Sewell, J. R. (1973). Assessment of Some Solid Filters for Removing Hydrogen Sulphide and Sulphur Dioxide From Air. Rep. CD.2168. Chemistry Div., DSIR, New Zealand.

Skinner, B. J., White, D. E., Rose, H. J., and Mays, R. E. (1967). *Econ. Geol.* **62**, 316.

Sourirajan, S., and Kennedy, G. C. (1962). *Am. J. Sci.* **260**, 115.

Stoker, A. K., and Kruger, P. (1975). *Proc. U.N. Symp. Develop. Use Geothermal Resources, 2nd, San Francisco, California, May* **3**, 1797.

Styrikovich, M. A., Khaibullin, I. K., and Tskhvirashvili, D. G. (1955). *Dokl. Akad. Nauk SSSR* **100**, 1123.

Styrikovich, M. A., Tskhvirashvili, D. G., and Nebieridze, D. P. (1960). *Dokl. Akad. Nauk SSSR* **134**, 615.

Tolivia, E. (1970). *Geothermics (Spec. Issue 2) 2(Pt 2)*, 1596.

Tolivia, E., Hoashi, H., and Miyazaki, M. (1975). *Proc. U.N. Symp. Develop. Use Geothermal Resources, 2nd, San Francisco, California, May* **3**, 1815.

Tonani, F. (1970). Geothermics (Spec. Issue 2) 2 (Pt 1), 492.

Tongiorgi, E. (1975). *U.N. Symp. Develop. Use Geothermal Resources, 2nd, San Francisco, California, May*, Abstract IV-18.

Von Stackelburg, M., Quatram, F., and Dressel, J. (1937). *Z. Elektrochem.* **43**, 14.

Weissberg, B. G. (1969). *Econ. Geol.* **64**, 95.

Weissberg, B. G., and Zobel, M. G. R. (1973). *Bull. Environ. Contam. Toxicol.* **9,** 148.

Werner, H. M. (1970). *Geothermics (Spec. Issue 2)* **2** *(Pt 2),* 1651.

White, D. E. (1967). *In* "Geochemistry of Hydrothermal Ore Deposits" (H. L. Barnes, ed.), pp. 575–631. Holt, New York.

White, D. E., Brannock, W. W., and Murata, K. J. (1956). *Geochim. Cosmochim. Acta* **10,** 27.

W.H.O. (1970). "European Standards for Drinking Water," 2nd ed., p. 52. World Health Organization, Geneva.

W.H.O. (1971). "International Standards for Drinking Water," 3rd ed., p. 70. World Health Organization, Geneva.

Wilson, J. S. (1974). *In* "Water 1973," pp. 782–787. Am. Inst. Chem. Eng. Symp. Ser. No. 136.

Yanagase, K., Yamaguchi, K., Yanagase, T., Suginohara, Y., Kozawa, S., and Yamazaki, H. (1970a). *Nippon Kagaku Zasshi* **91,** 1141.

Yanagase, T., Suginohara, Y., and Yanagase, K. (1970b). *Geothermics (Spec. Issue 2)* **2** *(Pt 2),* 1619.

Chapter 9

PHYSICAL AND CHEMICAL CHARACTERISTICS
OF NEW ZEALAND GEOTHERMAL AREAS

INTRODUCTION

This chapter discusses various scientific aspects and characteristics of different New Zealand geothermal systems. Particular attention is given to the geological and physical settings of the systems and the geochemistry of the hot fluids present. The approach and methods outlined in this chapter are applicable and pertinent to geothermal systems throughout the world.

9.1 LOCATION OF GEOTHERMAL AREAS IN NEW ZEALAND

The Taupo Volcanic Zone in the North Island of New Zealand is considered on geophysical evidence to occur on the southern extension of the Tonga–Kermadec Ridge, an active plate margin. Active volcanism and major tectonic movements have created many areas where geothermal gradients are unusually high. In the South Island of New Zealand a major transcurrent fault, the Alpine Fault, runs along the eastern margin of the Southern Alps. Rapid uprise of the country and tectonic activity have created a high geothermal gradient, and waters circulating within the metamorphic rocks give rise to many warm springs (up to 85°). Figure 1.4 gave the positions of the major North Island geothermal areas, while details of hot spring occurrences in the South Island were summarized by Ellis and Mahon (1964).

The Taupo Volcanic Zone extends some 250 km between the two active andesite volcanoes of Ruapehu in the center of the North Island

and White Island in the Bay of Plenty, and contains a large number of geothermal areas associated with rhyolitic and andesitic volcanism of Quaternary age. The areas contain large numbers of boiling springs, geysers, fumarolic activity, steaming ground, and extensive areas of hydrothermally altered rock. At Ngawha in North Auckland, high heat flows and hot springs occur in areas with basaltic volcanism of Pleistocene age. In the Hauraki Volcanic Province in the northeast of the North Island (Nathan, 1974), warm waters are associated with late Tertiary andesitic and rhyolitic volcanism.

During the last 30 years most geothermal areas in the Taupo Volcanic Zone of the North Island were surveyed to determine their potential for electricity generation or as industrial heat sources. Many were drilled to depths of about 700 m. These investigations led to the establishment of the Wairakei geothermal power station 8 km north of Lake Taupo, and to current plans to develop the Broadlands area. This chapter discusses the physical and chemical characteristics of some of the geothermal areas which have been investigated by drilling.

9.2 GEOLOGY OF THE VOLCANIC ZONE OF NEW ZEALAND

Nine of the ten geothermal areas investigated by deep drilling are located in the Taupo Volcanic Zone; the tenth, at Ngawha, is associated with Pleistocene basaltic volcanism of Northland (Grindley, 1961; Healy, 1961; Kear and Hay, 1961).

Over 16,000 km³ of predominantly acid volcanic lava and pyroclastics were erupted from the Taupo Volcanic Zone during late Pliocene and Quaternary times. The rhyolite volcanic areas include the Taupo–Reporoa, Tarawera, and Whakatane depressions, which are strongly faulted, actively subsiding structural depressions with a thick filling of pumice pyroclastics that constitute aquifers in the geothermal areas. The hot springs of highest temperatures and water flows occur in the Taupo zone and some have been established for at least 500,000 years. The different geothermal areas are closely associated with active northeast-trending faults resulting from the subsidence of major structural depressions. The faults may provide ascent channels for hot waters and it is relevant that the geothermal areas in the present active zone of subsidence appear to be located in places where the northeast faults are crossed by minor northwest faults. Figure 9.1 shows the major structural features of the Taupo Volcanic Zone.

The rocks in the geothermal areas generally form layered sequences

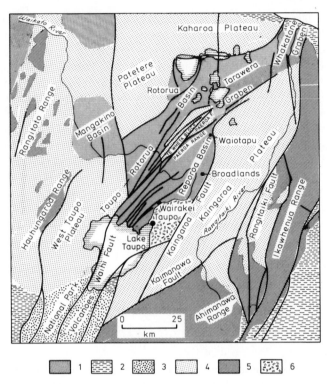

Fig. 9.1 Map of the Taupo Volcanic Zone showing surface geology and major structural features (adapted from Grindley, 1965a). 1, Quaternary freshwater sediments, lake beds, rhyolite domes, and pyroclastics; 2, Tertiary–Quaternary marine sediments; 3, Quaternary andesite (National Park); 4, Quaternary ignimbrites with minor rhyolites, andesites, and basalts; 5, Permian–Mesozoic greywackes and argillites; 6, Holocene pumice pyroclastics.

in which pumice breccias or tuffs and jointed andesite, rhyolite, or ignimbrite form permeable aquifers, and sandstones and mudstones form impermeable strata. In some geothermal areas, such as Wairakei and Waiotapu, a pumice aquifer is capped by lake sediment beds or ignimbrite sheets, whereas in others, such as Rotokaua and Orakeiko-rako, the pumice aquifer is open. On account of the generally high permeability of the strata, no major undetected reservoirs of hot water or steam are thought to exist within 1200 m of the surface.

The basement rocks underlying the structural depressions are Per-mian–Mesozoic, closely folded and indurated greywacke and argillite, phyllite, and subschists, similar to those exposed in the bordering ranges and plateaus. In the Kawerau and Broadlands geothermal areas, wells produce hot water from basement greywacke as well as from overlying volcanic rocks.

Steiner (1958) and Clark (1960) suggested a genetic relationship between the ignimbrites, rhyolites, and more basic rocks of the Taupo Volcanic Zone. They considered that all were formed from two primary magma types, basaltic and acidic, from which the intermediate types were produced by intermixing. Healy (1962a) pointed out that the rhyolites are concentrated in the widest part of the zone, where the structure is complicated by extensive folding and large-scale subsidence. Melting of the upper part of the crust has occurred to produce acid magma. Healy suggested that following large-scale eruptions of acid rocks, tongues of basaltic melt were drawn up to produce basaltic eruptions of a limited magnitude. Large-scale eruptions of basalt, or andesite produced by mixing of basalt with acid magma, would be expected to originate from lower levels than the rhyolitic magma. Support for this is found in the association of volcanoes of the andesitic-basaltic type with deep-seated fractures rather than with wide zones of superficial faulting. The alignment of the Tongariro volcanoes is significant in this respect.

The structure and volcanism of the Taupo Volcanic Zone were studied by Modriniak and Studt (1959) using geophysical techniques. A gravity survey outlined two major structural features: the Kaingaroa plateau, in which the Mesozoic basement rocks were above sea level; and the Taupo–White Island depression, in which the basement reached depths of at least 3700 m below sea level. Between the two were basement blocks at various stages of subsidence. The subsurface structure closely correlates with surface topography, indicating the recent structural evolution, and strong evidence exists of a genetic relationship between the geothermal areas and basement faults.

Magnetic surveys showed that apart from the ignimbrites, the volcanic rocks were concentrated in structural depressions. The dacites and andesites occurred over partly subsided basement blocks, whereas rhyolites appeared in the deepest depressions. This suggested an andesite–rhyolite sequence of volcanism, repeated as subsidence progressed from one part of the area to another.

A characteristic feature of volcanism in the Taupo Volcanic Zone is the small amount of basalt. Drilling to maximum depths of 2165 m in the various geothermal areas has seldom intercepted basalt. The predominance of acid rocks in the geothermal areas and their chemical uniformity has a considerable bearing on the chemical composition of the waters present in the various systems.

9.3 EXPLORED GEOTHERMAL AREAS

The geothermal areas of the North Island which have been investigated by deep drilling include, from south to north: Tauhara, Wairakei, Rotokaua, Broadlands, Reporoa, Orakeikorako, Te Kopia, Waiotapu, Kawerau, and Ngawha (see Fig. 1.4). Three other areas in the
central volcanic zone, Tokaanu-Waihi, Rotorua, and Tikitere, have
been investigated by surface scientific surveys and shallow wells. All
the areas contain hot chloride waters and most contain steam-heated
surface waters at shallow levels.

Investigation Well Data

A summary of data from the investigation wells drilled in the
various areas, taken from Smith (1958) and from records of the New
Zealand Ministry of Works and Development, is shown in Table 9.1.

Approximately 100 wells have been drilled in the Wairakei geothermal area by techniques outlined by Fooks (1961) and Stilwell (1970).
Sixty of these are operating production wells, 20 are perimeter investigation wells capable of production, and 20 are shallow, narrow casing,
or nonproducing wells. The average depth of production wells at
Wairakei is about 600 m and solid production casing is set to
approximately 460 m. Some investigation wells are deeper. The
maximum downhole temperature recorded at Wairakei was 274° at
2254 m in well 121, but the average maximum temperature in production wells was 250°–260° before extensive exploitation commenced.

TABLE 9.1

Physical and Drilling Data of Geothermal Investigation Wells in New Zealand

Area	Number of wells	Solid casing depths (m)	Total drilled depths (m)	Maximum downhole temp. (°C)
Tauhara	4	375–922	1047–1227	257–292
Rotokaua	2	494–612	888–1210	292–306
Broadlands	31	417–1547	760–2418	143–300
Waiotapu	7	257–309	455–1110	203–295
Reporoa	1	601	1351	234
Orakeikorako	4	486–646	1157–1411	224–266
Te Kopia	2	565–597	948–1257	233–241
Kawerau	16	306–613	433–1233	237–286

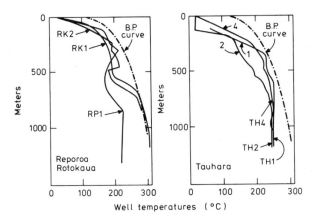

Fig. 9.2 Temperatures measured at depth in wells in the Rotokaua, Reporoa, and Tauhara fields, compared with the boiling point–depth curve (from Smith, 1970).

Maximum temperatures and temperature gradients in exploration wells in geothermal areas were shown by Smith (1970), and some of these are summarized in Fig. 9.2. The maximum temperatures ranged from about 230° to 306°. Downhole pressures throughout the drilled areas were generally several tens of bars above water vapor pressures, at least before production. In many cases gas pressures in the deep waters were significant and ranged from 1 bar at Wairakei up to tens of bars at Broadlands, Kawerau, and Ngawha.

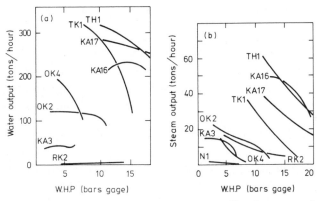

Fig. 9.3 The outputs of steam and water at various wellhead pressures from wells in the Tauhara (TH), Te Kopia (TK), Kawerau (KA), Orakeikorako (OK), and Rotokaua (RK) fields (from Smith, 1970).

Well Performance

Smith (1970) summarized the mass output data for New Zealand geothermal wells. Since the majority of the wells drilled had 20-cm production casing, individual outputs reflect rock permeability in the different areas. Figure 9.3 (from Smith, 1970) shows the outputs of some investigation wells.

9.4   CHEMISTRY OF STEAM AND WATER DISCHARGED FROM SPRINGS, FUMAROLES, AND WELLS

General

Chemical investigations of natural hot waters and steam in geothermal areas within the volcanic zone extend back at least 100 years. Many of the earlier chemical analyses and interpretations were summarized by Grange (1937) and later by Wilson (1955). More detailed investigations were commenced in the thermal areas about 1950 and were concerned with geothermal developments. Summaries of the regional chemistry of the natural hot fluids in the Taupo Volcanic Zone were given by Ellis and Mahon (1964, 1967), and that of the Ngawha area by Ellis and Mahon (1966). Tables 9.2 and 9.3 summarize the chemical compositions of water and steam discharged from a selection of hot springs, fumaroles, and wells. A high proportion of the surface hot waters and practically all the deep waters have many chemical similarities which may be related to the similar compositions of the rhyolite, dacite, and andesite volcanic rocks. The deep waters are all of the near-neutral-pH alkali chloride type.

Compared with geothermal waters in Chile, El Salvador, and the Imperial Valley of Southern California, New Zealand waters are relatively dilute, but they are more saline than most hot spring waters of Iceland and Turkey and a high proportion of those in the western United States.

The salinity of the New Zealand waters ranges from approximately 6000 ppm at Tokaanu in the extreme south of the volcanic zone to about 600 ppm at Orakeikorako and Tikitere. Figure 9.4 shows representative salinities as equivalent sodium chloride, and the total mineral contents of hot waters discharged from geothermal wells in various areas.

A variety of other chemical types of surface hot waters occur in the

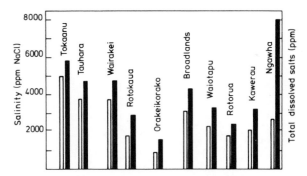

Fig. 9.4 Salinity (open columns) and total dissolved solids (black columns) in water from some New Zealand geothermal areas.

region; for example, White Island, an active andesite volcanic crater, has pools of acidic water high in mineral content. Steam temperatures of about 800° have been measured in fumaroles on the island and the steam has a high content of sulfur gases (H_2S and SO_2) and hydrogen chloride, together with ammonia, hydrogen fluoride, and boric acid. Fumarole condensates up to 1 *m* in hydrochloric acid have been collected. Wilson (1959), and Giggenbach (1975) gave details of the chemistry of hot pools on the island.

Acid sulfate–chloride waters, frequently containing comparable concentrations of chloride and sulfate, and acid sulfate waters low in chloride are present in many New Zealand geothermal areas (Ellis and Mahon, 1964). At Rotokaua there is some evidence that acid sulfate-chloride spring waters are formed when hot chloride water rises through sediments containing sulfur. However, the more usual cause of acidity is the oxidation of sulfide in steam-heated waters. The acid water compositions are dominated by rock leaching and have little bearing on the nature of the geothermal systems.

Neutral-pH sodium sulfate–bicarbonate waters with sulfate and bicarbonate present in approximately similar proportions occur in many areas. They are generally superimposed at shallow levels on the main chloride water aquifer, or located on the perimeter of the chloride systems. Water of this type can exist to a considerable depth and, for example, well 6 at Broadlands penetrated 600–700 m of aragonite-depositing sodium sulfate–bicarbonate water. These are groundwaters heated with steam from deeper chloride water. Acid sulfate, created by hydrogen sulfide oxidation, is neutralized by rock-water interaction, and carbon dioxide also reacts with rock to produce bicarbonate. Since calcium and magnesium sulfates and carbonates

TABLE 9.2

Geothermal area		Concentration (ppm)										
	pH	Li	Na	K	Rb	Cs	Ca	Mg	F	Cl	Br	I
Wairakei												
Well 4/1	7.9	12.6	1300	192	2.9	2.2	23	0.02	6.9	2140	5.7	0.2
Well 20	8.4	13.8	1300	220	3.1	2.6	18	0.005	8.2	2215	5.5	0.4
Well 44	8.6	14.2	1320	225	2.8	2.5	17	0.005	8.3	2260	6.0	0.3
Spring 97	6.8	6.8	665	68	0.7	1.7	45	4.2	4.4	1110	2.5	0.4
Spring 190	7.5	10.0	950	62	1.0	n.d.	20	0.05	5.8	1596	n.d.	n.d.
Waiotapu												
Well 6	8.9	6.6	860	155	2.4	0.8	10	0.06	7.5	1450	4.7	0.2
Well 7	8.8	6.4	790	90	0.7	0.9	10	0.05	5.3	1310	3.7	1.0
Spring 64 (Champagne Pool)	5.7	9.0	1220	160	2.3	1.7	35	n.d.	5.5	2000	7.2	0.4
Spring 20	8.6	4.0	450	22	0.3	0.6	9	0.08	5.2	688	2.0	0.8
Kawerau												
Well 7A	8.6	5.35	745	123	0.75	0.5	2.6	0.16	4.3	1234	5.05	0.2
Well 8	9.0	5.5	740	130	0.9	0.5	1.1	0.39	5.2	1262	6.2	0.2
Well 17	8.6	5.3	688	119	0.5	0.45	2.0	0.32	3.3	1220	5.1	0.2
Spring 2	6.2	2.7	330	49	n.d.	n.d.	13	0.9	1.4	445	1.4	0.7
Spring 4	7.6	3.3	398	53	0.27	0.25	13	1.0	1.9	544	1.6	0.8
Rotorua												
Well 219	9.4	2.5	375	35	0.27	0.23	1	0.06	n.d.	355	0.1	0.2
Well 137	9.4	1.4	565	31	0.26	0.31	1	0.22	4.0	632	2.1	0.7
Spring 83	9.05	4.7	485	58	0.5	0.4	1.2	n.d.	6.4	560	n.d.	n.d.
Orakeikorako												
Well 2	9.1	3.1	550	54	0.2	0.3	1.0	0.2	5.7	546	0.8	1.0
Well 3	9.0	3.1	254	40	0.18	0.27	1.0	0.15	17.0	301	0.6	0.2
Spring 98	8.3	4.0	280	42	0.15	0.22	2.5	0.5	8.5	284	0.6	0.2
Broadlands												
Well 2	8.5	11.4	1075	222	2.2	1.9	2.9	0.08	7.25	1844	6.4	0.3
Well 10	8.9	9.6	890	150	1.2	1.2	2.1	0.1	6.5	1262	3.9	0.4
Spring 1	7.0	7.4	860	82	0.1	1.2	2.6	0.48	5.2	1060	3.0	0.6
Tauhara												
Well 1	8.0	13.8	1275	223	2.5	1.9	14.0	0.07	6.8	2222	4.2	2.1
Taupo Spring	7.4	4.6	405	47	0.23	0.13	11.0	2.3	1.1	537	n.d.	n.d.
Rotokaua												
Well 1	8.45	2.4	695	120	0.5	0.5	3.7	0.15	9.6	1100	n.d.	0.2
Spring 6	2.5	7.8	990	102	1.7	2.0	12	10	1.0	1433	4.0	0.5
Ngawha												
Well 1	8.1	10.7	900	78	n.d.	n.d.	29	1.4	n.d.	1658	n.d.	n.d.
Jubilee Spring	6.4	8.0	830	63	0.3	0.55	7.8	2.5	0.3	1250	2.6	1.0
Tokaanu												
Spring 6	7.25	22.3	1750	165	1.1	2.8	36	0.5	1.5	3064	5.0	2.4
Spring 23A	7.3	17.3	1330	142	0.9	1.5	50	0.6	0.85	2319	4.1	1.6

[a] Concentrations are those present at atmospheric pressure and boiling point temperature. Springs listed have temperatures between 98°-100°. Temperatures in the production zones of the wells listed are as follows: Wairakei, 4/1-240°, 20-245°, 44-244°; Waiotapu, 6-240°, 7-238°; Kawerau, 7A-258°, 8-262°, 17-256°; Rotorua, 219-150°, 137-160°; Orakeikorako, 2-225°, 3-234°; Broadlands, 2-260°, 10-250°; Tauhara, 1-255°; Rotokaua, 1-258°; Ngawha, 1-225°.

Analysis of Waters from Springs and Wells in New Zealand Geothermal Areas[a]

		Concentration (ppm)					Ratios			
SO$_4$	As	Total SiO$_2$	Total HBO$_2$	Total NH$_3$	Total carbonate as CO$_2$	Total H$_2$S	Na/K	Na/Ca	Cl/B	Cl/F
33	4.3	590	115.5	0.25	25	1.5	11.5	100	29.9	165
35	4.7	640	118.5	0.20	17	1.7	10.0	125	23.1	145
36	4.8	640	117	0.15	19	1.4	10.0	135	23.9	145
72	1.8	235	57	0.22	88	1.9	16.6	26	24.1	135
56	2.8	245	82	0.37	38	2.0	26	83	24.0	145
52	3.1	620	56	0.9	65	2	9.4	150	32	105
86	2.9	600	63	0.8	90	2	14.9	140	25.7	130
145	4.9	490	117	11.5	235	6	13.0	60	21.1	195
93	1.1	380	27	0.4	58	5	35	87	31	71
9	—	780	255	0.75	68	6	10.3	500	6.0	154
10.5	—	815	248	0.5	17	4	9.7	1170	6.3	130
11.0	—	750	233	1.2	57	4	9.8	600	6.5	198
158	—	245	85	4.0	85	10	11.5	44	6.5	170
96	—	240	102	4.0	110	6	12.8	53	6.6	150
12	0.08	405	25.4	0.05	206	36	18.2	700	17.2	—
30	0.30	314	32.3	0.05	143	74	31	1000	24.2	85
88	n.d.	490	21.6	0.2	167	14	14	700	32	47
142	0.8	480	31.5	0.1	295	6	17.3	955	21.4	51.3
64	0.6	546	14.4	0.1	50	5	10.7	440	25.8	9.5
220	0.3	280	13.6	0.55	190	1.3	11.3	190	25.8	17.9
3.5	5.6	848	216	1.6	100	1.0	8.2	480	10.6	136
6.0	3.1	695	2205	1.2	400	2.0	10.1	737	6.9	104
100.0	1.0	338	130	3.8	490	1.0	17.8	575	10.1	109
30	4.1	726	153	0.1	13	1.0	9.7	160	18.0	175
101	0.4	235	38	0.1	180	0.4	14.7	64	17.5	260
46	n.d.	760	134	1.6	62	1.0	9.8	326	10.1	61
520	n.d.	340	183	1.6	144	0.2	16.5	140	9.7	800
20	n.d.	475	4800	55	14.5	1.0	19.6	53	0.43	—
347	0.2	178	3690	140	490	n.d.	22.0	185	0.42	2220
65	5.6	305	364	1.2	1.5	1.0	18.0	845	10.3	1100
39	3.7	240	279	1.7	95	1.0	15.9	46	10.3	1450

TABLE 9.3

Composition of Steam and Gas in Some New Zealand Geothermal Areas

	Moles of gas per	Composition of gas (% by volume)				
Area (well number)	kg water in total discharge	CO_2	H_2S	$\overline{HC}$ [a]	H_2	N_2
Ngawha (1)	0.5	93.9	0.7	3.9	0.5	1.0
Kawerau (8)	0.11	94.0	2.6	2.1	0.3	1.0
Wairakei (27)	0.012	90.0	4.1	2.1	0.5	2.4
Waiotapu (6)	0.026	88.0	10.3	0.2	1.0	0.5
Broadlands (2)	0.25	94.4	₁.6	2.2	0.2	1.5
Karapiti fumarole, Wairakei	0.012	94.5	2.3	1.1	1.0	1.1
Tauhara fumarole	0.01	96.4	1.3	0.5	0.4	1.4

[a] $\overline{HC}$ = total hydrocarbons.

have low solubilities at high temperatures, sodium is the predominant cation.

The water from Ngawha springs and from a 600-m well in the area comes from an aquifer in sedimentary rocks. In some respects this water differs little in constituent concentrations from those in the Taupo Volcanic Zone. However, the Ngawha waters have high concentrations of boron, ammonia, and carbon dioxide derived from the local sediments. Cinnabar and native mercury also occur at Ngawha in association with the spring sinters (Ellis and Mahon, 1966).

Geochemistry of Constituents

Sodium and potassium concentrations in the waters reaching the surface range from about 100 to 1800 ppm and from 20 to about 300 ppm, respectively. Highest concentrations of potassium are present in the waters from the Wairakei, Tauhara, and Broadlands production wells. The atomic ratios of Na to K range from less than 10 for high-temperature wells to over 30 for major spring waters. Lithium ranges from less than 1 ppm up to the highest concentration of 25 ppm in Tokaanu spring waters. Ratios of Na to Li range from 20 to 50 with one or two exceptions and a large number of areas have Na/Li ratios in the range of 20–30. Rubidium and cesium are present at a level of 0.01– 0.001 of the potassium molal concentration (Ellis and Wilson, 1960; Golding and Speer, 1961). The highest concentrations of both these constituents are found in well waters at Wairakei, Broadlands, and Waiotapu.

The Na/K and Na/Rb ratios are controlled by temperature-dependent mineral solution equilibria. There appears to be no simple relationship, however, between Na/Li ratios and temperature. Table 9.4 shows the temperatures and the atomic Na/K, Na/Rb, and Na/Li ratios for a selection of deep waters at Broadlands.

Calcium concentrations in the deep well waters rarely exceed 50 ppm (Mahon, 1965a) and are influenced by water temperature, pH, and carbon dioxide concentrations. For example, the low calcium concentrations at Orakeikorako result mainly from the low salinity and consequently high pH of the water (Chapter 4).

Magnesium concentrations are so low in the deep waters (0.001–1.0 ppm) that in the past they have been difficult to determine. The lowest concentrations appear to occur in the dominantly rhyolitic areas such as Tauhara, Wairakei, Waiotapu, and Broadlands. Higher concentrations in the spring waters are generally due to mixing with local meteoric waters which frequently contain magnesium leached from surface rocks (Mahon, 1965a).

Appreciable concentrations of arsenic, up to approximately 6 ppm, occur in the deep hot waters (Ritchie, 1961), and the atomic ratio of Cl to As is frequently around 1000. At Rotorua the ratios are considerably higher than the average.

The Cl/B ratio in the waters of different areas is frequently similar to that in the major aquifer rocks (Mahon, 1970a). Thus, in areas of andesite such as Tokaanu and Kawerau, Cl/B ratios of less than 10 are common; in areas where rhyolite and ignimbrite predominate, as for example, Wairakei, Waiotapu, Orakeikorako, and western Broadlands,

TABLE 9.4

Relationship between Na/K, Na/Rb, and Na/Li Ratios and Maximum Downhole Temperatures for Waters Discharged from Broadlands Wells

Well number	Temp. (°C)	Na/K	Na/Rb	Na/Li
19	267	7.8	1605	25
2	262	8.0	1775	27
17	255	8.2	2000	30.2
18	240	8.5	2240	29.9
16	184	11.4	4155	30.2

the values are often between 10 and 30; while in areas where sedimentary rocks influence the water composition, such as at Ngawha and in the eastern Broadlands field, the ratios are between 0.1 and 5.

Figure 9.5, taken from Ellis (1970), gives the molecular proportions of chloride, bicarbonate, and boron in a range of waters in the New Zealand geothermal areas, including both hot spring and well waters. The homogeneity of the waters in individual areas is evident, except for Waiotapu and Broadlands.

Fluoride concentrations in the deep waters are in the range of 1–12 ppm (Mahon, 1964). Maximum concentrations are influenced by underground temperatures and the solubility of calcium fluoride. Spring waters in an area often have lower Cl/F ratios than water tapped from the deep reservoirs, due to leaching of fluoride from the rocks by water as it cools during ascent to the surface.

With few exceptions the Cl/Br atomic ratios in the waters are in the range of 600–1100, in comparison with 660 for seawater (Ellis and Anderson, 1961). Iodine concentrations are uniformly low—values of 0.1–1 ppm occur frequently. Higher values of both bromide and iodide are sometimes found in springs where contamination from surface vegetation has occurred.

Sulfate concentrations are rather erratic but values above 100 ppm

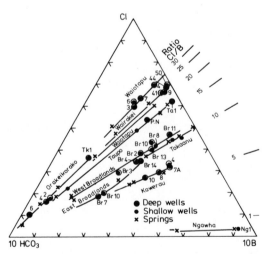

Fig. 9.5 The interrelationships between molal concentrations of chloride, boron, and bicarbonate in spring and well waters from various New Zealand areas (from Ellis, 1970).

are rare in the deep waters. Lowest concentrations occur in the highest-temperature waters, and for example at Kawerau and Broadlands, at temperatures close to 300°, concentrations are in the range of 5–20 ppm. It is of interest that values of around 15 ppm would correspond approximately to the sulfate that would be formed by quantitative oxidation of sulfide by the oxygen dissolved in meteoric water at air temperatures. Sulfate concentrations in surface springs of small flow are frequently high due to the oxidation of hydrogen sulfide.

After a detailed investigation of the silica concentrations in the waters from deep Wairakei wells, Mahon (1966) concluded that the waters were in equilibrium with quartz at the measured downhole temperatures. In areas of highest underground temperatures (e.g., Broadlands and Kawerau), the waters discharged from wells contain up to 1050 ppm silica at atmospheric pressure and boiling point temperature.

Boiling surface springs of good flow ($\approx$0.5 l/sec) usually give reasonably representative samples of the upper portions of the deep water in a geothermal system. However, as the water cools in rising through the country, the silica concentrations may decrease by deposition in the rock, so that the "silica temperature" may be on the low side. Below 150°–200° the quartz–solution reaction becomes too slow to maintain equilibrium in ascending water and further deposition in the rocks takes place only when the solubility of the amorphous silica is exceeded, close to or at the surface. Silica temperatures of this magnitude are therefore common.

The concentrations of minor constituents in the deep neutral-pH chloride waters have been determined in a number of areas. Some results were given in Table 3.3, and for the deep waters at Wairakei further results were as follows: strontium, 0.1; barium, 0.01; phosphorus, 0.05; selenium, 0.01; and molybdenum, 0.02 ppm.

Chloride water springs frequently contain concentrations of minor constituents similar to those in the deep water of an area. Arsenic, antimony, mercury, gold, and thallium, however, may be precipitated at shallow levels or in the rocks (Chapter 3). Re-solution of the elements may also occur in some hot waters at the surface. Acid spring waters contain appreciable quantities of iron, aluminum, and other metals from rocks as a result of leaching.

Although the deep waters in the Taupo Volcanic Zone contain only minor concentrations of trace metals, sinters surrounding the springs in a number of areas contain appreciable concentrations of these constituents. Also, in at least some areas (Broadlands and Waiotapu) mineralized bands of sphalerite, galena, and chalcopyrite were intersected in silicified fracture zones during drilling (Chapter 3).

There is a limited amount of published data on the isotopic composition of New Zealand hot waters. A preliminary survey of deuterium ratios was carried out by Hulston (1962), and Banwell (1963) reported oxygen and hydrogen isotope ratios for a representative number of hot waters from the Taupo Volcanic Zone. Additional unpublished values for Wairakei were measured by Stewart (personal communication) and a resurvey was recently completed (Hulston, 1975). An investigation of the hydrogen and oxygen ratios in waters of the Ngawha area was described by Macdonald (1966) and for Broadlands by Giggenbach (1971). Figure 3.1 included δD and δ¹⁸O relationships for Broadlands and Wairakei. A feature of the results is the small ¹⁸O shift observed for Wairakei waters in comparison with that for many other geothermal areas.

Many other features of isotopic work relating to hydrology or geothermometry in New Zealand geothermal areas were discussed in Chapters 3 and 4, while changes with time for Wairakei well discharges were summarized in Chapter 8.

The steam discharged from fumaroles in geothermal areas of the Taupo zone is derived from the boiling of hot water beneath the areas. The major gases in the steam are carbon dioxide and hydrogen sulfide, with smaller quantities of hydrogen, nitrogen, ammonia, hydrocarbons, and inert gases. The composition of major fumarole steam is usually very similar to that of steam in well discharges in an area.

The gas concentrations in the deep waters vary from area to area, being highest in sedimentary rock environments, as at Ngawha, and low in areas of predominantly volcanic rock, as at Wairakei. The highest concentrations of ammonia also occur in waters of sedimentary areas, and the lowest in waters within rhyolitic rocks.

9.5 ELECTRICAL PROSPECTING AND HEAT SURVEYS

Electrical Prospecting

Macdonald (1967) presented data from electrical resistivity measurements from Taupo to Waiotapu, using a Wenner electrode array. Results of similar measurements made in the Kawerau, Tikitere, and Rotorua areas are available from Macdonald *et al.* (1970) and from Macdonald (1974). Electrical measurements have also been made in less detail at Tokaanu, Orakeikorako, and Ngawha.

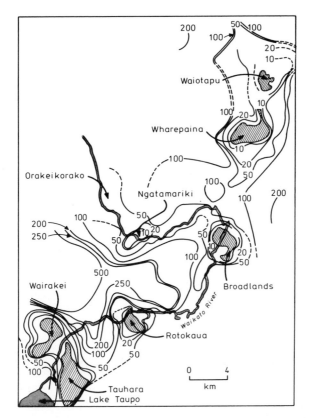

Fig. 9.6 Resistivity contours (in ohm meters) in the central part of the Taupo
Volcanic Zone from Taupo to Waiotapu (adapted from Macdonald, 1967).

Figure 9.6, taken from Macdonald (1967), shows resistivity contours
over the Taupo–Waiotapu area, the shaded zones representing areas
where the resistivity is 5 ohm m and less. The salinities of the deep
waters in a number of areas in the Taupo zone commonly lie in the
range of 0.01–0,1 mole of NaCl per kilogram, the rock porosity is
frequently about 10–20%, and temperatures are usually in the range of
220°–300°. The resistivity of rocks under these chemical and physical
conditions is 5 ohm m or less (see Chapter 4) but areas where the
resistivity is 10 ohm m or less were generally considered as significant
during the exploration.

Macdonald (1967) listed the size of the area where the apparent
electrical resistivity was 10 ohm m or less for different geothermal
fields. This list, together with similar results from other fields, is

TABLE 9.5

Areas of Low Resistivities and Natural Heat Flow in Some New Zealand Geothermal Areas

Geothermal area	Area (km²) within		Natural heat flow	
	5 ohm m contour	10 ohm m contour	(kcal/sec)	Year of measurement
Tauhara	14	⎫32	25,000	1960
Wairakei	8	⎭	101,000	1958
Rotokaua	4	7.5	50,000	1951–1952
Broadlands	6.5	11.5	20,000–45,000	1967–1971
Reporoa	10	14.5	—	—
Orakeikorako	—	—	82,000	1960
Waiotapu	3	14	134,000	1957
Tikitere	0.7[a]	4.4[a]	28,800	1972
Kawerau	6[b]	10[b]	18,000	1962

[a] Macdonald and Dawson, penetration depth approximately 460 m.
[b] Macdonald, Muffler, and Dawson, penetration depth approximately 460 m.

shown in Table 9.5. The total area within the 5-ohm m contours was 51.1 km² and within the 10-ohm m contours 93.9 km². On many of the field margins there was only a small separation between 5-, 10-, and 20-ohm m contours, suggesting a steep temperature gradient between hot and cold areas. Less steep boundaries occurred where there was runoff of hot water from the main systems.

A series of electrical soundings were made at Orakeikorako and Te Kopia using a Wenner electrode configuration with a maximum interelectrode spacing of 1097 m (Macdonald, 1966). The lowest values of 15–25 ohm m were considerably higher than for many of the other geothermal areas due to the low salinity of the deep hot waters.

On the basis of electrical resistivity measurements Hochstein (1975) considered that the shallower parts of the ten most extensively investigated geothermal areas in New Zealand had a total volume of between 15 and 40 km³ and that the electrical energy-generating potential of this volume was of the order of 2000 MW.

Keller (1969) carried out electromagnetic surveys to deep levels across the Taupo Volcanic Zone to check on the possible existence of active magma chambers. Surveys were made on such a scale that electrical properties might be determined to depths of 35 km beneath the thermal areas. Preliminary interpretations suggested that no large volume of molten rock lay within the reach of the surveys, but further work on the properties of molten rock was required to substantiate this.

The electrical structure of the volcanic zone was summarized by Keller as follows:

(1)   The main geothermal areas (Tauhara, Wairakei, Rotokaua, Broadlands, Waiotapu) appeared to lie along a conductive vertical slab having a width of 1-2 km and a depth of at least 5 km. The resistivities within this slab were in the range of from 5 to 10 ohm m, considerably higher than would be anticipated for molten rock, and very probably representative of porous rock filled with warm or hot water.

(2)   Greywacke rocks to the east of the zone had a moderately high resistivity (1500-2000 ohm-m) to a depth of 5 km. The resistivity beneath these rocks may be as high as 10,000 ohm m.

(3)   Volcanic rocks to the west of the zone exhibited a moderately high resistivity, similar to that of the eastern greywackes, to considerable depths.

(4)   The rocks in the lower part of the crust, below a depth of about 20-30 km, had relatively low resistivities, ranging from 50 to 200 ohm m.

Heat Flow

The natural heat flows from the majority of geothermal areas in the volcanic zone have been measured and the results are included in Table 9.5. Results are given to a base temperature of 12°. Values are from Dawson and Dickinson (1970) except the Tikitere result (from Dickinson, 1972). The results in this table can be compared with the figure of 7.3×10^7 kcal/sec quoted by Bolton (1975) as the heat output above 0°C for the total geothermal resources of New Zealand.

9.6   GEOLOGICAL AND CHEMICAL CHARACTERISTICS OF INDIVIDUAL AREAS

Wairakei

Geology and Geophysics

The geology of the Wairakei geothermal area was described by Grindley (1965a), while the petrology and palynology are available from the work of Steiner (1955) and Harris (1965), respectively. Figures

9.7 and 9.8 show a geological map and a geological section through the Wairakei area (Grindley, 1965a).

The area consists of a rhyolite pumice breccia aquifer (Waiora formation) varying from 450 to over 900 m in thickness. This aquifer is capped by relatively impermeable lacustrine mudstones (Huka formation) between 60 and 150 m thick, at a depth of between 180 and 300 m below surface. The Huka formation is overlain by a younger sheet of pumice breccia or lapilli tuff (Waiora breccia) which is exposed at the ground surface over the hydrothermal area.

The Waiora formation is underlain by the Wairakei ignimbrites, a relatively impermeable sequence of dense rhyolite ignimbrite sheets known, from the drilling of a 2286-m well, to be at least 1020 m thick in the west of the area. In the western section of the area an andesite flow (Waiora Valley andesite) was penetrated by a number of wells. The thickness of the flow varies from 3 to 183 m and it occurs in most areas immediately above the Wairakei ignimbrite. Below the ignim-

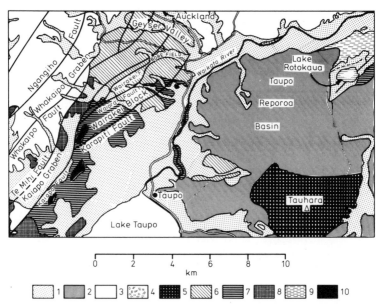

Fig. 9.7 The surface geology and major fault patterns in the Taupo–Wairakei–Rotokaua district (from Grindley, 1965a). 1, Taupo pumice alluvium; 2, Waitahanui pumice breccia; 3, Haparangi rhyolitic pumice; 4, Haparangi younger rhyolite domes and sills; 5, Tauhara dacite; 7, Huka Falls lacustrine mudstones and sandstones; 8, Haparangi older rhyolite flows and sills; 9, Wairoa formation pumice breccias, tuffs, sandstones, and siltstones; 10, K-trig basalt.

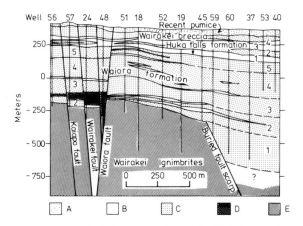

Fig. 9.8 An east–west geological cross section of the Wairakei geothermal field (from Grindley, 1965a). A, ash showers and alluvium; B, siltstone and pumiceous sandstone, diatomite; C, pumice breccia and vitric tuff; D, vesicular hypersthene andesites; E, dense quartz ignimbrite.

brite a second pumice breccia aquifer (Ohakuri formation) occurs and this is underlain by a sequence of andesite flows and tuffs.

The geothermal area is situated on the crest of a major structural elevation known as the Wairakei High, originally discovered by gravity surveys (Beck and Robertson, 1955). The Wairakei High is a complex faulted horst or fault block, step faulted down to depressions on either side, which continues south for 8 km from the central part of the area. High gravity values to the south are principally due to a thick sill of rhyolite, up to 305 m thick, intruded into the Waiora aquifer above the Wairakei ignimbrites. The center of the rhyolite intrusion lies 8 km south of the Wairakei area, where it is probably responsible for a large positive magnetic anomaly (Grindley, 1965a).

Faults occur throughout the Wairakei area and are frequently recognized at the surface as small scarplets or lineaments on air photographs. Many production wells are drilled on three major faults, the Wairakei, Waiora, and Kaiapo faults. The wells were drilled to intersect the faults at approximately 400–500 m.

Figure 9.9 shows isotherms at 600 m below sea level, while Fig. 9.10 shows resistivity contours across the same general area. Of particular significance in Fig. 9.10 is the small distance separating the resistivity contours in the north and south of the area, suggesting very abrupt edges to the system. Wells drilled close to these boundaries have shown that the deep chloride water disappears rapidly over a small

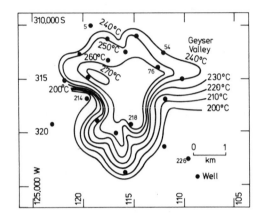

Fig. 9.9 Isotherms within the wes Wairakei field 600 m below sea level du August 1963 (from Banwell, 1965a).

lateral distance. A more gradual trend in chloride concentrations occurs when the contours are widely spread.

Chemistry

The geochemistry of the Wairakei area has been described in many papers, including those of Wilson (1955), Ellis, (1962, 1970), Ellis and Wilson (1960), Ellis and Mahon (1964), Mahon (1962a, 1965b, 1966, 1970b); Mahon and Glover (1965); Glover (1970); Hulston (1962) and Banwell (1963).

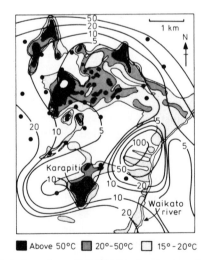

■ Above 50°C ▓ 20°-50°C □ 15°-20°C

Fig. 9.10 The Wairakei geothermal field showing temperatures at 1-m depth (1960) and resistivity contours in ohm meters (from Banwell, 1965a).

The fluid underlying Wairakei at depths of 200–2290 m is a rather homogeneous body of hot water (200°–270°) that contains approximately 2500 ppm of sodium chloride, is saturated with quartz, and is slightly undersaturated with calcium carbonate. Below the Huka formation, at depths varying between 150 and 245 m, to a depth of 2286 m, the salinity of the hot water is almost constant. Above 150 m there is superficial dilution of the hot water with local cold meteoric water to produce sodium chloride concentrations of approximately 1500 ppm. The lateral extent of the hot chloride water is a minimum of 8 km², but evidence from chemistry, geophysics, and aquifer pressure changes during exploitation suggests that the Wairakei and Tauhara geothermal areas are connected at depth. The same reservoir of hot water could underlie a maximum area of 22–25 km² to a depth of around 2 km.

Before development, hot spring activity occurred in two parts of the area. In the north in Geyser Valley, near-neutral-pH chloride water was discharged from hot springs along the edges of the Wairakei stream. Waiora Valley, situated in the center of the area and at a higher altitude, contained many springs of the acid sulfate and acid sulfate–chloride water types. The concentration of sodium chloride in the Geyser Valley springs was variable but the maximum concentration of approximately 3300 ppm indicated that the springs obtained a reasonably direct supply of water from depths of at least 200–300 m in the deep reservoir. After Wairakei wells had produced for several years, the level of the hot water in the aquifer fell and most of the springs in the valley ceased discharging. The area was converted into a fumarolic area and local hot waters changed from a neutral-pH chloride composition into acid sulfate and acid sulfate–chloride waters.

The Karapiti area in the west of the Wairakei system is an extensive zone (1 km²) of fumarolic activity and steaming ground. The largest fumarole in the area for many years was Karapiti, which discharged 46 tons/h of superheated steam at a temperature of 115°–120° (Table 9.3). Other fumaroles and considerable areas of steaming ground occur in many elevated parts of the area.

Table 9.6 shows analyses of waters discharged from wells and springs at atmospheric pressure and boiling point temperature. The chemical homogeneity of the chloride water is illustrated by the constancy of molecular and atomic ratios of the constituents in spring and well waters. The Cl/B ratio varies between 19 and 26 throughout the area, but most values lie between 22 and 24. The Cl/As, Cl/Br, Cl/Cs, Cl/F, and Na/Li ratios are nearly constant.

TABLE 9.6

Typical Analyses of Waters Discharged from Wells and Springs at Wairakei[a]

Concentration (ppm)

	Date	pH	Li	Na	K	Rb	Cs	Ca	Mg	F	Cl	Br	I	SO₄	Total HBO₂	Total SiO₂	As	Total NH₃	Total CO₂	Total H₂S
Well																				
27	1964	8.5	13.2	1200	200	3.0	2.7	17.5	0.005	8.1	2156	5.9	0.6	25	115	660	4.7	0.15	23	—
28	1964	8.4	13.4	1200	200	2.6	2.5	20.0	0.005	7.7	2170	5.0	0.6	27	116.5	680	4.5	0.15	31	2
41	1964	8.5	12.4	1175	172	2.4	2.2	24.0	0.005	7.5	2113	5.8	0.3	23	114.5	540	4.5	0.15	20	3
43	1964	8.2	12.6	1320	150	2.5	2.5	27.0	—	7.7	2106	5.0	—	22	113	460	4.1	0.15	16	—
44	1964	8.6	14.2		225	2.8	2.5	17.5	0.004	8.3	2260	6.0	0.3	36	117	650	—	0.15	19	1
Spring																				
97	1958	6.8	6.8	665	68	0.7	1.7	45.0	4.2	4.4	1110	2.5	0.4	72	57	235	1.8	0.22	88	1.9
190	1958	7.5	10.0	950	62	1.0	—	20.0	0.05	5.8	1596	—	—	56	82	245	2.8	0.37	38	2.0

	Molecular ratio of chloride to					Molecular ratio of sodium to				
	B	F	As	Br	SO₄	Li	K	Rb	Cs	Ca
Well										
27	23.2	145	970	820	235	27.4	10.2	1490	2600	120
28	23.0	150	1020	980	220	27.0	10.2	1710	2770	105
41	22.8	150	990	820	250	28.6	11.6	1820	3080	85
43	23.0	140	1080	950	260	28.1	13.3	1750	2720	75
44	23.8	145	—	850	170	28.0	10.0	1750	3100	130
Spring										
97	24.1	135	1300	1000	42	28.6	16.6	3535	2270	26
190	24.0	145	1200	—	77	28.6	26.0	3535	—	83

[a] Concentrations are those present at atmospheric pressure and boiling temperature. Downhole temperatures in the wells listed are as follows: 27–247°, 28–249°, 41–233°, 43–222°, 44–245°.

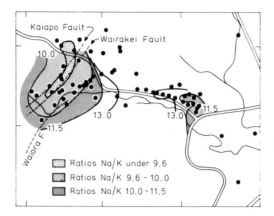

Fig. 9.11 The Na/K atomic ratios in waters from wells in different parts of the Wairakei geothermal field in 1960 (from Ellis and Wilson, 1961).

The Na/K ratios in the deep water vary between 9 and 20 and are in line with values expected for measured downhole temperatures (Chapter 4). Figure 9.11 shows the distribution of Na/K ratios in major wells at Wairakei during 1960.

The distribution of Cl/B atomic ratios in Wairakei well waters is shown in Fig. 9.12. The lowest ratios are present in the east of the area

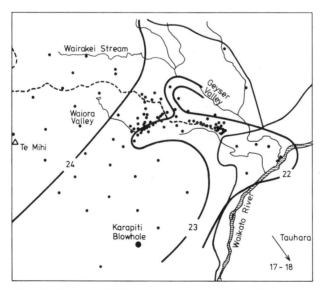

Fig. 9.12 The distribution of Cl/B atomic ratios in the Wairakei field in 1970.

($\approx$22), the ratios decreasing to the south to values of 17–18 in both deep and shallow waters of the Tauhara area. The changes in the Cl/B ratio in the main Wairakei area are a reflection of variations in chloride rather than in boron concentrations, while the low ratios in the south at Tauhara are the result of increasing boron concentrations. The ratio for water from 2000 m discharged from well 121 was 22, and similar to values in shallow wells (<1000 m) nearby.

Rock–water interaction experiments (Chapter 3) have demonstrated that chloride is removed more rapidly from rocks than boron, and ratios in solution tend to decrease to the values present in the rocks as alteration proceeds. Assuming that the rocks at Wairakei had similar Cl/B ratios, the lowest ratios in hot waters of the area would occur in the zones of highest permeability and most extensive rock alteration. The higher ratios in the northwest of the area may represent a zone of lower permeability or an area which has been invaded by hot water in more recent geological times.

Ratios of Cl to As in the Wairakei well waters vary little and have not changed significantly over 15 years. Small amounts of arsenic are lost from the waters in migrating from deep levels to the springs, especially where the waters pass through clay sediments. The variation of As/Sb ratios in the well waters is possibly related to the deposition of antimony with silica, since the ratio in a silica deposit from several hundred meters in a well casing was much lower than that in the hot waters (Ritchie, 1961). Antimony concentrations in Geyser Valley springs were double those for deep well waters, relative to arsenic and chloride. This could be due to the reworking by hot spring waters of earlier shallow hydrothermal deposits enriched in antimony.

The fluoride concentrations and the Cl/F ratios are also nearly constant throughout the area, lowest ratios generally appearing in the shallower hot waters. Laboratory experiments demonstrated that fluoride is readily leached from rhyolite pumice (Chapter 3) and leaching of near-surface pumice deposits probably explains the lower Cl/F ratios originally found in some of the Geyser Valley spring waters.

The Cl/Br and Cl/I ratios in the Wairakei waters vary between 855–863 and 12,500–25,000 respectively, and have shown little change with time. The small variation in the Cl/Br ratio provides evidence that the Wairakei waters have a common source. The variation in the Cl/I ratio, particularly between the deep and shallow waters, is probably due to leaching of shallow pumice horizons or mudstones containing iodine-rich organic materials, or to absorption of iodide into clay minerals. Koga (1967a) showed that readily soluble iodide in Wairakei rocks was highest in mudstones, followed by rhyolite pumice, rhyolite breccia, andesite, and ignimbrite, in that order.

The concentrations of various trace metals in the Wairakei waters were shown in Table 3.3. Iron and aluminum are low in both well and neutral-pH chloride spring waters, concentrations of 0.008 and 0.35 ppm, respectively, being present (Goguel, 1976). In acid spring waters in the Waiora Valley, concentrations of both constituents are 10–50 times higher. Manganese concentrations are also low in the neutral-pH waters, values of 0.001 ppm being present in the deep well waters. The Cl/Mo and Cl/Ge ratios are relatively constant for wells throughout the area (Koga, 1967b) and there is little distinction between deep well waters and spring waters. Copper and zinc concentrations in the deep well waters average 1.3 and 1.5 ppb, respectively, but in acid spring waters the concentrations are considerably higher (40–100 ppb).

Studies of deuterium and ^{18}O concentrations in Wairakei waters have shown that the deep hot water originates from local precipitation which has circulated to deep levels (Chapter 3). The isotope ratios in Lake Taupo, lying 8 km to the south, are similar to but slightly higher than those in the deep Wairakei water, but regional pressure gradients indicate that recharge to Wairakei is from the western catchment area of Lake Taupo rather than from the lake itself. As discussed in Chapter 8, the consistent near-zero tritium concentrations over 20 years for major wells show that there is no fresh meteoric water entering the central hot system. This was also indicated by the small changes in the chemistry of the deep waters during this period.

Table 9.7 shows the carbon dioxide, hydrogen sulfide, and ammonia concentrations in a selection of well discharges at Wairakei, while Table 9.3 gave the total gas composition of a representative deep water. Carbon dioxide and hydrogen sulfide are the major gases in the deep waters and well discharges, and together contribute 97% of the total gases present. Ammonia, hydrogen, nitrogen, saturated hydrocarbons, and noble gases make up the remaining 3%. Before exploitation of the area, the deep water contained in solution 0.011 mole/kg of carbon dioxide and 0.0036 mole/kg of hydrogen sulfide. The gas concentrations in discharges from the production wells, the integrated gas, steam, and water outputs of the area, and the absence of very high gas concentrations in major fumaroles indicated that little steam existed in the reservoir before exploitation.

The distribution of some of the minor gases in fumarole discharges in the area have been studied (Mahon and Glover, 1965). It was found that an inverse relationship existed between the concentrations of nitrogen and hydrogen. This relationship could be produced by either

TABLE 9.7

Gas Composition of Steam from Wells and Fumaroles at Wairakei

		Gas in total discharge (mmole/100 moles water)			Gas ratios	
	Date	CO_2	H_2S	NH_3	CO_2/H_2S	CO_2/NH_3
Well						
18	9/60	31	1.1	0.2	28.2	155
27	9/60	16.5	0.55	0.14	30.0	118
43	9/60	3.9	0.2	0.13	19.5	30
46	3/64	2.1	0.24	0.13	8.7	16.1
47	9/60	6.0	0.35	0.12	17.1	50
203	10/60	418.0	9.2	0.53	45.4	789
Fumarole						
Karapiti	10/61	165	4.0	0.41	41	400
F22/1	10/61	1210	13.0	0.39	93	3100

subsurface oxidation of the gases, possibly in the groundwater where steam is partly condensed, or from the reaction $2NH_3 = N_2 + 3H_2$. The methane concentrations in the discharges were random, suggesting that a portion of this gas was derived from shallow organic material.

During the exploitation of Wairakei, the temperatures and pressures in deeper parts of the system have decreased (Chapter 7). In some of the production areas the chloride concentrations in waters from deep wells have decreased as lower-chloride water from shallow level aquifers has been drawn into the wells. Hot water levels have subsided in many parts of the area, causing the water output from wells, the salinity of the waters, and the local heat transfer processes to gradually change (Mahon, 1975). Some wells which discharged steam–water mixtures are now discharging only steam. There is no evidence to indicate that the cold water of low salinity which surrounds the area has penetrated laterally into the hot system.

Changes in the gas content of the deeper well waters at Wairakei were discussed in Chapter 7. As the hot water level has fallen in the aquifer rocks, the empty pore spaces have filled with steam and gas boiled from the water. The gas content of the water remaining in the central system at the depths of major production (700 m) had decreased in 1976 to about 0.02–0.03 mole/kg. Deep drilling (>2000 m) in the central western production area after exploitation had proceeded for several years showed that the gas concentration in water below the main production area appeared to have been unaffected by shallow production and was similar to that originally present in waters at 600 m (about 0.011 mole/kg).

Systematic downhole pressure measurements have been made in Tauhara and Wairakei wells during the 18 years of production from Wairakei. A deep connection between the two areas with little impedance to water movement is indicated, and this together with the similarity in chemistry between the two areas presents strong evidence that Tauhara and Wairakei are part of the same hot water system.

Rotokaua

Geology and Geophysics

The main thermal area of Rotokaua is 9 km east of Wairakei. The area is dominated by Lake Rotokaua, which occupies 0.7 km² and contains highly mineralized water (Fig. 9.13). The thermal area is

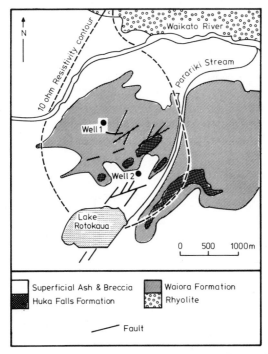

Fig. 9.13 Outline of surface geology and fault patterns in the Rotokaua geothermal field (from Browne, 1974).

characterized by at least 25 hydrothermal eruption craters. Widespread surface and near-surface sulfur deposits surround the lake and occur over the entire hot area. Recent shallow drilling has indicated sulfur reserves of around 4.3×10^9 kg.

With only two deep wells drilled in the area, knowledge of the subsurface geology is limited. Browne (1974) summarized the stratigraphy (Table 9.8). Most of the area is covered by superficial ash, sand, and sulfur. East of the thermal area, thinly bedded lacustrine strata have been correlated with the Huka Falls formation, while north of the Waikato River, Haparangi rhyolite is exposed. Inside the 10-ohm m resistivity contour (Macdonald, 1966) most of the area is covered by volcanic ash of the Waiora formation.

Browne (1974) discussed the structure of Rotokaua, which lies at the eastern edge of the Taupo–Reporoa Basin and is close to the Rotokaua fault zone, which extends west of Mount Tauhara northeast to the Broadlands area. The Huka pumice breccias, which appear to dip gently westward, are cut by two large faults, downthrown to the east, while most of the area is crossed by a swarm of small cracks and lineaments. Some of the small faults link explosion craters and may be deep seated.

Chemistry

The hot springs at Rotokaua were described by Gregg (1958) and the chemistry of the area by Mahon (1960, 1965b) and Ellis and Wilson (1961). The area is of particular chemical interest due to the sulfur deposits which are actively forming in the sinter surrounds of many of

TABLE 9.8

Rock Stratigraphy at Rotokaua

Unit	Lithology
Superficial deposits Taupo pumice Rotokaua breccia Waitahanui breccia	Sulfur, pumice breccia, grit, sandstone, conglomerate
Huka Falls formation	Lacustrine siltstone and sandstone
Waiora formation	Pumice tuff and breccia, ignimbrite, tuffaceous sediment
Rhyolite	Lithic, but locally spherulitic, banded quartz–andesine rhyolite
Waiora formation?	Sandy water-laid tuff

the springs and outlets of fumaroles and to the large subsurface deposits of sulfur.

A high proportion of the hot springs in the area are situated on or near the shores of Lake Rotokaua and along the Parariki stream, which drains the lake (Fig. 9.13). These are waters of the acid sulfate–chloride type (Table 9.2). Small hot water seepages also occur along the banks of the Waikato River 1.2 km to the west and are thought to be associated with the activity at Rotokaua. From the grouping of chloride and sulfate concentrations in the surface hot waters, Ellis and Wilson (1961) suggested that there were two distinct types of water, having chloride concentrations of 1600 and 750 ppm and sulfate contents of 400 and 1000 ppm, respectively. If the sulfate were of surface origin, a random distribution of ratios would be expected but ratios cluster about values of either 10 or 2.

The waters discharged from wells RK1 and RK2 (the only wells drilled in the area; Table 9.2) are of the usual neutral-pH chloride type, with low concentrations of sulfate. The acid sulfate–chloride spring water results from the hydrolysis of near-surface buried sulfur deposits by the hot chloride water (Chapter 4) or from oxidation of hydrogen sulfide to sulfate by ferric iron in buried weathered volcanic rocks.

The deep chloride concentrations in RK1 and RK2 are 700 and 1500 ppm, respectively, the lower value being of the same order as that present in many of the flowing springs. The higher concentration is similar to that found in the deep Wairakei waters. The Cl/B ratio in the springs and well waters lies between 8 and 12 and is similar to the ratio of 9.3 in the rhyolite rocks penetrated by the wells. This suggests that considerable rock–water reaction has occurred in the past and that rock permeability, which now appears to be low, was once considerably higher. The low Na/K ratio and high silica concentration in the discharge from well RK1 reflect the high temperatures, which reach a maximum of 306°.

The gas content of the deep water is not accurately known but the available data suggest that carbon dioxide concentrations are 0.15–0.2 mole/kg and hydrogen sulfide concentrations are 0.01–0.06 mole/kg. The gas composition is similar to that at Wairakei and Tauhara, with CO_2 making up over 90% by volume of total gas.

Broadlands

Geology and Geophysics

The Broadlands geothermal area is located 20 km northeast of Wairakei and 3–4 km west of the Kaingaroa escarpment. The subsur-

face geology of the area is well established through the drilling of 31
exploration wells (Grindley and Browne, 1968; Grindley, 1970; Hochstein and Hunt, 1970; and Browne, 1973). Figure 9.14 shows the major
structural features at Broadlands inferred from photogeology, seismic
surveys, and drilling. A section of the stratigraphy of the area was
shown in Fig. 3.5.

The system is dominated by buried rhyolite domes, rhyolite and
dacite flows, and sills. Twin domes of Ohaki rhyolite and a major flow
of rhyolite (the Broadlands rhyolite) occur in the west of the area.
Rangitaiki ignimbrite, which outcrops on the Kaingaroa block to the
east at 305–457 m above sea level, is found in the Broadlands wells at
610–762 m below sea level, demonstrating a vertical displacement of
over 900 m between the Kaingaroa block and the Taupo–Reporoa
Basin. The displacement is assumed to take place along the Kaingaroa

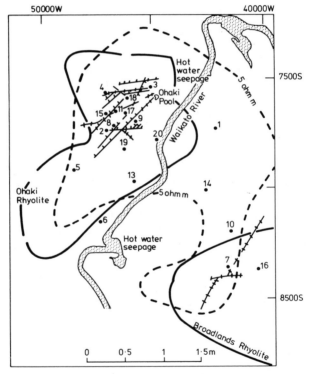

Fig. 9.14 The Broadlands geothermal area (approximately 300 m above sea level)
showing contours on the Ohaki rhyolite V_4 and Broadlands rhyolite V_3 seismic layers at
sea level. The 5-ohm m resistivity contour is shown as a dashed line; also fault traces
(modified from Hochstein and Hunt, 1970).

fault and some evidence for this is obtained from the steep gravity gradient westward from the Kaingaroa block into the Taupo-Reporoa Basin.

Basement Mesozoic greywacke and argillite were penetrated by wells in the east of the area at approximately 975 m (701 m below sea level), although it was not detectable from seismic measurements. In the west of the area the greywacke was encountered by a deep well at approximately 2286 m (−1981 m), and across the area a displacement of the basement of some 1280 m was indicated.

In the north of the area the surface escape of heat is related to a pattern of small faults striking northeast which are also associated with the northernmost of the two rhyolite domes outlined by seismic surveys. Heat escape in the south of the area appears to be associated with an intersection of faults trending northeast and east, respectively (Grindley, 1970).

Extensive geophysical studies at Broadlands have included heat flow, magnetic, electromagnetic, microearthquake, seismic, gravity, and resistivity measurements. The boundaries of the system have been accurately located using a modified Schlumberger resistivity array (Risk, 1975a) and buried zones of fractured rock have been recognized by resistivity anisotropy measurements (Risk, 1975b); Chapter 5).

Chemistry

The chemistry of the Broadlands geothermal system was discussed by Mahon and Finlayson (1972) and Finlayson and Mahon (1976), and the relationship between the chemistry and mineralogy by Browne and Ellis (1970) (Chapter 3). Surface activity in the area is minor and apart from a large pool, Ohaki Pool (Fig. 9.14), some 850 m^2 in area and formerly discharging 10 l/sec of hot water at 98°, only a small number of flowing springs are scattered around the hot area. In the north of the area, just east of Ohaki Pool, hot water seepage occurs along the banks of the Waikato River. There are no fumaroles in the area, although considerable cold gas emissions occur.

Deep-water temperatures range up to 300° but waters tapped by wells are usually at 260°-270°. After separation at atmospheric pressure, the waters contain as major constituents NaCl, 2500 ppm; KCl, 400 ppm; boron, 50 ppm; silica, up to 1050 ppm; and bicarbonate, 150 ppm. Constituent ratios such as Cl/B, Cl/F, Cl/Cs, and Na/Li are not as constant over the area as they are at Wairakei and Tauhara. However, the overall similarity in the chemistry of the deep waters suggests that they may have a common origin.

TABLE 9.9

Analysis of Waters Discharged from Broadlands Wells[a]

Well	Date	pH (20°C)	Li	Na	K	Rb	Cs	Ca	Mg	F	Cl	Br	I	SO4
1	1.2.66	8.3	10.9	1065	152	0.6	0.6	4.6	0.08	6.7	1701	—	1.4	43
2	23.8.66	8.3	11.7	1050	224	2.2	1.7	2.2	0.80	7.3	1743	6.4	0.8	8
3	28.4.67	8.0	12.2	1045	213	1.8	1.8	3.0	0.02	6.1	1801	6.6	—	6
4	6.9.67	8.35	11.8	1075	218	2.0	1.4	3.0	—	4.5	1851	6.7	1.3	29
5	14.11.67	9.1	4.9	1400	146	—	—	—	—	—	1142	—	—	—
6	4.12.67	7.4	1.2	435	39	—	—	260	—	—	28	—	0.95	15
7	14.5.68	8.55	15.0	1300	234	2.4	1.8	1.4	0.013	8.4	1823	5.2	0.35	6
8	29.2.68	7.9	11.4	975	232	2.9	1.6	3.0	0.1	8.0	1858	6.3	0.6	3.5
9	14.5.68	8.45	12.7	930	203	2.4	1.5	2.0	0.35	6.2	1709	5.9	0.4	4.0
10	18.10.68	8.6	9.5	910	142	1.1	1.4	1.1	0.05	6.4	1244	3.6	0.1	11.5
11	22.8.68	8.25	12.2	1020	218	1.7	1.4	7.3	0.92	6.4	1794	7.0	0.1	10
12	13.12.68	8.51	10.0	1510	198	0.95	1.25	2.8	0.13	3.2	1546	5.1	0.5	25
13	10.10.68	8.6	12.6	980	200	2.2	1.3	2.4	0.015	4.5	1668	5.25	0.1	6.5
14	6.2.69	8.5	10.5	880	175	0.9	1.5	2.0	0.7	3.2	1482	4.2	0.7	6.0
15	3.4.70	8.25	11.5	1060	207	1.8	1.55	5.7	0.13	6.8	1750	5.8	0.4	7.1
16	16.1.70	8.25	6.2	570	85	0.51	0.47	8.55	7.46	6.05	528	1.6	0.30	45
17	24.4.70	8.15	10.0	1000	224	1.86	1.88	4.82	0.007	8.6	1778	4.5	0.25	5.5
18	24.4.70	7.83	11.2	1110	237	1.84	1.80	9.0	0.18	3.9	1985	6.4	0.35	4.0
19	2.4.70	8.25	11.8	950	206	1.6	1.7	2.5	0.05	7.4	1720	4.6	0.65	20.6

Well	B	SiO$_2$	As	Total NH$_4$	Total HCO$_3$	Cl in deep water	Molecular ratio of chloride to							Molecular ratio of sodium to		
							B	F	As	Br	SO$_4$	Cs	Li	K	Rb	Ca
1	48.1	565	—	12.0	230	1233	10.8	136	—	—	109	10,625	30	11.9	6595	405
2	48.4	805	5.7	2.1	178	1177	11.0	128	650	615	590	3840	27	8.0	1775	830
3	55.3	725	4.0	4.0	174	1245	9.9	160	951	615	800	3750	26	8.3	2155	605
4	56.5	880	5.7	2.3	282	1150	10.0	220	685	622	170	4955	28	8.4	1995	625
5	23.4	—	—	1.5	1410	1000	14.8	—	—	—	—	—	86	16.3	—	—
6	1.2	180	—	1.1	2284	25	6.9	—	—	—	5.1	—	110	19.0	—	2.9
7	82.7	1000	—	1.0	910	1002	6.7	—	—	790	825	3795	26.2	9.4	2010	1620
8	52.6	796	5.7	2.3	157	1220	11.0	114	690	665	1440	4350	25.8	7.1	1250	565
9	43.95	805	4.5	—	134	1147	11.9	114	800	650	1155	4270	22	7.8	1140	810
10	54.8	635	4.1	1.2	553	891	6.9	107	640	780	295	3330	28.9	10.85	3075	1440
11	48.9	805	5.5	1.35	78	1213	11.2	150	690	575	485	4800	25.2	7.9	2230	245
12	43.7	400	1.0	2.4	1820	1196	10.8	258	3265	680	165	4635	45.5	13.0	5905	940
13	48.0	750	3.2	1.9	162	1144	10.6	199	1100	715	695	4810	23.5	8.3	1655	710
14	47.4	844	5.7	1.45	218	961	9.6	248	550	795	670	3700	25.3	8.5	3635	765
15	37.5	1126	—	2.0	310	1150	14.2	138	—	680	665	4230	27.8	8.7	2190	325
16	40.6	265	—	12.0	555	441	4.0	47	—	744	32	4210	30.2	11.4	4155	115
17	46.0	765	—	1.9	115	1215	11.8	111	—	890	875	3544	30.2	7.6	2000	360
18	58.5	635	—	3.05	250	1367	10.3	273	—	699	1345	5135	29.9	8.0	2240	215
19	48.2	875	—	1.9	175	1150	10.9	124	—	844	225	3775	25.0	7.8	2205	660

^{a} Well samples were separated at atmospheric pressure.

Table 9.9 shows analyses of waters discharged from wells at Broad-lands. The chloride concentrations in the deep waters over a wide area are of similar magnitude (1150 ppm) but in the southeast of the area dilution lowers concentrations to between 400 and 1000 ppm. In the southwest of the area near well 6, a low-chloride, high-bicarbonate-sulfate water, at about 160°, occurs to considerable depths (500–600 m). The high-temperature chloride waters are saturated with silica and calcium carbonate, the latter depositing as calcite in some of the wells during discharge.

Only in the Ohaki Pool does hot water from depth reach the surface without major dilution. The pool appears to be supplied with hot water from a depth of approximately 300 m, where temperatures are of the order of 200°. In the Wairakei and Tauhara areas there was little chloride gradient below a depth of approximately 300 m but at Broadlands there is a gradient in some parts of the field. In well 20 a significant temperature inversion at 1965 m corresponded to the interception of a water layer of temperature 227° containing 720 ppm chloride. Similar inversions also occurred at about this depth in two other wells. However, both the temperature inversions and the lower chloride concentrations in the discharges disappeared after some weeks of production and the quantities of lower-temperature lower-chloride water in the system appeared to be small.

The high silica concentrations, up to 1050 ppm, low Na/K ratios (7–9), and low calcium concentrations in the deep Broadlands waters result from high subsurface water temperatures. The water tempera-tures calculated from the silica concentrations in the wells (Mahon and Finlayson, 1972) were often 20°–30° lower than the measured tempera-tures and temperatures estimated from Na/K ratios (Table 6.4), and there is evidence that waters of temperature between 300° and 310° ascend vertically through the area and lose heat, mainly by boiling, before reaching the production zones of the various wells. Boiling is accompanied by the rapid degassing of the waters on reaching lower-pressure zones.

The Cl/B ratio in waters on the west side of the system, in zones where hot water occurs in ignimbrite, is approximately 14, but values of approximately 10 are found in waters in contact with rhyolite. Ratios in wells which penetrate greywacke and argillites at shallow depths on the eastern side of the area are considerably lower (4–6), while waters discharged from wells in the center of the area are also partly influenced by the greywacke and have intermediate Cl/B ratios between 6 and 10.

Carbon dioxide and hydrogen sulfide concentrations in the deep

Broadlands waters are ten and five times higher, respectively, than at Wairakei (Table 9.10). Ammonia concentrations are also quite high (5–20 ppm). The high gas concentrations appear to be related to the basement greywackes and argillites and it is notable that the gas concentrations in wells penetrating these basement rocks are some of the highest in the area. There are no major differences, however, between the compositions of gases at Broadlands and those in other geothermal areas in the volcanic zone.

The average carbon dioxide concentration in the water at depths of 400–800 m is approximately 0.6% by weight, but at deeper levels concentrations as high as 6% by weight could be present. The partial pressure of carbon dioxide at the average well depth ranges from 8 to 14 bars but at greater depths the partial pressure could rise to 100 bars. The major effect of high gas pressures is to depress isotherms below the boiling point–depth relationship for water (Chapter 3).

Closed wells frequently develop very high gas pressures at the wellheads (55 atm) as a result of the high gas concentrations. Condensate forming on the inside of the steel casing, close to the wellhead,

TABLE 9.10

Gas Concentrations in Broadlands Wells

Well	Date	Gas in total discharge (mmoles/100 moles)		Gas in total discharge (wt %)		
		CO_2	H_2S	CO_2	H_2S	CO_2/H_2S
1	28.2.66	2770	7	6.7	0.013	395
2	30.9.66	256	7.1	0.62	0.013	36.0
3	28.4.67	1213	14.3	2.96	0.027	85.0
4	29.2.68	1240	19.3	3.02	0.036	64.0
7	2.4.68	2060	16.5	5.03	0.031	124
8	15.1.68	795	15.6	1.94	0.029	50.8
9	22.4.68	270	5.4	0.66	0.010	50.0
10	11.5.68	522	4.1	4.27	0.008	127
11	29.8.68	208	4.7	0.51	0.009	44.7
12	5.12.68	1563	7.3	3.81	0.014	214
13	18.10.68	261	4.65	0.64	0.009	55.9
14	6.2.69	1769	20.8	4.32	0.039	84.9
15	15.12.69	108	2.8	0.26	0.005	38.4
17	2.2.70	452	9.7	1.10	0.018	46.5
18	31.3.70	1808	20.8	4.41	0.039	86.7
19	2.4.70	239	4.4	0.58	0.008	54.5

becomes saturated with ammonia and carbon dioxide, and ammonium bicarbonate is deposited.

Waiotapu

Geology and Geophysics

The Waiotapu geothermal area, with an areal extent of about 18 km², is located at the north end of the Reporoa Valley approximately 60 km north of Wairakei. Seven investigation wells to depths of up to 1000 m revealed the subsurface geology and structure of the Waiotapu area (Grindley, 1957; Studt, 1957; Healy 1974). Figure 9.15 shows a geological cross section.

The dacite domes of Maungaongaonga and Maungakakaramea form the northwest and northeast corners of the system, and a rhyolite dome the southwest margin. Subsurface rock units appear to dip toward the southwest and to rise to the northwest, where they are cut off by the northwest-striking Ngapouri fault.

The following rock formations were penetrated by wells 2–7. A sequence of siltstone and sandstone 53–91 m thick (Onuku breccia); Rangitaiki ignimbrite, 31–94 m; vitric tuff and pumice breccia, 15–99 m; and Waiotapu ignimbrite, 236–358 m. Well 1 passed from Maungaongaonga dacite at the surface directly into the tuff and pumice breccia. The deep wells 4, 6, and 7 penetrated three units of Paeroa ignimbrite separated by tuffs and breccias. Wells 6 and 7 passed through andesite above the lowest Paeroa ignimbrite unit, and well 4 encountered rhyolite beneath it.

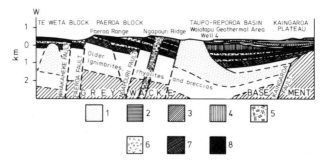

Fig. 9.15 Cross section of the Waiotapu geothermal field: 1, Haparangi rhyolite; 2, Huka group lake beds, vitric tuff, pumice breccia; 3, Rangitaiki ignimbrites; 4, Waiotapu ignimbrite; 5, Waiotapu rhyolite; 6, Ohakuri group—pumice breccia, tuff, ignimbrite, sandstone; 7, Te Kopia ignimbrites; 8, Paeroa ignimbrites (from Grindley, 1965c).

Most of the surface fault traces are discontinuous because of recent sedimentation and ash showers. Surface evidence indicates a fairly continuous fault zone with branch faults crossing the area in a north-northeast direction from west of Lake Ngakoro (Green Lake) to north of the Lady Knox geyser. Much of the heat flow at Waiotapu may be related to this fault zone.

Resistivity measurements using a fixed electrode spacing of 550 m (Hatherton *et al.*, 1966) showed that the 5-ohm m contour enclosed an area extending from Waiotapu to Reporoa and it was suggested by Healy and Hochstein (1973) that this low represented a flow of hot water south between the areas.

Chemistry

The hot springs at Waiotapu are scattered throughout the area, and extend to the west along the Ngapouri fault toward the Waikite and Te Kopia areas located on the Paeroa fault. The chemistry of the hot springs was discussed by Wilson (1957) and Lloyd (1958). The compositions of the spring and well waters were later interpreted by Ellis and Mahon (1964) and Mahon (1965b).

The degree of mineralization of the Waiotapu waters is intermediate between Wairakei and Broadlands (Table 9.2), with variable deep chloride concentrations ranging from about 900 to 1400 ppm. Below 600 m the chloride concentration is probably around 1200-1400 ppm, while at shallower depths concentrations are lower due to dilution. Lloyd (1958) plotted isochloride lines within the area (Fig. 6.3), outlining two zones where the deep chloride water ascends to the surface without suffering major dilution. These zones are located at Champagne Pool in the central part of the area and along the shore of Lake Ngakoro in the south. The highest chloride concentration in the surface springs is approximately 2000 ppm.

The springs at Waiotapu are more variable in pH than at Wairakei or Tauhara. High-chloride springs frequently have pHs between 4 and 6 and this is thought to be due to steam rising through the waters or to the effect of subsurface sulfur deposits.

Oxidation of the hydrogen sulfide in surface waters has produced a wide range of acid sulfate-chloride waters together with deposits of elemental sulfur, which is frequently found around springs and fumaroles.

Antimony and arsenic sulfides, elemental sulfur, alums, and silica sinter are the major precipitates found around the Waiotapu springs and fumaroles. The occurrence of natural black sulfur is rare in New Zealand but it is formed at Waiotapu in large globules in an area of mainly acid springs (Lloyd, 1958). The presence of black sulfur and the large number of acid springs bears a rather close resemblance to the conditions at Rotokaua, where buried sulfur deposits exist.

Major fumarolic areas occur in the north of Waiotapu, on the flanks of Maungaongaonga and Maungakakaramea. A large number of acid sulfate springs are present, formed by hydrogen sulfide from steam flows oxidizing in surface water.

Constituent ratios in the waters are more variable than in most other areas and the lack of chemical homogeneity suggests the presence of more than one deep aquifer. The Cl/B ratio in wells and springs, for example, ranges from approximately 25 to 32. Lower values frequently appear in springs which are slightly acid, although in wells 6 and 7 similar differences occur in alkaline waters. The variations in constituent ratios may simply reflect the variations in acidity of the waters and reaction with rocks at different positions within the system.

The carbon dioxide and hydrogen sulfide concentrations in the well waters are approximately 0.024 and 0.0025 mole/kg, respectively, intermediate between Wairakei and Broadlands. The CO_2 concentrations at Waiotapu were sufficient for calcite to be deposited in the casings of all the wells which were discharged.

Orakeikorako

Geology and Geophysics

The Orakeikorako geothermal area is located on the Paeroa fault 50 km north-northwest of Wairakei. It was formerly one of the most active hot spring areas of the Taupo Volcanic Zone. The stratigraphy of the Orakeikorako area is shown in Fig. 9.16.

Four investigation wells drilled in the area started at various levels in the Waiora formation, which overlies a series of ignimbrites and rhyolites. The oldest ignimbrite (Akatarewa ignimbrite) is overlain by two other ignimbrites, the Te Kopia ignimbrite and the younger Paeroa ignimbrite. A rhyolite (Haparangi rhyolite) and pumiceous pyroclastics (Ohakuri group) occur between the Akatarewa and Te Kopia ignimbrites. The Ohakuri group, the Te Kopia ignimbrites, and the Akatarewa ignimbrites, do not outcrop at the surface (Grindley, 1965b).

The structure of Orakeikorako was outlined by Modriniak and Studt (1959) and Grindley (1965c). The northeast-trending Paeroa range

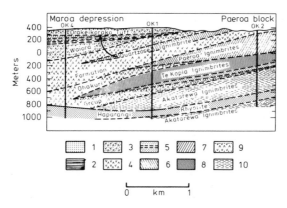

Fig. 9.16 Geological cross section of the Orakeikorako field (from Grindley, 1965b). 1, Haparangi rhyolite; 2, Kaingaroa ignimbrites; 3, subaqueous tuff and pumice breccia; 4, subaerial tuff and pumice breccia; 5, lacustrine siltstone and sandstone; 6, soft quartzose ignimbrites; 7, Paeroa ignimbrites (quartz–biotite–plagioclase); 8, Te Kopia ignimbrites (glassy quartzose); 9, Ohakuri sandstone, tuff, pumice breccia; 10, Akatarewa ignimbrites (dense, quartzose, lenticular).

separates the Taupo–White Island depression from the Reporoa depression and is bounded to the southwest by the northwest-trending Maroa Graben (Fig. 9.1). A section through the Orakeikorako area obtained from drilling confirmed Grindley's suggestion that the Taupo–White Island depression is dominantly an early Pleistocene basin of volcano-tectonic subsidence filled with thick ignimbrites, rhyolites, and pumice pyroclastics. Tectonic and hydrothermal activity has now shifted eastward into the Taupo–Reporoa Basin to the east of the Paeroa range.

Gravity values decrease southwest from the Paeroa range and rise again at Wairakei. A zone of low gravity values in the intervening region was given the name "Maroa Graben" by Modriniak and Studt (1959). The most noticeable feature is the thinning and disappearance of the ignimbrites on the Paeroa block and their replacement by thick pumiceous pyroclastics of the Waiora formation and Ohakuri group, intruded by thick sills of rhyolite. The Maroa depression appears to be a basin of subsidence filled with less dense products of acid volcanism, notably pumice breccias and rhyolite (Grindley, 1965c).

Chemistry

The chemistry of the Orakeikorako springs was discussed by Mahon (1972). The thermal activity covers an area of 1.8 km² and is notable for the large number of boiling springs and geysers, and the beautiful multicolored silica sinter terraces. Table 9.2 gave examples of analyses of well and spring waters.

The majority of springs in the area discharge near-neutral-pH chloride water containing between 250 and 400 ppm chloride, 50 and 250 ppm sulfate, and 100 and 300 ppm bicarbonate. Acid sulfate and acid sulfate-chloride waters are present, but to a lesser extent than at Waiotapu. A high proportion of the chloride springs are on the banks, and close to water level of the Waikato River, while acid waters are located at higher altitudes.

The general chemistry of the hot water at Orakeikorako is similar to that of a diluted Wairakei water (e.g., similar Cl/B ratios). The main differences are the higher proportions of fluoride, bicarbonate, and sulfate relative to chloride at Orakeikorako. Constituent ratios in the spring waters are rather variable, indicating a complex interaction between rising hot water, steam heating, and cold water dilution. Chemical and geological data similarly indicate the presence of a number of different hot water aquifers or permeable horizons in the system. The chemistry and degree of mineralization of waters discharged from the deep wells (Table 9.2) is, with one exception, similar to that of the springs.

The low mineral content of the waters is thought to reflect rapid water movement through the system, limited rock-water interaction, and dilution. Orakeikorako does not have a well-defined "capping" rock, and convective recirculation at shallow levels is unlikely to occur.

Due to the low rock permeability, the deep wells were unable to sustain discharges long enough for deep gas concentrations to be accurately assessed. The data collected suggested that the carbon dioxide and hydrogen sulfide concentrations were of the order of 0.07 and 0.001 mole/kg, respectively.

The similarity in the chemistry of waters in geothermal areas along the Paeroa fault (Orakeikorako, Te Kopia, Waikite, and the western zones of Waiotapu) demonstrates the importance of the fault in controlling the movement of hot water in this region. The deep hot waters along the fault exist in rocks of rather similar chemical composition and individual areas appear to be at a similar stage of development.

Kawerau

Geology and Geophysics

The major geothermal areas in the northern part of the Taupo Volcanic Zone are Rotorua, Tikitere, and Kawerau. The Kawerau geothermal system has been utilized by a major pulp and paper company since 1958 to supply steam for processing and for the generation of electricity. Figure 9.17, from Healy (1974), shows a geological map of Kawerau.

Healy (1962b) and Grindley (1965c) summarized subsurface geology at Kawerau revealed by drilling. Later information from wells was correlated by Macdonald *et al.* (1970), who divided the stratigraphic sequence at Kawerau into five major units. From top to bottom these are:

Unit 1.   Alluvium and pumice breccia (43-99 m thick).
Unit 2.   Rhyolite, with subordinate tuff breccia, ash, sandstone, and siltstone (533-722 m thick).
Unit 3.   Andesite (84 to more than 251 m thick).

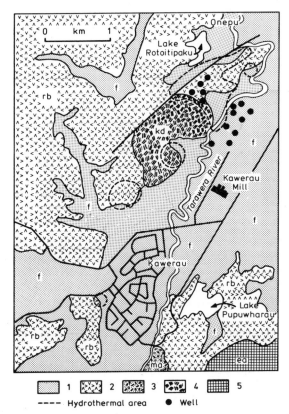

Fig. 9.17 Outline surface geology of the Kawerau geothermal area showing major centers of activity: 1, pumice alluvium; 2, Rotoiti breccia; 3, Matahina ignimbrite; 4, Kawerau dacite; 5, Edgecumbe andesite (from Healy, 1974).

Unit 4.   Ignimbrite, with intercolated tuff breccia and sandstone.
Unit 5.   Rhyolite.

The structure of the Kawerau area was discussed by Studt (1958) and Healy (1974). The area lies near the axis of a northeast-trending complex graben, the Whakatane Graben, that continues northeastward out to sea to form the White Island trench. Seismic refraction shows extensive bodies of rhyolite, which are of low permeability and act as cap rocks for the hydrothermal system. Interpretation of gravity data by Studt (1958) indicates that the minimum depth to the nonvolcanic basement beneath the area is approximately 1500 m and greywacke was intersected by recent drilling.

In the vicinity of Kawerau the graben is filled with volcanic rocks and associated sedimentary rocks composed of volcanic detritus. The volcanic rocks are primarily of rhyolitic composition and include flows, tuff breccias, pumice breccias, and ignimbrites. There are a few andesitic rocks (at Mount Edgecumbe, Monauake, and at depth in the Kawerau area). Young dacite extrusions form the Onepu hills nearby.

The initial wells in the area produced from rhyolite and breccias in Unit 2, in which temperatures fell with time. The later deeper wells produce from the andesite (Unit 3), which has permeability due to fractures. Maximum temperatures tend to occur in the andesite, suggesting that the hot water spreads horizontally through this unit from a deeper source. The highest temperature at 285° was measured in andesite in well 8, and both above and within the greywacke in well 21.

Chemistry

The Onepu springs are located on the left bank of the Tarawera River and cover 0.6 km². Hot water seepages and scattered hot pools extend up the Tarawera River and Pururanga stream for about 1.5 km above their confluence (Studt, 1958).

Mahon (1962b, 1968) discussed the chemistry of the springs and wells at Kawerau. Spring waters are of the neutral-pH chloride and acid sulfate types (Table 9.2). Chloride concentrations increase from approximately 300 ppm in springs located near the Tarawera River to 600 ppm in springs located at higher elevations above the cold water table. The chloride concentrations in the well waters increase with depth. At 500 m the hot water contains approximately 600 ppm, a concentration similar to that in the more concentrated springs, which are apparently supplied by an aquifer at this depth. Chloride concentrations increase to between 800 and 850 ppm at 820 m, then remain constant with increasing depth.

The Cl/B ratio in the spring and well waters is approximately 6.5. The constancy of this and other ratios, such as Cl/F and Na/Li, indicate that the area has a single hot water reservoir. The Cl/B ratio in the local andesites is 2.6 and in the basement greywackes somewhat lower. The water compositions in the area are apparently influenced mainly by these rocks.

The chemistry of the Kawerau waters is generally similar to that of the Wairakei waters, and constituent ratios such as Na/Li and Cl/SO$_4$ are of the same order (Table 9.2). The higher temperatures at Kawerau are reflected in the higher silica concentrations (1000 ppm) and low calcium and fluoride concentrations and Na/K ratios. The low concentrations of rubidium and cesium and relatively high concentrations of magnesium, compared with concentrations in hot water more directly associated with rhyolite, are an indication of compositional control by andesite.

The carbon dioxide and hydrogen sulfide concentrations in the well discharges of 0.1 and 0.003 mole/kg, respectively, are similar to those at Broadlands and considerably higher than at Wairakei. Ammonia concentrations (10–12 ppm) are also high, while total gas compositions are similar to those in other areas. The high carbon dioxide concentrations are conducive to calcite formation and this mineral forms in the casing of many of the wells.

Ngawha

Geology and Geophysics

The Ngawha geothermal area in Northland is 6 km east of Kaikohe. A geological map of the Kaikohe–Ngawha area from Brothers (1965) is shown in Fig. 9.18. The area is situated in an ancient river valley eroded in claystones, part of the Mangakahia series shales and sandstones of Kear and Hay (1961). Kerikeri series olivine basalts of Pleistocene age flowed into this valley and later blocked the northern end, causing pondage of the drainage and lake formation. In-silting of the lake then occurred. Thermal activity is considered to be associated with recent subsurface basalt intrusions, which are represented in the surrounding northern country by surface flows. The hot springs seep from the Ngawha lake beds, which in places have become silicified and mineralized.

The stratigraphic sequence found in a 600-m exploratory well drilled

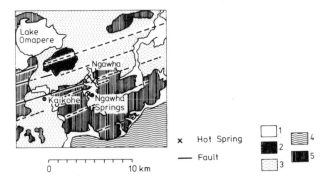

Fig. 9.18 Surface geology and fault patterns of the Ngawha geothermal area: 1, alluvium; 2, rhyolite; 3, basalt; 4, sandstones; 5, claystone, sandstone, siltstone (modified from Brothers, 1965).

in the area is as follows. From 0 to 29 m, the rocks are grey siltstones, shales, and sandstones with Upper Cretaceous to Paleocene microfossils. These are followed, at from approximately 229 to 472 m, by greenish-grey sheared and brecciated calcareous fine-grained sediments with Upper Cretaceous, Paleocene, and Eocene microfossils and with rare pyrite. From 472 to 533 m there are hard, relatively unsheared calcareous siltstones and sandstones of Eocene age, with some slickensiding, and these overlie argillites of Permian to Lower Cretaceous age containing iron sulfide. The argillites extend from 533 to 586 m.

Kear (1959) suggested that volcanic events in the Taupo Volcanic Zone may have paralleled those which occurred in the far north Cenozoic volcanism. Kear found evidence in the north of a general sequence from andesite through dacite and rhyolite to basalt, with deformation occurring in the rhyolite phase.

Jones (1939), in a magnetic survey of the Ngawha area, correlated a ridge of high vertical magnetic intensities on the eastern border of the main thermal area with Kerikeri series basalts reaching almost to the surface. West of the ridge of magnetic highs lay a zone of low magnetic intensities followed by normal magnetic values in the main area of old cinnabar workings. The magnetic lows could reflect deep hydrothermal alteration of magnetic minerals in zones of hot water upflow. The magnetic results suggest a pattern of hot water upflow deflected along the western margin of a basaltic intrusion.

Heat flow and rock resistivity measurements at Ngawha were discussed by Macdonald and Banwell (1965). Figure 9.19 shows a geophysical map of the Ngawha area. In a series of nine wells drilled to depths between 100 and 120 m, temperatures were erratic down to 30 m but below this the temperatures increased almost linearly with depth. The area included within the outermost isogradient line (50°/km) was 180 km^2 and the conductive heat flow was approximately 9000 kcal/sec, giving an average heat flow of 5 μcal cm^{-2} sec^{-1}, a little over four times normal. The heat flow in the central part of the area was above 20 times normal. Heat discharge as steam and hot water from the area appears to be negligible in comparison with the conductive flow. If a heat flow corresponding to the terrestrial average for an area of this size (1700 kcal/sec) is subtracted from this figure, an anomalous heat flow of 7300 kcal/sec is obtained (approximately 7% of the natural heat flow from Wairakei).

Resistivity soundings were carried out using a Wenner system with interelectrode spacings increasing from 3 to 1000 m. A conspicuous feature was a deep 200-ohm m layer at increasing depth toward the southwest, considered to be due to sedimentary rocks of the Waipapa group. Above the 200-ohm m layer a zone of lower resistivities,

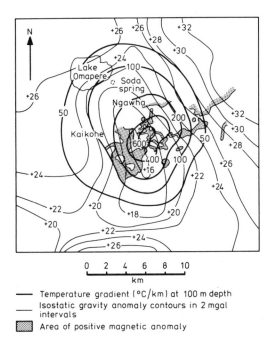

    0 2 4 6 8 10
    |__|__|__|__|__|
           km

— Temperature gradient (°C/km) at 100 m depth
--- Isostatic gravity anomaly contours in 2 mgal intervals
▨ Area of positive magnetic anomaly

Fig. 9.19 The Ngawha geothermal area, showing temperature gradients (°C/km) at 90 m depth (heavy contour lines), isostatic gravity anomalies contours (thin lines) in 2 m gal intervals and areas of positive magnetic anomaly (shaded) (from Banwell, 1965b).

ranging from 1 to 4 ohm m, occurred in a broad belt across the hot area. It was bordered laterally on either side with resistivities of about 10 ohm m, which may represent the same formation, but at a lower temperature, less affected by thermal alteration, or permeated by water of lower mineral content. The lowest resistivities (1 ohm m) were found adjacent to the highest (1500 ohm m), and the high-resistance layer may represent either a zone in which water is replaced by steam and gas or, more probably, an impervious compact rock.

Chemistry

The chemistry of the area was summarized by Ellis and Mahon (1966). In its chemical and geological characteristics Ngawha resembles the hydrothermal area of Sulphur Bank, California, which was discussed in detail by White and Roberson (1962). Both areas are sites of mercury mineralization and have hot waters with high boron, ammonia, and bicarbonate concentrations. In both cases there is a rock sequence of thick sandstones and shales, followed by shallow lake beds with included carbonaceous material and late basic volcanics.

Cinnabar deposits occur at Ngawha and were discussed by Brothers (1965). The mineral is sparsely distributed throughout old lake beds in the vicinity of hot springs, as an impregnation in slightly silicified grit, sand, and carbonaceous mud. Some hundreds of prospect holes showed that the maximum concentration of ore was within 3 m of the surface of the lake sediments. There was no overlying impervious trapping bed. The richest patch was about 0.3 m thick and fairly solid, covering an area of 16 m². The ore consisted of cinnabar in crusts with marcasite, a little siliceous sinter, and native sulfur. The order of deposition was not consistent, but most commonly sulfur or marcasite occupied the interior of ore lumps, the cinnabar occurring as a crust on the outside. Elsewhere in the lake beds small cavities were coated with sulfur or alum, along with realgar and crystals of stibnite.

The surface flows of the springs and pools at Ngawha are very small. The flowing springs are slightly acidic, whereas the stagnant pools are appreciably acidic due to bisulfate formation. The major features of the waters (Table 9.2) are the high degree of mineralization, the high concentrations of boron, ammonia, and bicarbonate relative to chloride, and the significant concentrations of lithium, rubidium, and cesium. The concentrations of carbon dioxide in the spring waters often approach saturation for 1 atm gas pressure.

In waters of appreciable chloride concentration (higher than about 250 ppm) in the main hot spring area, the atomic ratios of Cl to B are almost constant between 0.4 and 0.5. For the well water the ratio varied between 0.41 and 0.46. Chloride concentrations in the more concentrated spring waters range from 400 to 1500 ppm compared with the concentration of 1600 ppm in water discharged from the well. In a number of zones within the system the water obviously migrates from depth to the surface with only minor dilution. The chloride concentration in the deep system is 1200 ppm.

The deep water is saturated with quartz at the measured downhole temperature. This is of particular interest because it was the first evidence that geothermal waters associated with sedimentary rocks are in equilibrium with quartz. Calcium and sulfate concentrations are approximately the same as found in the Taupo Volcanic Zone hot waters. Fluoride, bromide, rubidium, and cesium are lower, whereas magnesium and iodide are considerably higher.

The gases discharged from springs, fumaroles, and the deep well are predominantly carbon dioxide (88% and 94%) and 4–8% of methane. The high percentage of hydrocarbons in the gas probably represents steam reaction with the local organic-rich sediments. The proportion of gas to steam in the well discharge is 1000 times higher than at Wairakei and 100 times higher than at Kawerau. The CO_2/H_2S ratio is also five times higher than at Wairakei.

The chemistry of the hot waters from Ngawha and Sulphur Bank, California, is compared in Table 9.11. White and Roberson's 1962 results for the hottest water (77°C) in the latter area are compared with an analysis for Jubilee Pool, Ngawha. There are striking similarities between concentrations of key elements such as lithium, ammonia, fluoride, chloride, and boron. The more stagnant waters in both areas contain the highest sulfate and bicarbonate concentrations. The neu-

TABLE 9.11

Comparison of the Chemistry of Ngawha, New Zealand, and Sulphur Bank, California

	pH	Li	Na	K	Ca	Mg	NH_3
Herman Pit, Sulphur Bank	7.5	4.8	1100	33	7	22	450
Jubilee Pool, Ngawha	6.4	8.0	830	63	7.8	2.5	140

	F	Cl	SO_4	HCO_3	HBO_2	H_2S	SiO_2
Herman Pit, Sulphur Bank	1.0	690	454	2960	2680	3.6	72
Jubilee Pool, Ngawha	0.3	1250	347	340	3692	6	178

tral-pH Sulphur Bank waters also contain about 0.02 ppm of mercury, and the acid pools up to 0.2 ppm.

The gases at Sulphur Bank consist mainly of carbon dioxide and methane, again probably derived from organic material in the surface sediments. The organic-rich layer appears to be important in providing suitable conditions for the deposition of mercury from the rising waters, since there is a strong association between organic material and mercury in both areas.

The rocks underlying Sulphur Bank, the Franciscan formation of greywacke and shales, are rather different in texture from the Mangakahia sediments beneath Ngawha but could be similar to the argillites found in the Ngawha well at depths over 533 m.

REFERENCES

Banwell, C. J. (1963). *In* "Nuclear Geology on Geothermal Areas" (E. Tongiorgi, ed.), pp. 95-138. Consiglio Nazionale della Richerche, Lab. di Geol. Nucl., Pisa.

Banwell, C. J. (1965a). *In* "New Zealand Volcanology—Central Volcanic Region" (B. N. Thompson, L. O. Kermode, and A. Ewart, eds.), pp. 161-171. New Zealand Dept. Sci. Ind. Res., Inform. Series No. 50, Wellington.

Banwell, C. J. (1965b). *In* "New Zealand Volcanology—Northland-Coromandel-Auckland" (B. N. Thompson and L. O. Kermode, eds.), pp. 38-41. New Zealand Dept. Sci. Ind. Res., Inform. Series No. 49, Wellington.

Beck, A. C., and Robertson, E. I. (1955). *In* "Geothermal Steam for Power in New Zealand" (L. I. Grange, ed.), pp. 15-19. New Zealand Dept. Sci. Ind. Res., Bull. 117, Wellington.

Bolton, R. S. (1975). *Proc. U.N. Symp. Develop. Use Geothermal Resources, 2nd, San Francisco, California, May* **1**, 37.

Brothers, R. N. (1965). *In* "New Zealand Volcanology—Northland-Auckland-Coromandel" (B. N. Thompson and L. O. Kermode, eds.), pp. 33-37. New Zealand Dept. Sci. and Ind. Res. Informat. Ser. No. 49, Wellington.

Browne, P. R. L. (1973). The geology, mineralogy and geothermometry of the Broadlands geothermal field, Taupo Volcanic Zone, New Zealand. PhD thesis, Victoria Univ. of Wellington, Wellington.

Browne, P. R. L. (1974). *In* "Minerals of New Zealand," Rep. N.Z.G.S. No. 38D (Part D Geothermal). Rotokaua. New Zealand Dept. Sci. and Ind. Res., Wellington.

Browne, P. R. L. and Ellis, A. J. (1970). *Am. J. Sci.* **269**, 97.

Clark, R. H. (1960). *In* "The Geology of Tongariro Subdivision" (D. R. Gregg, ed.), pp. 107-123. N.Z.G.S. Bull. 40, New Zealand Dept. Sci. and Ind. Res., Wellington.

Dawson, G. B., and Dickinson, D. J. (1970). *Geothermics (Spec. Issue 2)* **2** *(Pt 1)*, 466.

Dickinson, D. J. (1972). Geophys. Div. Rep. No. 75. New Zealand Dept. Sci. and Ind. Res., Wellington.

Ellis, A. J. (1962). *N.Z. J. Sci.* **5**, 434.

Ellis, A. J. (1970). *Geothermics (Spec. Issue 2)* **2**(*Pt 1*), 516.

Ellis, A. J. and Anderson, D. W. (1961). *N.Z. J. Sci.* **4,** 415.

Ellis, A. J., and Mahon, W. A. J. (1964). *Geochim. Cosmochim. Acta* **28,** 1323.

Ellis, A. J., and Mahon, W. A. J. (1966). *N.Z. J. Sci.* **9,** 440.

Ellis, A. J., and Mahon, W. A. J. (1967). *Geochim. Cosmochim. Acta* **31,** 519.

Ellis, A. J., and Wilson, S. H. (1960). *N.Z. J. Geol. Geophys.* **3,** 593.

Ellis, A. J., and Wilson, S. H. (1961). *Nature (London)* **191,** 696.

Finlayson, J. B., and Mahon, W. A. J. (1976). Geochemical aspects of the Broadlands geothermal system. Paper presented at the *N.Z. Volcanol. Geothermal Conf., Auckland Univ., Feb. Auckland.*

Fooks, A. C. L. *(1961). Proc. U.N. Conf. New Sources of Energy Rome* **3,** 170.

Giggenbach, W. (1971). *N.Z. J. Sci.* **14,** 959.

Giggenbach, W. (1975). *Bull. Volcanol.* **39,** 15.

Glover, R. B. (1970). *Geothermics (Spec. Issue 2)* **2** *(Pt 2),* 1355.

Goguel, R. L. (1976). *N.Z. J. Sci.* **19,** 359.

Golding, R. M., and Speer, M. G. (1961). *N.Z. J. Sci.* **4,** 203.

Grange, L. I. (1937). "The Geology of the Rotorua-Taupo Subdivision" (L. I. Grange, ed.). N.Z.G.S. Bull. 37 New Zealand Dept. Sci. Ind. Res., Wellington.

Gregg, D. R. (1958). *N.Z. J. Geol. Geophys.* **1,** 65.

Grindley, G. W. (1957). *In* "Report on Geothermal Survey at Waiotapu 1957" (D.S.I.R., ed.), pp. 10–25. New Zealand Dept. of Sci. Ind. Res., Geothermal Rep. No. 4, Wellington.

Grindley, G. W. (1961). *Proc. U.N. Conf. New Sources Energy Rome* **2,** 237.

Grindley, G. W. (1965a). "The Geology, Structure, and Exploitation of the Wairakei Geothermal Field, Taupo, New Zealand." N.Z.G.S. Bull. 75, New Zealand Dept. Sci. Ind. Res., Wellington.

Grindley, G. W. (1965b). N.Z.G.S. Geothermal Rep. 3. New Zealand Dept. Sci. and Ind. Res., Wellington.

Grindley, G. W. (1965c). *In* "New Zealand Volcanology—Central Volcanic Region" (B. N. Thompson, L. O. Kermode, and A. Ewart, eds.), pp. 185–186. New Zealand Dept. Sci. and Ind. Res., Informat. Ser. No. 50, Wellington.

Grindley, G. W. (1970). *Geothermics (Spec. Issue 2)* **2** *(Pt 1),* 248.

Grindley, G. W., and Browne, P. R. L. (1968). N.Z.G.S. Geothermal Rep. No. 5. New Zealand Dept. Sci. and Ind. Res., Wellington.

Harris, W. F. (1965). *In* "The Geology, Structure and Exploitation of the Wairakei Geothermal Field, Taupo, New Zealand" (G. W. Grindley, ed.). N.Z.G.S. Bull. 75, New Zealand Dept. Sci. and Ind. Res., Wellington.

Hatherton, T., Macdonald, W. J. P., and Thompson, G. E. K. (1966). *Bull. Volcanol.* **29,** 485.

Healy, J. (1961). *Proc. U. N. Conf. New Sources Energy Rome* **2,** 250.

Healy, J. (1962a). *In* "Crust of the Pacific Basin," Geophys. Monogr. No. 6, pp. 151–157. Am. Geophys. Un. Nat. Acad. of Sci.

Healy, J. (1962b). *In* "Report for Tasman Pulp and Paper Co. Ltd. on Geothermal Survey at Kawerau 1962" (D.S.I.R., ed.), pp. 6–15. New Zealand Dept. Sci. and Ind. Res., Wellington.

Healy, J. (1974). *In* "Minerals of New Zealand," Rep. N.Z.G.S. No. 38D (Part D. Geothermal) Waiotapu. New Zealand Dept. Sci. and Ind. Res., Wellington.

Healy, J., and Hochstein, M. P. (1973). *J. Hydrol. (N.Z.)* **12,** 71.

Hochstein, M. P., and Hunt, T. M. (1970). *Geothermics (Spec. Issue 2)* **2** (Pt 1), 333.

Hochstein, M. P. (1975). *Proc. U.N. Symp. Develop. Use Geothermal Resources, 2nd, San Francisco, California, May* **2,** 1049.

Hulston, J. R. (1962). N.Z. Inst. of Nucl. Sci. Rep. No. 15. New Zealand Dept. Sci. and Ind. Res., Wellington.

Hulston, J. R. (1975). *Proc. Int. At. Energy Agency Advisory Groups Meeting Appl. Nucl. Tech. Geothermal Stud., Pisa, Italy, Sept.* (in press).

Jones, W. M. (1939). *N.Z. J. Sci.* **20B,** 272.

Kear, D. (1959). *N.Z. J. Geol. Geophys.* **2,** 578.

Kear, D., and Hay, R. F. (1961). N.Z.G.S. Geological map of New Zealand 1:250000, Sheet 1. New Zealand Dept. Sci. and Ind. Res., Wellington.

Keller, G. V. (1969). Geophys. Div. Rep. No. 55. New Zealand Dept. Sci. and Ind. Res., Wellington.

Koga, A. (1967a). *N.Z. J. Sci.* **10,** 979.

Koga, A. (1967b). *N.Z. J. Sci.* **10,** 428.

Lloyd, E. F. (1958). *N.Z. J. Geol. Geophys.* **2,** 141.

Lloyd, E. F. (1972). "Geology and Hot Springs of Orakeikorako" (E. F. Lloyd, ed.). N.Z.G.S. Bull. 85, New Zealand Dept. Sci. and Ind. Res., Wellington.

Macdonald, D. C. (1966). *Bull. volcanol.* **24,** 691.

Macdonald, W. J. P. (1966). Geophys. Div. Rep. No. 41. New Zealand Dept. Sci. and Ind. Res., Wellington.

Macdonald, W. J. P. (1967). Geophys. Div. Rep. No. 46. New Zealand Dept. Sci. and Ind. Res., Wellington.

Macdonald, W. J. P. (1974). *In* "Geothermal Resources Survey—Rotorua Geothermal District" (D.S.I.R., ed.), pp. 53–78. New Zealand Dept. Sci. and Ind. Res., Geothermal Rep. No. 6, Wellington.

Macdonald, W. J. P., and Banwell, C. J. (1965). *In* "New Zealand Volcanology— Northland-Auckland-Coromandel" (B. N. Thompson and L. O. Kermode, eds.), pp. 41–43. New Zealand Dept. Sci. and Ind. Res., Informat. Ser. 49, Wellington.

Macdonald, W. J. P., Muffler, L. J. P., and Dawson, B. G. (1970). Geophys. Div. Rep. No. 62. New Zealand Dept. Sci. and Ind. Res., Wellington.

Mahon, W. A. J. (1960). Chem. Div. open file Rep. No. DL 118/12—WAJM 4. New Zealand Dept. Sci. and Ind. Res., Wellington.

Mahon, W. A. J. (1962a). *N.Z. J. Sci.* **5,** 85.

Mahon, W. A. J. (1962b). *N.Z. J. Sci.* **5,** 417.

Mahon, W. A. J. (1964). *N.Z. J. Sci.* **7,** 3.

Mahon, W. A. J. (1965a). *N.Z. J. Sci.* **8,** 66.

Mahon, W. A. J. (1965b). *In* "New Zealand Volcanology—Central Volcanic Region" (B. N. Thompson, L. O. Kermode, and A. Ewart, eds.), pp. 148–150. New Zealand Dept. Sci. and Ind. Res., Informat. Ser. No. 50, Wellington.

Mahon, W. A. J. (1966). *N.Z. J. Sci.* **9,** 135.

Mahon, W. A. J. (1968). Chem. Div. open file rep. CD118/22-WAJM/48. New Zealand Dept. Sci. and Ind. Res., Wellington.

Mahon, W. A. J. (1970a). *In Proc. Symp. Hydrogeochem. Biogeochem., Tokyo* **1,** 196–210. Clarke Co., Washington, D.C.

Mahon, W. A. J. (1970b). *Geothermics (Spec. Issue 2)* **2** (Pt. 2), 1310.

Mahon, W. A. J. (1972). *In* "Geology and Hot Springs of Orakeikorako" (E. F. Lloyd, ed.), pp. 104–112. N.Z.G.S. Bull. 85, New Zealand Dept. Sci. and Ind. Res., Wellington.

Mahon, W. A. J. (1975). *Proc. U.N. Symp. Develop. Use Geothermal Resources, 2nd, San Francisco, California, May* **1,** 775.

Mahon, W. A. J., and Finlayson, J. B. (1972). *Am. J. Sci.* **272,** 48.

Mahon, W. A. J., and Glover, R. B. (1965). *Proc. Commonwealth Min. Metall. Congr., 8th, Aust—N.Z.* Paper No. 209.

Modriniak, N., and Studt, F. E. (1959). *N.Z. J. Geol. Geophys.* **2,** 654.

Nathan, S. (1974). *In* "Minerals of New Zealand," Rep. N.Z.G.S. No. 38D (Part D Geothermal). New Zealand Dept. Sci. and Ind. Res., Wellington.

Risk, G. F. (1975a). *Proc. U.N. Symp. Develop. Use Geothermal Resources, 2nd, San Francisco, California, May* **2,** 1185.

Risk, G. F. (1975b). *Proc. U.N. Symp. Develop. Use Geothermal Resources, 2nd, San Francisco, California, May* **2,** 1191.

Ritchie, J. A. (1961). *N.Z. J. Sci.* **4,** 218.

Smith, J. H. (1958). *N.Z. Eng.* **13,** 354.

Smith, J. H. (1970). *Geothermics (Spec. Issue 2)* **2** (Pt 1), 232.

Steiner, A. (1955). *In* "Geothermal Steam for Power in New Zealand" (L. I. Grange, ed.), pp. 21–26. New Zealand Dept. Sci. and Ind. Res. Bull. 117, Wellington.

Steiner, A. (1958). *N.Z. J. Geol. Geophys.* **1,** 325.

Stilwell, W. B. (1970). *Geothermics (Spec. Issue 2)* **2** (Pt 1), 714.

Studt, F. E. (1957). *In* "Report on Geothermal Survey at Waiotapu 1957" (D.S.I.R., ed.), pp. 17–25. New Zealand Dept. of Sci. and Ind. Res., Geothermal Rep. No. 4. Wellington.

Studt, F. E. (1958). *N.Z. J. Sci.* **1,** 219.

White, D. E., and Roberson, C. E. (1962). *Geol.. Soc. Am. Buddington Volume*, 397.

Wilson, S. H. (1955). *In* "Geothermal Steam for Power in New Zealand" (L. I. Grange, ed.), pp. 27–42. New Zealand Dept. Sci. and Ind. Res., Bull. 117, Wellington.

Wilson, S. H. (1957). *In* "Report on Geothermal Survey at Waiotapu 1975" (D.S.I.R., ed.), pp. 95–102. New Zealand Dept. Sci. and Ind. Res., Geothermal Rep. No. 4, Wellington.

Wilson, S. H. (1959). *In* "White Island" (W. M. Hamilton and I. L. Baumgart, eds.), pp. 32–50. New Zealand Dept. Sci. and Ind. Res., Bull. 127, Wellington.

APPENDIX

TABLE A.1

Properties of Liquid Water and Steam at Saturated Water Vapor Pressures

Temperature (°C)	Liquid[a,b] density (gm/cm³)	Steam density[b] (mgm/cm³)	Steam pressure[a] (bars)	Specific enthalpy liquid[b] (cal/gm)	Specific enthalpy steam[b] (cal/gm)	Dielectric constant[a]
0	0.9998	0.00485	0.006	0.00	597.4	87.79
25	0.9971	0.02306	0.032	25.05	608.4	78.47
50	0.9880	0.08311	0.123	50.00	619.1	69.96
75	0.9748	0.2421	0.386	74.98	629.4	62.30
100	0.9583	0.5978	1.013	100.09	639.2	55.47
125	0.9390	1.298	2.321	125.39	648.1	49.37
150	0.9170	2.546	4.758	151.00	656.0	43.91
175	0.8923	4.612	8.920	177.03	662.5	39.02
200	0.8647	7.852	15.537	203.60	667.1	34.60
225	0.8339	12.74	25.476	230.91	669.6	30.58
250	0.7992	19.95	39.728	259.23	669.2	26.87
275	0.7594	30.50	59.415	289.02	665.2	23.38
300	0.7125	46.15	85.805	321.01	656.6	19.99
325	0.6543	70.41	120.387	353.52	642.7	16.58
350	0.5746	113.47	165.125	399.02	612.4	12.87

[a] Helgeson and Kirkham (1974a).
[b] Keenan *et al. (1969).*
Refer to Chapter 4 reference list for above references (see pages 159–162).

TABLE A.2

Densities of Sodium Chloride Solutions[a]

Tempera-ture (°C)	0	0.1	0.25	0.50	1.0	2.0	3.0	4.0
					m NaCl			
0	0.9998	1.0041	1.0107	1.0216	1.0418	1.0800	1.1154	1.1480
25	0.9971	1.0012	1.0091	1.0171	1.0362	1.0721	1.1057	1.1372
50	0.9881	0.9921	0.9998	1.0075	1.0258	1.0610	1.0935	1.1242
75	0.9749	0.9788	0.9846	0.9940	1.0122	1.047	1.080	1.111
100	0.9583	0.9623	0.9683	0.9777	0.9951	1.032	1.065	1.096
125	0.9391	0.9431	0.9490	0.9589	0.9774	1.014	1.047	1.078
150	0.9169	0.9211	0.9274	0.9374	0.9567	0.994	1.027	1.058
175	0.8922	0.8966	0.8949	0.9137	0.9338	0.972	1.006	1.037
200	0.8647	0.8696	0.8765	0.8877	0.9087	0.948	0.983	1.015
225	0.8339	0.839	0.846	0.858	0.880	0.921	0.958	0.992
250	0.7992	0.804	0.812	0.825	0.849	0.893	0.932	0.967
275	0.7594	0.765	0.773	0.787	0.813	0.861	0.904	0.942
300	0.7125	0.718	0.728	0.743	0.772	0.825	0.873	0.916
325	0.6541	0.662	0.672	0.689	0.722	0.783	0.839	0.888

Data:

Water from R. W. Bain (1964), "Steam Tables," HMSO, London.

NaCl 0.1–1.0 m, 25°–200°; A. J. Ellis (1966), *J. Chem. Soc. A*, 1579, adjusted to vapor saturation pressures using data from A. M. Rowe and J. C. S. Chou (1970). *J. Chem. Eng. Data* **15**, 61.

NaCl at 0°C from F. J. Millero (1970). *J. Phys. Chem.* **74**, 356.

1.0–4.0 m NaCl, 0°–50°, interpolated from *Int. Crit. Tables* (1928) **3**, 79.

0.1–1 m NaCl over 200° and 2.0–4.0 m over 75°, from J. L. Haas (1970). *Am. J. Sci.* **269**, 489–493.

[a] At pressure of 1 bar for temperatures below 100°, and at saturated vapor pressures for temperatures over 100°.

TABLE A.3

Values of Weak Acid Dissociation Constants for Various Temperatures at Saturated Water Vapor Pressures[a]

	25°	50°	75°	100°	125°	150°	200°	250°	300°	Source[b]
H_2O	14.00	13.26	12.70	12.26	11.91	11.64	11.26	11.05	11.04	1
H_2CO_3	6.36	6.31	6.33	6.42	6.55	6.76	7.24	7.74	8.40	2
H_2S	6.98	6.70	6.61	6.61	6.68	6.80	7.17	7.60	8.05	3
NH_4^+	9.25	8.56	7.95	7.41	6.94	6.54	5.73	5.05	4.29	4
H_4SiO_4	9.63	9.35	9.15	9.01	8.90	8.86	8.86	9.04	9.48	1, 5
H_3BO_3	9.23	9.06	8.98	8.94	8.92	8.93	8.99	9.07	9.24	1
HF	3.18	3.40	3.64	3.85	4.09	4.34	4.89	5.8	6.8	4
HSO_4^-	1.99	2.27	2.63	2.99	3.37	3.74	4.49	5.41	7.06	6
HCl	−6.1	−5.0	−3.9	−2.90	−2.02	−1.23	−0.06	0.67	1.24	6

[a] As values of $-\log K_a = pK_a$.
[b] 1. Kharaka and Barnes (1973).
2. Read ((1975).
3. Ellis and Giggenbach (1971).
4. Truesdell and Singers (1971).
5. Seward ((1974).
6. Helgeson (1969).
Refer to Chapter 4 reference list for above references (see pages 159–162).

INDEX

A

Acid-base equilibria, computer calculation, 259
Acid sulfate waters, *see also* Hot waters
  compositions of, 60
  formation of, 60
Acid sulfate-chloride waters, *see also* Hot waters
  compositions of, 60
  formation of, 60
  in New Zealand, 331
  at Rotokaua, 353
  in Taiwan wells, 66
Acid waters, association with sulfur deposits, 72
Activity coefficients
  in ionic equilibria, 129
  for NaCl solutions, 130-131
Aerial photography, *see* Photography
Afyon area, 19
Agricultural uses, *see specific areas*
Ahuachapan area, 5
Albite, *see* Hydrothermal alteration
Aleutian Islands areas, 22
Alkali chloride waters, *see also* Hot waters
  compositions of, 60, 66-69, 217, 345-349, 355-358
  composition, New Zealand, 332-333
  salinity *v* trace metal concentrations, 67
Alkali metals, *see* Lithium; Sodium; etc.
Alloys, *see* Corrosion chemistry
Alpine fault, New Zealand, 324
Aluminum, method of analysis, 196
Alunite, *see* Hydrothermal alteration

Ammonia
  analysis in steam, 201
  methods of analysis, 191
  steam/water distribution, 287
  in Wairakei wells, 349
Ammonium bicarbonate, from Wairakei steam, 290
Analysis, *see* Water analysis; Steam analysis; etc.
Analytical techniques, 188-202
Anatolia, western, *see* Kizildere
Anhydrite, *see* Calcium sulfate
Ankara area, 19
Antimony, method of analysis, 192
Aragonite, *see* Deposits from waters
Argon, *see* Inert gases
Arsenic
  in alkali chloride waters, 66
  in El Tatio water, 236
  methods of analysis, 192
  in New Zealand waters, 335
  removal from geothermal waters, 318
  in Wairakei waters, 348
Atomic absorption, *see specific element*
Austenitic stainless steels, *see* Corrosion chemistry

B

Beowawe area, 22
Beppu area, 12
Bicarbonate, method of analysis, 193
Bicarbonate waters, *see also* Hot waters
  compositions of, 61
  formation of, 61
  in New Zealand, 331

A
B 7
C 8
D 9
E 0
F 1
G 2
H 3
I 4
J 5